Microfluidics in Detection Science
Lab-on-a-chip Technologies

RSC Detection Science Series

Editor-in-Chief:
Professor Michael Thompson, *University of Toronto, Canada*

Series Editors:
Dr Sub Reddy, *University of Surrey, Guildford, UK*
Dr Damien Arrigan, *Curtin University, Australia*

Titles in the Series:
1: Sensor Technology in Neuroscience
2: Detection Challenges in Clinical Diagnostics
3: Advanced Synthetic Materials in Detection Science
4: Principles and Practice of Analytical Techniques in Geosciences
5: Microfluidics in Detection Science: Lab-on-a-chip Technologies

How to obtain future titles on publication:
A standing order plan is available for this series. A standing order will bring delivery of each new volume immediately on publication.

For further information please contact:
Book Sales Department, Royal Society of Chemistry, Thomas Graham House, Science Park, Milton Road, Cambridge, CB4 0WF, UK
Telephone: +44 (0)1223 420066, Fax: +44 (0)1223 420247
Email: booksales@rsc.org
Visit our website at www.rsc.org/books

Microfluidics in Detection Science
Lab-on-a-chip Technologies

Edited by

Fatima H Labeed
University of Surrey, Guildford, Surrey, UK
Email: f.labeed@surrey.ac.uk

Henry O Fatoyinbo
University of Surrey, Guildford, Surrey, UK
Email: h.fatoyinbo@surrey.ac.uk

THE QUEEN'S AWARDS
FOR ENTERPRISE:
INTERNATIONAL TRADE
2013

RSC Detection Science Series No. 5

Print ISBN: 978-1-84973-638-1
PDF eISBN: 978-1-84973-760-9
ISSN: 2052-3068

A catalogue record for this book is available from the British Library

Published by The Royal Society of Chemistry,
Thomas Graham House, Science Park, Milton Road,
Cambridge CB4 0WF, UK

Registered Charity Number 207890

For further information see our web site at www.rsc.org

Printed and bound by CPI Group (UK) Ltd, Croydon, CR0 4YY

Preface

Microfluidics is the name given to a family of techniques that manipulate, detect and process fluids and associated particles at the micro-scale; its name derives from a fluid analogy to microelectronics, and many of the technologies have both parallels and, in some cases, origins in that field. There are many reasons why we would want to explore miniaturisation of detection, such as minimising amounts of sample use. This conserves small quantities of potentially rare or precious sample whilst still providing a high level of speed and accuracy at relatively low cost when compared to conventional bench-top techniques.

Microfluidic technologies have formed the basis of lab-on-a-chip devices, and advanced an extensive profile of applications in the areas of chemistry, biological sciences and engineering. The concept of an integrated, miniaturised laboratory on a disposable chip has gone from hypothetical device to academic study to use in industry, medicine and defence application over the last twenty years. However, as these devices enter circulation and become relevant to industrial applications, so there is a need for a simple guide to the design and manufacture of laboratory on a chip devices. In this book, we present for the first time a modular approach to the construction and integration of lab-on-a-chip components. With application chapters divided into applications for sample preparation and detection methods, and with unique chapters on the integration of lab-on-a-chip components with each other and with supporting technology, such as dielectrophoresis. This book offers the reader a convenient guide to this emerging field of technology. This book offers insights into lab-on-a-chip applications from biosensing to molecular and chemical analysis through a variety of methods, from scaled-down versions of existing technology to unique approaches exploiting the physics of the micro and nano scales.

RSC Detection Science Series No. 5
Microfluidics in Detection Science: Lab-on-a-chip Technologies
Edited by Fatima H Labeed and Henry O Fatoyinbo
© The Royal Society of Chemistry 2015
Published by the Royal Society of Chemistry, www.rsc.org

The book will give insight information in areas including; material fabrication- encompassing a range of existing and new materials to create microfluidic devices along with factors determining the choice of material are covered; an overview of fluid hydrodynamics-providing solid background in the limitations present when operating at small scale; novel microsystems used for particle sorting-this encompasses electrokinetic techniques applied for rare cell detection; digital microfluidics-where confined microenvironments are used for a range of chemical, biochemical and biological screening applications with novel architectures to avoid contaminations discussed; surface acoustic wave based micro and nano scale lab-on-a-chip applications particle manipulations; optofluidics- this discusses the integration of discrete optofluidic technologies to create all-optical lab-on-a-chip devices capable of delivering compact and inexpensive routes for sample pre-processing; dielectrophoresis in microfluidic applications-a leading micro-fluidic technique exhibiting flexible operation modes that can be electrode-or insulator-based and used for a number of biological samples; novel trans-ducer strategies for lab-on-a-chip biosensing-it discusses platforms that allow easy use, low cost and automation of sensors.

The editors have strived to ensure that his handbook is a valuable resource both to those new to the technology, and practitioners requiring a convenient reference to a comprehensive coverage of microfluidics (in its various forms of methods and applications) in detection science. The editors would like to express their sincere gratitude to the authors who have made this aim achievable.

Contents

RSC Detection Science Series No. 5
Microfluidics in Detection Science: Lab-on-a-chip Technologies
Edited by Fatima H Labeed and Henry O Fatoyinbo
© The Royal Society of Chemistry 2015
Published by the Royal Society of Chemistry, www.rsc.org

CHAPTER 1

Materials and Fabrication Techniques for Nano- and Microfluidic Devices

KIN FONG LEI

Department of Mechanical Engineering, Graduate Institute of Medical Mechatronics, Chang Gung University, Tao-Yuan, Taiwan, Republic of China
E-mail: kflei@mail.cgu.edu.tw

1.1   Introduction

Microfluidic technology has enabled the realisation of a vast range of miniaturised analytical devices. Microfluidic devices are commonly associated with lab-on-chip (LOC) systems or micrototal-analysis system (µTAS), when scaled-down operations are performed on miniaturised versions of conventional laboratory bench top instruments.[1] One of the main objectives of microfluidic technologies is to provide a total solution, from sample input to display of the analysed results. Complete analytical protocols, from sample pretreatment through to sample/reagent manipulation, separation, reaction, and detection, can be performed automatically on well-designed and integrated miniaturised devices.

Historically, developmental advances of microfluidic devices originated from the microelectronics manufacturing sector. Silicon has been used as the base substrate material for fabricating microfluidic devices for various applications.[2–4] Well-established silicon processing and extensive studies of

RSC Detection Science Series No. 5
Microfluidics in Detection Science: Lab-on-a-chip Technologies
Edited by Fatima H Labeed and Henry O Fatoyinbo
© The Royal Society of Chemistry 2015
Published by the Royal Society of Chemistry, www.rsc.org

silicon properties have contributed to the rapid evolution of microfluidic technologies. The fabrication process for silicon-based microfluidic devices involve substrate cleaning, photolithography, metal deposition, and wet/dry etching.[5,6] However, silicon substrate is relatively expensive and optically opaque to certain electromagnetic wavelengths, limiting its applications in optical detection. To combat these shortcomings, glass and polymeric materials have been used to fabricate microfluidic devices.[7,8] Compared with silicon, glass and polymer materials are inexpensive and optically transparent. Polymer materials include polymethylmethacrylate (PMMA), polystyrene (PS), polycarbonate (PC), and polydimethylsiloxane (PDMS). Amongst these polymer materials, PDMS has been one of the most widely used materials for fabricating microfluidic devices in recent years due to its flexibility in moulding and stamping, optical transparency, and biocompatibility. Recently, paper has been proposed to be an alternative material used as a substrate of microfluidic devices.[9] Paper is inexpensive, lightweight, available in a wide range of thickness, and is disposable. Aqueous solutions can be transported by wicking, thus realising passive pumping. In addition, well-defined pore sizes in paper can be manufactured and suspended solids within samples can be separated based on size exclusion before an assay is performed. Paper is biocompatible with various biological samples and can thus be modified with a wide range of functional groups to enable covalent bonding of proteins, DNA, or small molecules creating bespoke biochemical sensing systems.[10–12]

In this chapter, materials used in the fabrication of microfluidic devices are grouped for discussing microfabrication techniques and applications. Moreover, the ability of system integration, cost of processing, and suitability for specific applications will be highlighted. An up-to-date and systematic approach for fabricating nano- and microfluidic devices will be presented.

1.2   Traditional Silicon-Based Microfluidic Devices

From the beginning of the 20th century, continual rapid development of microelectronic technologies made computing processors fast and inexpensive. In 1965, Gordon Moore observed that the number of transistors per unit area would double every two years.[13] This extraordinary growth rate led to the realisation of current personal computers that run on the computing power of millions of transistors within a centimetre scale environment. In the 1980s, microelectromechanical systems (MEMS) were inspired by microelectronic technologies and were developed from microelectronic fabrication processes to build machines on the order of micrometres.[14] The majority of MEMS devices are made from single-crystal silicon wafers and their fabrication processes include deposition of polycrystalline silicon for resistive elements, metal deposition for conductors, silicon oxide for insulation and as a sacrificial layer, and silicon nitride and titanium nitride for electrical insulation and passivation. Sensors, actuators, and control functions can also be cofabricated on standard silicon wafers. There has since been remarkable progress in research in MEMS technologies, under strong capital promotions from both national governments and industry.

Microfluidic technology is one of the branches of MEMS that handles fluids within submillimetre environments, *i.e.* typically microlitres, nano-litres, or even picolitres. Fluids are manipulated, mixed, or separated on a compact platform for various biomedical, biochemical and chemical analytical applications. One of the objectives of the development of micro-fluidic devices is to provide a total solution (*i.e.* sample-to-answer) in low-cost and rapid systems. For instance, point-of-care (POC) diagnostic applications can be realised based on the advantages of miniaturisation, integration, and automation of the microfluidic system. Microfluidic devices can, and are often modelled as miniaturised versions of conventional laboratory devices, with early developments of microfluidic technologies being based predomi-nantly on silicon as the substrate of choice for many microfluidic devices.

1.2.1 Microfabrication with Silicon

Silicon microfabrication is the process for the production of devices on silicon wafers in the submicrometre to millimetre range. Normally, struc-tures in microfluidic devices have relatively high aspect ratios compared to those in microelectronic devices, which are fabricated to within the top few micrometres of the substrate material. Microfluidic devices may require the whole substrate thickness, utilise both sides of the substrate, or require bonding multiple substrates together. Besides the conventional microelec-tronic fabrication techniques, such as photolithography, thin-film deposi-tion, and etching, some newer processes were introduced to fulfill the fabrication requirement of microfluidic devices. Since there is a plethora of silicon microfabrication techniques, only the important processes in fabri-cating microfluidic devices are discussed. For a more comprehensive over-view of further techniques refer to ref. 15.

1.2.1.1 Photolithography

Photolithography is the transfer of a pattern on to a material and is arguably the most important step in the microfabrication process. It predominantly utilises ultraviolet (UV) light to transfer a geometric pattern from a photo-mask to a light-sensitive chemical photoresist, *e.g.*, AZ1500-series resists, on the substrate. For higher-resolution patterns, expensive technologies such as X-ray, electron beams, or ion beams are used in the photolithographic process. Generally, a series of steps are included in the photolithographic process, such as photomask creation, wafer cleaning, photoresist applica-tion, UV exposure, and development (for exact steps and specification/tolerances, refer to the photoresist manufacturer's datasheet). A brief over-view of the steps involved is discussed below.

A photomask is a glass or quartz plate with a chromium geometric pattern, which can be designed by computer software, *e.g.*, Tanner EDA L-Edit. The creation of the photomask begins from a square glass or quartz plate covered with a full layer of chromium. A laser beam or electron beam is used to travel over the photoresist on the chromium surface for exposure of the pattern defined by the computer software. Where the photoresist is exposed, the chromium can be etched away. A transparent path is left for the illuminating light to penetrate through. The glass or quartz plate is transparent to UV light and chromium blocks the light. Therefore, a photomask can define the geometric pattern over the entire wafer in a single step of UV exposure.

To obtain highly reliable devices and improve the yield rate, contaminants present on the surface of silicon wafer must be removed before microfabrication. RCA clean is the industrial standard cleaning procedure. Werner Kern developed the basic procedure in 1965 while working for RCA (Radio Corporation of America).[16] The RCA clean procedure has three major sequential steps: (1) Mixing one part NH_4OH with five parts deionised water, heating to 80 °C, then adding one part H_2O_2, and immersing the wafer for 10 min to remove organic contaminants (Note: A thin silicon dioxide layer along with metallic contamination on the silicon surface is formed and has to be removed in subsequent steps). (2) A short immersion in a 1 : 50 solution of HF and H_2O at 25 °C in order to remove the thin oxide layer. (3) Mixing one part HCl and six parts deionised water, heating to 80 °C, then adding one part H_2O_2, and immersing the wafer for 10 min to remove metal ions.

Photoresist application is followed to create a thin-film photoresist on the wafer surface. Photoresist is a solution of a light-sensitive polymer. Either positive or negative resists can be selected, depending on whether it is desirable to have the opaque regions of the photomask for the protection of the resist during UV exposure, or *vice versa*. Application of the photoresist layer on the silicon wafer is normally accomplished *via* a spin-coating process. The photoresist is carefully dispensed onto the wafer surface, avoiding bubble formations, and spun at high speeds, *i.e.* 3000 rpm. The spinning speed determines the photoresist film thickness. The wafer is then "soft baked" at 75–100 °C for about 1 min to remove solvents and improve adhesion. This process can create a uniform and smooth photoresist film on the wafer surface. The exact values of the parameters including spinning speed, baking temperature, and baking duration can be determined from the manufacturer's datasheet.

After photoresist application, the photoresist is exposed under UV light for the definition of the geometric pattern based on the photomask. The photomask and the photoresist-coated wafer are respectively placed into the UV exposure machine. The machine has micrometre manipulators and a microscope that allow for precise alignment of the fiducial markers located on the photomask and the wafer. The alignment process is critical for fabricating multilayer structures. The photomask is placed in direct contact or proximity contact (10–20 μm) above the photoresist-coated surface.

The photoresist is then exposed to UV light under programmed time duration, which is determined by the energy adsorption of the photoresist.

The photoresist is then developed to remove the undesired regions. With positive photoresist, the regions that are exposed by UV light become soluble in the developer. The converse is true for negative photoresist. Control of development time is critical for fabricating high-resolution patterns. Photoresist stripping can be accomplished by etching. Both an oxygen plasma etching process or Piranha solution, *i.e.* 3 : 1 mixture of H_2SO_4 and H_2O_2, at around 120 °C can remove photoresist.

An example of the photolithography process to fabricate metal patterns on a silicon wafer is illustrated in Figure 1.1. It shows the fabrication process from wafer cleaning, photoresist application through to UV exposure and development. The patterned photoresist on the wafer is thus created. Then, thin-film metal is deposited on the entire wafer surface (see Section 1.2.1.2). Photoresist stripping is performed and the metal pattern is left on the wafer surface. This process is called lift-off and is commonly used for the fabrication of metal electrodes on wafers.

1.2.1.2   Thin-Film Deposition

Thin-film deposition process is an "additive" process that adds a thin film on the entire surface, such as silicon dioxide, polysilicon, silicon nitride, and metal. Basically, physical vapour deposition (PVD) is a common approach for fabricating microfluidic devices, where a thin film is deposited onto a substrate surface by the condensable vapour of the desired material, through a vacuum or low-pressure gaseous environment. The vapour is physically transported from a target source by various techniques, *e.g.*, evaporation, sputtering, cathodic arc, and pulsed laser. Evaporative deposition and sputter deposition are discussed and compared below.

In general, evaporation is used to deposit metals and some compounds with low fusion temperature, *e.g.*, Au, Al, Ti, Cr, and SiO. Resistive and electron beam heating are the major types of evaporation techniques. In resistive heating, the desired metal is placed in a refractory tungsten "boat"

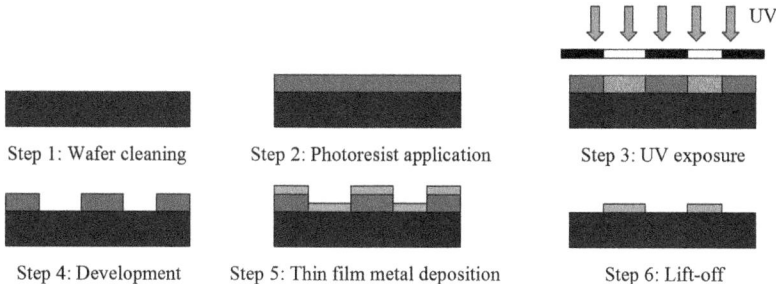

Figure 1.1   Fabrication of metal pattern on silicon wafer.

working as an electrically heated filament. When current is applied to the filament, the metal is evaporated and deposited onto the substrate. An illustration of the resistive evaporation is shown in Figure 1.2. In electron-beam heating, an electron beam is scanned over the metal to generate a vapour for deposition. It has high deposition rates and lower substrate heating than resistive evaporation, thus generating less tensile stress in the deposited film on the unheated substrate.

Sputter deposition is based on the bombardment of a sputtering target by accelerated inert ions, *e.g.*, Ar^+. Ejected clusters of the target material become condensable vapour for deposition. There are several methods to obtain the necessary plasma, such as direct current (DC), radio frequency (RF), magnetron, and reactive sputtering techniques. Most materials can be sputtered if sufficiently high-energy plasma can be generated. Sputtering provides a flux of more energetic atoms than in evaporation, thus, it has better step coverage and stress control. Because sputter deposition is from a spatially distributed source, the molecules maintain some directionality for only roughly their mean free path. Sidewall coating is nearly equal to surface coating because the substrate is immersed in the plasma. Evaporation deposits directionally from the source, and thus has poor step coverage. However, this is very useful for lift-off processes. Illustrations of the step coverage between evaporation and sputtering are shown in Figure 1.3.

1.2.1.3   Etching

Once the substrate has been protected with a photoresist or mask, either a thin-film deposition process or an etching process can be performed to add or remove material. The etching process is a "subtractive" process and is

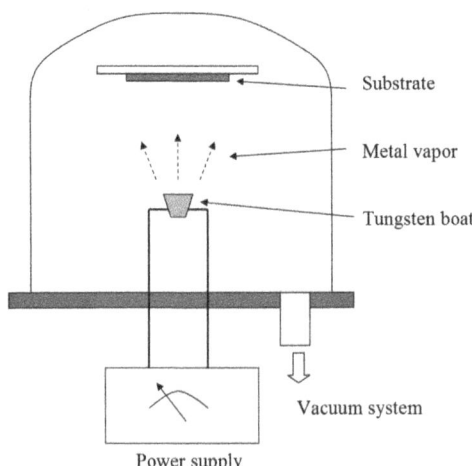

Figure 1.2   Illustration of resistive evaporation.

generally divided into wet etching with chemical solutions and dry etching with plasma methods.

In order to fabricate subtractive structures, *e.g.*, holes, trenches, and mesas, on silicon wafer, wet silicon etching is performed with appropriate choice of substrate type, thin-film masking layer, and etchant. Depending on the selection of etchants, silicon wafer can be etched in all directions at nearly the same rate, *i.e.* isotropic etching, or unequal etch rates for different crystal planes, *i.e.* anisotropic etching. The most common isotropic etchant is HNA, which is a mixture of hydrofluoric acid (HF), nitric acid (HNO_3), and acetic acid (CH_3COOH). With the SiO_2 mask, microfluidic channels can be etched in the silicon wafer. However, the etch rate is slowed down when the channels are long and narrow due to the diffusion limits of the reaction. Good agitation of the etchant can enhance the etch rate and the etched channel surface (a near perfect hemispherical shape). For anisotropic etching, hydroxides of alkali metals, *i.e.* KOH, NaOH, CeOH, RbOH, *etc.*, can be used as crystal-orientation-dependent etchants of silicon. They etch much slower on the (111) plane of silicon than other planes. It is also important to note that etching at concave corners on (100) silicon stops at (111) inter-sections, but convex corners are undercut. Illustrations of isotropic and anisotropic etching are shown in Figure 1.4.

Reactive ion etching (RIE) can be accomplished by chemical reactions under low pressure (a few millitorr) with an RF-generated plasma. Etching at low temperatures (from room temperature to 250 °C) can be achieved by such

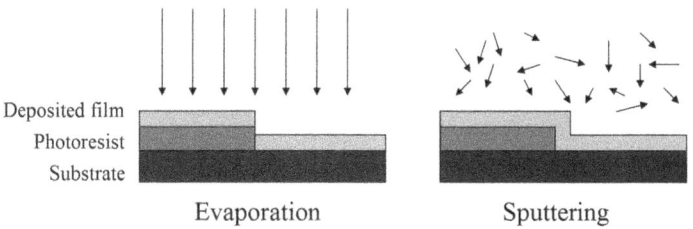

Figure 1.3 Illustration of the step coverage between evaporation and sputtering.

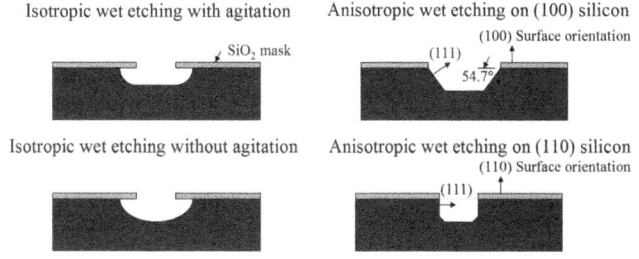

Figure 1.4 Illustration of the etched profiles using isotropic and anisotropic etching.

energetic ions. The reactant gas SF_6 is commonly used for silicon etching. The bulk silicon etching is carried out by the release of the fluorine free radicals through dissociation, ionisation, and attachment reactions. Since the reaction proceeds spontaneously, the etch profiles are nearly isotropic. For fabricating very high-aspect ratio subtractive structures, deep reactive ion etching (DRIE) was introduced.[17,18] A high-density inductively coupled plasma and an alternating process of etching and protective polymer deposition are used to achieve an aspect ratio of up to 30 : 1. The practical maximum etch depth capability is of the order of 1 mm. The etching step uses SF_6 with a substrate bias of −5 to −30 V. The cations generated in the plasma are accelerated nearly vertically into the substrate being etched. After etching for a period of time, polymer deposition using C_4F_8 is started for coating with a protective polymer layer on all exposed surfaces. Then, the etching step is repeated and the polymer layer on the horizontal surfaces is rapidly removed due to the ion bombardment and the presence of reactive fluorine radicals. With the protective polymer layer on the vertical surfaces, this alternating process can enhance the aspect ratio of the silicon etching.

1.2.1.4   LIGA

LIGA is a German acronym that translates into English as lithography, electroplating, and moulding. These are the major processing steps of LIGA for fabricating high-aspect ratio microstructures. In the process, there are two main lithography technologies, *i.e.* X-ray and UV, to create structures. PMMA is an X-ray sensitive photoresist that creates high-resolution structures and SU-8 is exposed by an inexpensive UV-light source. Both photoresists are thick resists and can create high aspect ratio microstructures for the electroplating process. Metals, *e.g.*, nickel, copper, or gold, are plated upward from the conductive seed substrate into the volume that is left by the developed photoresist. Typically, DC electroplating is used and a regulated current is set to obtain the desired current density. Reference recipes of plating solutions[15] are listed in Table 1.1. An alternative is to use pulsed currents at a specified duty cycle. The benefit of pulsed electroplating is that

Table 1.1   Reference recipes of plating solutions

Nickel (at 60 °C with a current density of 100 mA cm^{-2})	
Nickel sulfate ($NiSO_4 \cdot 6H_2O$)	330 g l^{-1}
Nickel chloride ($NiCl_2 \cdot 6H_2O$)	45 g l^{-1}
Boric acid (H_3BO_3)	38 g l^{-1}
Copper (with a current density of 10 mA cm^{-2})	
Copper sulfate ($CuSO_4 \cdot 5H_2O$)	120 g l^{-1}
Sulfuric acid (H_2SO_4)	100 g l^{-1}
Gold (with a current density of 100 mA cm^{-2})	
Potassium gold cyanide (dicyanoaurate) ($KAu(CN)_2$)	20 g l^{-1}
Potassium citrate ($K_3C_6H_5O_7 \cdot H_2O$)	150 g l^{-1}
Potassium phosphate (dibasic) (HK_2O_4P)	40 g l^{-1}

diffusion of reactant species from the bulk solution in between current pulses can recharge the region closest to the cathode where the reactants are depleted by each plating pulse. This can raise the reactant concentration at the interface for plating finer-grained metal with less stress. After lithography and electroplating, the photoresist is stripped and the metal microstructures are left on the substrate. The metal part then serves as a either a final product, *e.g.*, gears, or replication mould for stamping or injection moulding.

1.2.1.5 Substrate Bonding

Most microfluidic devices require bonding of substrates to create enclosed volumes for fluid handling. Silicon-to-silicon bonding can be employed by fusion bonding and silicon-to-glass is bonded by anodic bonding techniques. Fusion bonding brings two cleansed silicon surfaces tightly together and subsequent high-temperature annealing (300–800 °C) in oxygen or nitrogen is performed to strengthen the bonds. The resulting composite has almost no thermal stress because the thermal expansion coefficients of the two substrates are identical. Anodic bonding provides an excellent bond between silicon and glass substrates. The two substrates are compressed and heated to 300–400 °C. Then, 1000 V direct current is applied across the composite with the glass as the cathode and the silicon as the anode. An illustration of anodic bonding procedure is shown in Figure 1.5.

1.2.2 Application Examples

Based on the above silicon microfabrication techniques, a number of silicon-based microfluidic devices have been fabricated and demonstrated for many applications. For example, a piezoelectrically actuated microdiaphragm pump with check valves was reported.[19-22] Three silicon substrates with subtractive structures are fabricated by wet etching respectively. Hence, they are bonded together and a piezodisk is attached on the pump diaphragm. The working principle of the pump is based on the movement of the diaphragm driven by the piezodisk. When the diaphragm is moved upward and enlarges the pump chamber volume, a negative pressure is generated in the pump chamber and fluid flows through the inlet valve into the pump chamber. Conversely, when the pump chamber volume is reduced, a positive

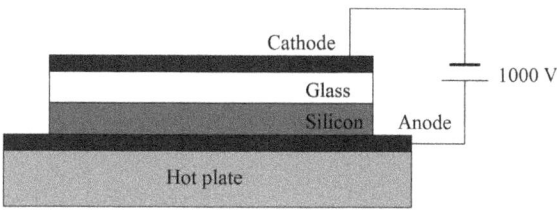

Figure 1.5 Illustration of anodic bonding procedure.

pressure is generated in the pump chamber and fluid flows through the outlet valve. By repeating this cycle process, fluid can be pumped continuously. The total volume of the pump chamber defines the effective dead volume and the volume changes generated by the movement of the diaphragm define the stroke volume. The ratio between stroke volume and dead volume defines the compression ratio. Three types of micropumps with different compression ratios are illustrated in Figure 1.6. Based on these mature developments, a microfluidic device for DNA amplification has been developed based on peristaltic pumping driven by three diaphragm pumps.[23] Reactant droplets can be pumped back and forth between three reaction chambers, stabilising at 90 °C, 72 °C, and 55 °C. A maximum flow rate of 3.14 μl s^{-1} at an operating frequency of 10 Hz was obtained. Polymerase chain reaction (PCR) was achieved after 20–30 thermal cycles. The PCR products were pumped into the reservoir to be collected and analysed by gel electrophoresis. This diaphragm pump is a typical example of a microfluidic device with several silicon substrates bonded together.

Because silicon substrate is poorly transparent, it is generally not used in biological detection applications where optical interrogations are performed. Microfluidic devices can thus be fabricated by bonding of silicon and glass substrates. A silicon–glass microfluidic device has been developed for simultaneous DNA amplification and detection.[24] The device was fabricated by bonding of a silicon substrate and an indium tin oxide (ITO)-coated glass substrate, as shown in Figure 1.7. Thin-film Pt heaters and temperature sensors were deposited on the front side of the silicon substrate to control thermal cycling of the PCR. On the back side of the silicon substrate, a chamber etched by DRIE process was used for the solution-phase PCR. The glass substrate had thin-film Pt reference and counter electrodes on the centre and four ITO circular working electrodes on the surrounding. Oligonucleotide capture probes were immobilised on the working electrodes. The PCR master mix was pipetted into the chamber and on-chip thermal cycling was performed. Finally, the PCR products were detected by differential pulse voltammetric measurement.

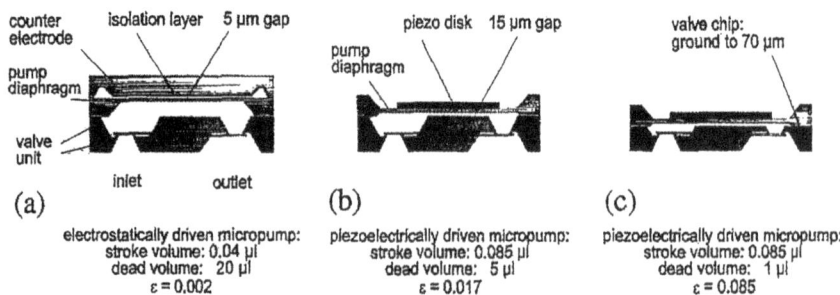

Figure 1.6 Structure of the piezoelectrically actuated microdiaphragm pumps with different compression ratios of (a) 0.002, (b) 0.017, and (c) 0.085. [Reprinted with permission from ref. 21. Copyright (1998) Elsevier.]

1.3 Glass/Polymer-Based Microfluidic Devices

Although silicon microfabrication is well-established, silicon substrates are being overlooked for microfluidic device fabrication in favour of better quality and low-cost materials, particularly in biological detection applications. Microfluidic devices may be hybrids of silicon, glass, and polymer materials. Polymer materials have the advantages of easy replication, optical transparency, biocompatibility, and acceptable thermal and electrical properties. These are very important characteristics for developing microfluidic devices in biological detection applications.

1.3.1 Soft Microfabrication

In the late 20th century, soft lithography was proposed to represent a non-photolithographic strategy, based on self-assembly and replica moulding, for carrying out micro- and nanofabrication.[25] Its advantages include the capability for rapid prototyping and easy fabrication without expensive capital equipment. In the past decade, a number of different methods have been proposed to fabricate patterns and structures with micro- and nanofeature

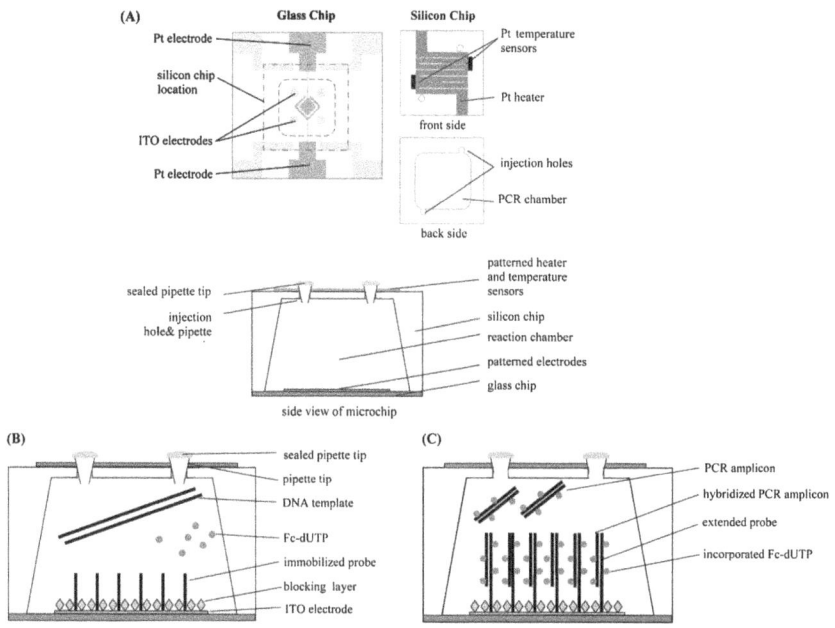

Figure 1.7 Schematic illustration of the silicon–glass microfluidic device. (A) The layout of the silicon and glass substrates and the side view of the device. (B) The initial status of the PCR cycle. (C) The final status of PCR cycle. [Reprinted with permission from ref. 24. Copyright (2008) American Chemical Society.]

sizes, such as micromoulding,[26] microcontact printing (µCP),[27] microtransfer moulding (µTM),[28] and micromoulding in capillaries (MIMIC).[29] Here, micromoulding and microcontact printing processes will be discussed and the substrate bonding techniques applied will also be addressed as it is represents an important process in the fabrication route for microfluidic device creation.

1.3.1.1 Micromoulding

Micromoulding is a technique that replicates microstructures on polymer substrates and generally includes PDMS replication, injection moulding, and hot embossing processes. In recent years, the most widely used method to fabricate microfluidic devices in research laboratories is the PDMS replication process. PDMS is an elastomer that is optically clear, inert, nontoxic, and nonflammable. Figure 1.8 shows the fabrication PDMS-based microstructures. A moulding template is fabricated by either patterning negative photoresist, *e.g.*, SU-8, on silicon wafer for precise microstructures (*i.e.* <50 µm) or direct micromachining PMMA substrate for larger micro-structures (*i.e.* >50 µm). A PDMS mixture is applied over the moulding template after thoroughly mixing the PDMS prepolymer and curing agent in a weight ratio of 10 : 1 according to the manufacturer's instruction.[30] Hence, it is cured at 70 °C for 1 h and then peeled off from the template. The desired microstructures allowing high aspect ratios can be obtained in the PDMS substrate. This process is simple and inexpensive. However, the entire protocol takes several hours and it is not practical for mass production. Alternatively, for fabricating microstructures in thermoplastics, injection moulding or a hot embossing process can be used in industry because of mass-production capabilities. In injection moulding, thermoplastic pellets are poured into a hopper, melted, and injected into the steel or aluminium mould with the desired microstructures. Molten plastic enters the mould under high pressure and travels to the cavity of the mould. After cooling, the plastic substrate with the microstructures are formed and released from the mould. An illustration of an injection moulding machine is shown in Figure 1.9. In hot embossing, the thermoplastic substrate is inserted into a moulding machine. The microstructures in the mould are transferred to the plastic substrate under heat and pressure. Temperatures above the glass-transition temperature of the specific plastic are required to

Step 1: Fabrication of   Step 2: PDMS application and   Step 3: Release from template
molding template   curing

Figure 1.8   Illustration of the PDMS replication process.

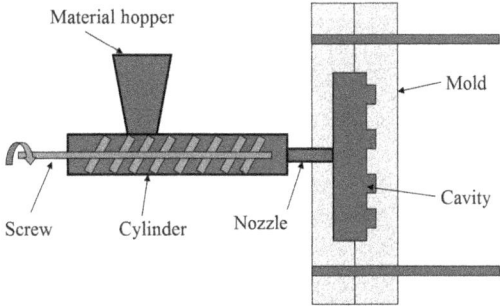

Figure 1.9   Illustration of the injection moulding machine.

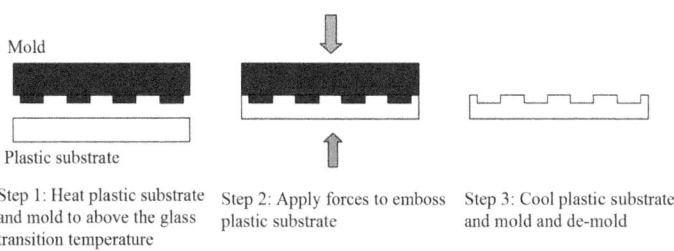

Step 1: Heat plastic substrate and mold to above the glass transition temperature

Step 2: Apply forces to emboss plastic substrate

Step 3: Cool plastic substrate and mold and de-mold

Figure 1.10   Illustration of the hot embossing process.

soften the plastic substrate and pressure is normally from 5 to 10 tons. Figure 1.10 shows the process steps. Since the plastic substrate is not required to be molten, high molecular weight polymers are suitable for this process and provide better mechanical and thermal properties. However, microstructures with high aspect ratios are difficult to form and have high residual stresses.

1.3.1.2   Microcontact Printing

Microcontact printing is the transfer of materials, *e.g.*, organic molecules and biologically active molecules, to a surface in a well-defined pattern. The process is illustrated in Figure 1.11. A PDMS stamp with the desired convex pattern is first wetted with the material to be transferred suspended in a solution, which puts the material on the raised part of the PDMS stamp. Then the stamp is pressed physically on the substrate surface and the material is area-selectively transferred to the substrate based on the pattern of the stamp. Once the stamp is removed, the material is patterned on the substrate in a well-defined pattern. This process is widely applied to microelectronics, surface chemistry, and cell biology.

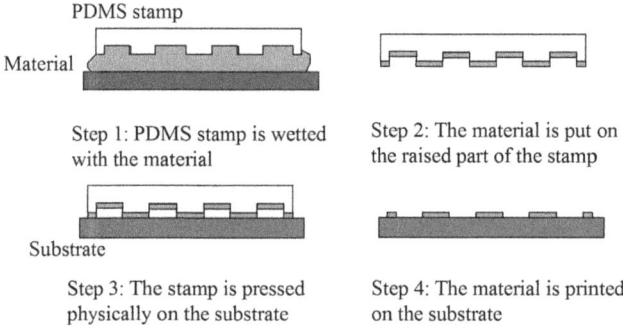

Figure 1.11   Illustration of the microcontact printing.

1.3.1.3   *Substrate Bonding*

As anodic bonding and fusion bonding are not applicable for polymeric materials, alternative substrate bonding processes operated in low temperatures are required. For bonding PDMS layers, oxygen plasma bonding is a widely used method to fabricate multilayer PDMS microfluidic devices.[31] An example of fabricating enclosed microfluidic channels in oxidised PDMS is shown in Figure 1.12. To form enclosed channels, oxidising PDMS surfaces in a plasma discharge for 1 min and bringing them into conformal contact can achieve an irreversible seal. The bonding is sufficiently strong that the two PDMS substrates cannot be peeled apart without failure in cohesion of the bulk PDMS.[32] For bonding thermoplastic layers, thermal compression and gluing are widely used for polymer-to-polymer substrate bonding. The thermal compression technique bonds polymer substrates by compression at temperatures elevated to the glass transition. The gluing can be achieved by application of a layer of glue, *e.g.*, resins or UV-curable materials, between the polymer substrates. These bonding techniques are simple and do not require sophisticated apparatus. The drawbacks are that they may induce global and localised geometric deformation of the substrates or leave an interfacial layer with significant thickness variation. For microchannels of a few hundred micrometres width, these drawbacks are tolerable. However, constructing micro- or nanosized channels is implausible since significant global and local deformation may distort the channel geometries. Alternatively, localised welding can be adopted for bonding thermoplastic substrates. The mating parts beside the microchannels are melted by either ultrasonics,[33] microwaves,[34] or an infrared laser.[35] An advantage of these methods is that energy is applied for targeted specific regions locally or uniform bonding of all mating interfaces, therefore the channel geometries are maintained and high-resolution local bonding can be achieved. However, these techniques have not been widely adopted for fabricating microfluidic devices because the requirement of special designs of the mating parts complicates the use of these techniques for efficient chip prototyping.

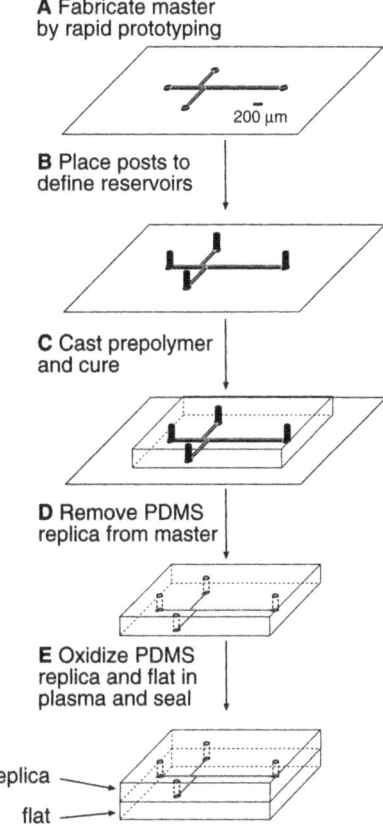

Figure 1.12   Description of fabricating enclose microfluidic channel in oxidised PDMS. (A) A high-resolution transparency containing the design of the channels was used as the mask in photolithography to produce a positive relief of photoresist on a silicon wafer. (B) Glass posts were placed on the wafer to define reservoirs. (C) PDMS was then cast onto the silicon wafer and cured at 65 °C for 1 h. (D) The polymer replica of the master containing a negative relief of channels was peeled away from the silicon wafer, and the glass posts were removed. (E) The PDMS replica and a flat slab of PDMS were oxidised in a plasma discharge for 1 min. Irreversible seal was formed between PDMS layers. [Reprinted with permission from ref. 31. Copyright (1998) American Chemical Society.]

1.3.2   Application Examples

1.3.2.1   PDMS-Based Microfluidic Devices

By bonding multilayers of elastomeric materials, *e.g.*, PDMS, microfluidic actuators including on–off valves, switching valves and pumps can be fabricated on-chip.[36,37] Crossed-channel architectures are used to build the

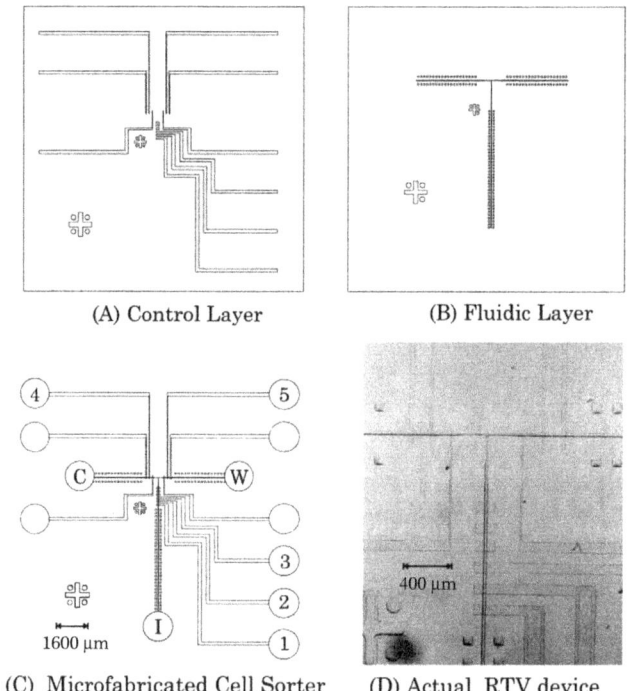

(A) Control Layer   (B) Fluidic Layer

(C) Microfabricated Cell Sorter   (D) Actual RTV device

Figure 1.13   Integrated cell sorter made of two different elastomeric layers bonded together. (A) The layout of the control layer. (B) The layout of the fluidic layer. (C) The integrated cell sorter. Valves 1–3 worked as a peristaltic pump and valves 4 and 5 worked as switch valves. Holes labelled as I, C, and W were the input, collection, and waste wells, respectively. (D) A snapshot of the integrated cell sorter. [Reprinted with permission from ref. 38. Copyright (2002) American Chemical Society.]

basic element of the actuators. The polymer membrane between the channels is engineered to be relatively thin, *e.g.*, typically 30 μm. When pressure is applied to the upper channel, the membrane deflects downward and closes the lower channel. Based on this architecture, an integrated microfabricated cell sorter was developed.[38] Switching valves, dampers, and peristaltic pumps were integrated for sorting, sample dispensing, flushing, recovery, and adsorption of any fluidic perturbation. The cell sorter is shown in Figure 1.13. Two different elastomeric layers, *i.e.* control layer and fluidic layer, were bonded together to build the cell sorter. The control layer contained lines where pressurised nitrogen and vacuum were introduced to actuate the closing and opening of the valves, respectively. The fluidic layer contained lines where the sample was injected and the cells were collected. Valves 1–3 work as a peristaltic pump and valves 4 and 5 work as switch valves. Holes labelled as I, C, and W were the input, collection, and waste wells, respectively. Since the pumping rate and the valve switching rate can

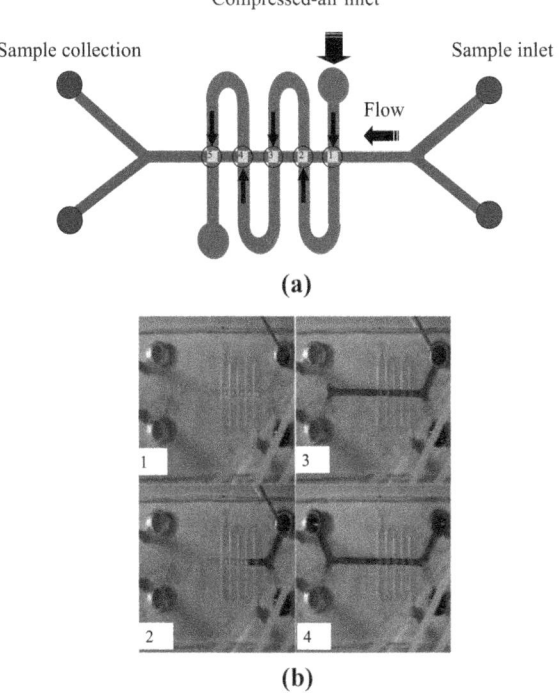

Figure 1.14   The S-shaped pneumatic micropump. (a) Illustration of the double-layer PDMS structure. (b) Photographs of the pumping operation. [Reprinted with permission from ref. 39. Copyright (2007) Springer.]

be changed at any time, cell sorting using this device was demonstrated in a variety of ways. Moreover, one S-shaped air channel across the fluidic channel could accomplish the same peristaltic pumping, as shown in Figure 1.14. The peristaltic effect driven by the time-phased deflection of PDMS membranes along the fluidic channel can be used to deliver liquid. The S-shaped pneumatic micropump has been used to drive samples for the study of cell separation and nucleus collection using dielectrophoretic (DEP) forces.[39] The pumping rate can achieve 39.8 µl min^{-1} at a driving frequency of 28 Hz under a pressure of 25 psi. The viable and nonviable cells can be separated by DEP forces and collected respectively in their specific reservoir. In addition, after cell lysis, the nucleus can also be collected using a similar scheme.

Simple PDMS–glass microfluidic devices are also widely used for detection science. An electrokinetic controlled DNA hybridisation micro-fluidic device has been developed and can perform all processes from sample dispensation to hybridisation detection within 5 min.[40] The device consists of a PDMS upper substrate that provides an H-type channel structure for fluidic transportation and a lower glass substrate that serves

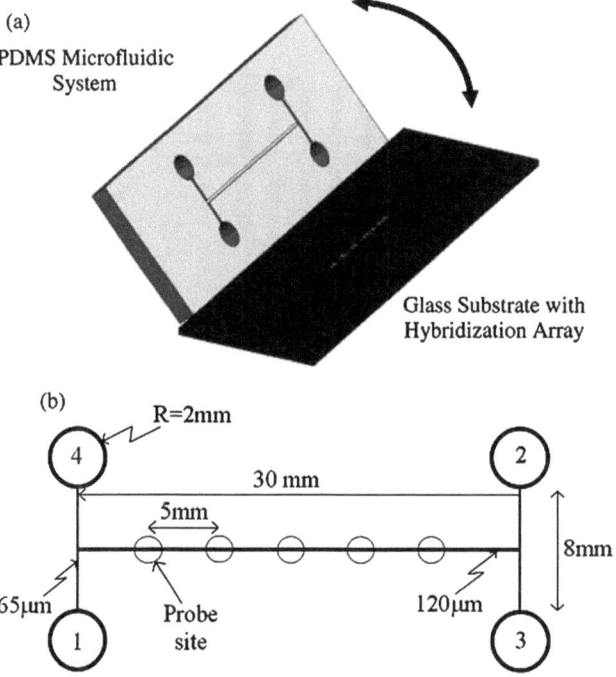

Figure 1.15 (a) Assembly procedure for PDMS H-type channel and immobilised
hybridisation array. (b) The location of the probe sites and the ports:
(1) sample port, (2) auxiliary port, (3) buffer port, and (4) wash port.
[Reprinted with permission from ref. 40. Copyright (2004) American
Chemical Society.]

as a solid support for the immobilised single-stranded oligonucleotide
probe, as shown in Figure 1.15. The electro-osmotic pumping driven by
the electrodes in four reservoirs can dispense controlled samples of
nanolitre volumes directly to the hybridisation array and remove
nonspecific adsorption. Hybridisation, washing, and scanning procedures
can be conducted simultaneously. Detection levels as low as 50 pM were
recorded using an epifluorescence microscope. Alternatively, a high aspect
ratio microfluidic device was developed for culturing cells inside an array
of microchambers with continuous perfusion of media.[41] The PDMS layer
shown in Figure 1.16 consists of circular microfluidic chambers of 40 μm
in height surrounded by multiple narrow perfusion channels of 2 μm in
height. This setting offered the advantage of a uniform flow profile for
medium perfusion and the creation of a homogeneous culture microenvironment. The PDMS layer was bonded to a glass substrate to build the
microfluidic cell-culture device, and the device was shown to culture
human carcinoma (HeLa) cells with continuous medium perfusion at
37 °C for 7.5 days, as shown in Figure 1.17.

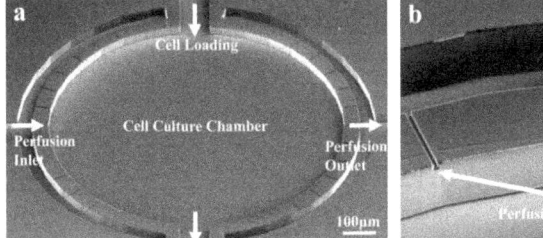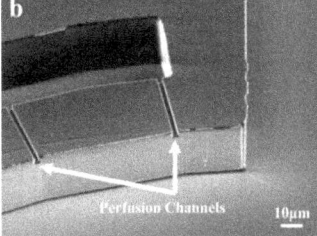

Figure 1.16 High aspect ratio microfluidic cell culture chamber. (a) Single unit of the cell culture chamber. (b) Multiple perfusion channels surrounding the main culture chamber. [Reprinted from ref. 41.]

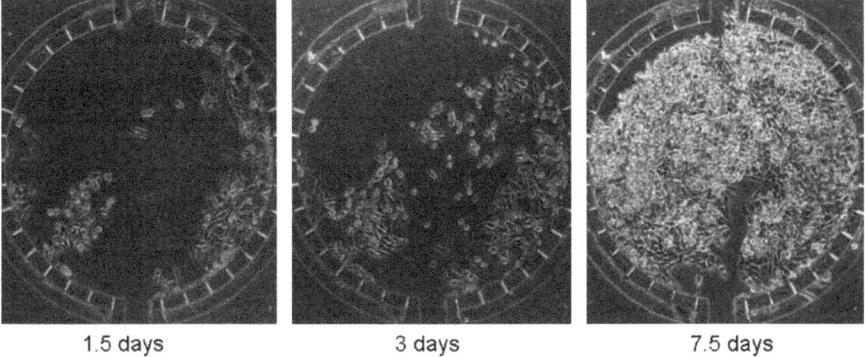

1.5 days   3 days   7.5 days

Figure 1.17 Cell growth inside the microfluidic chamber array. [Reprinted from ref. 41.]

1.3.2.2 Thermoplastic-Based Microfluidic Devices

Thermoplastic-based microfluidic devices can have great commercial potential with mass production being a critical factor in development. Immuno-assay on a compact disc (CD) has been reported and is expected to be a powerful platform for medical and clinical diagnostics.[42,43] Fluids on CD can be pumped by the centrifugal force controlled by the rotational speed of the CD. Several microfluidic functions, *e.g.*, capillary valving, centrifugal pumping, and flow sequencing can be performed on the CD-based microfluidic device. As shown in Figure 1.18, simultaneous and identical assays in parallel layouts can be fabricated on a single CD platform. Enzyme-linked immuno-sorbent assay (ELISA) was demonstrated and high-throughput screening (HTS) of analytes could be realised. Centrifugal and capillary forces were used to control the flow sequence of different solutions involved in the ELISA process. An analysis of rat IgG from hybridoma cell culture showed that the same detection range as the conventional method on the 96-well plate had been obtained with the advantages of lower reagent consumption and shorter assay time.[42] Another demonstration was to determine α-fetoprotein (AFP),

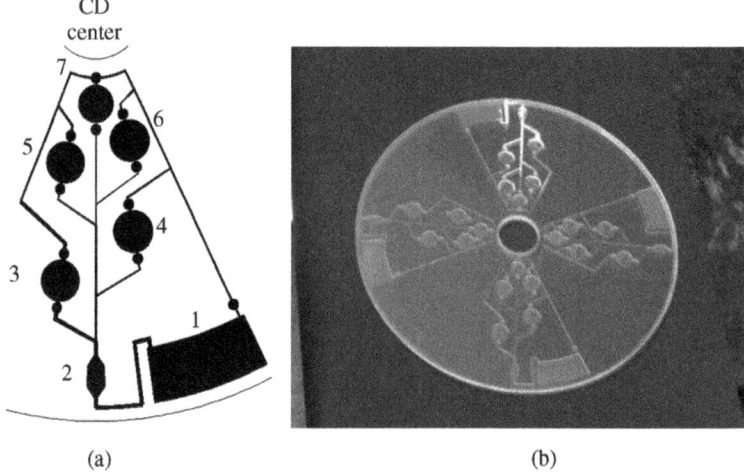

(a)   (b)

Figure 1.18   (a) The layout of the CD-based microfluidic device. (b) Photograph of a CNC-machined CD. [Reprinted with permission from ref. 42. Copyright (2004) American Chemical Society.]

interleukin-6 (IL-6), and carcinoembryonic antigen (CEA).[43] A 200-nl sample was applied to the CD-based microfluidic device and passed through the microchannel packed with antibodies against AFP, IL-6, and CEA. Alexa 647-labelled detection antibody was used for the detection. The flow rate was controlled by altering the rotational speed and the results were measured by a laser-induced fluorescent detector. The detection limits for AFP, IL-6, and CEA were 0.15, 1.25, and 1.31 pmol l^{-1}, respectively, with up to 104 sandwich-type immunoassays being completed within 50 min. Excellent analytical efficiencies within acceptable variations was demonstrated when compared with the traditional assays performed by 96-well ELISA plate.

1.4   Paper-Based Microfluidic Devices

Microfluidic devices made of paper have been proposed recently for a new class of diagnostic devices.[44] Most of the fluid functions including sample manipulation and detection can be completed within a sheet of paper. Paper is inexpensive, thin, light-weight, available in a wide range of thickness, and disposable, thus paper-based microfluidic devices are suitable for the development of diagnostic assays in developing countries and harsh environments.

1.4.1   Fabrication Techniques

In order to manipulate fluids along the desired direction in paper, hydrophobic barriers are patterned to realise the paper-based microfluidic device. Thus, the patterned barriers define the shape, *i.e.* width and length, of the channels and the thickness of the paper defines the height of the channels.

Aqueous solutions can be transported passively along the channels by wicking through the hydrophilic fibres of paper. There are several methods to pattern the hydrophobic barriers on paper, such as photolithography,[45,46] wax printing,[47,48] PDMS application,[49] and plasma treatment.[50] These fabrication techniques are described below.

Photolithography was demonstrated to pattern filter paper, thus defining the microchannels.[45] SU-8 3025 photoresist was utilised and spin-coated over the paper. The photoresist-coated paper was then baked at 95 °C for 10 min. After baking, UV exposure was performed with a photomask to define the pattern. Following the baking at 95 °C for 1–3 min, unpolymerised photoresist was removed from the paper by submergence in acetone for 1 min. Finally, the paper was dried under ambient conditions for 1 h. Areas covered with photoresist remain hydrophobic, while areas without photoresist were hydrophilic. The paper-based microfluidic device patterned by photolithography is shown in Figure 1.19.

Wax printing was reported to make features for paper-based microfluidic devices.[47,48] Figure 1.20 shows the patterning process of hydrophobic barriers in paper by wax printing. After designing the patterns of the hydrophobic barriers by computer-aided design (CAD) software, filter paper is printed on by a solid ink printer. The printed paper is then placed on a digital hot plate set at 150 °C for 120 s. When the wax on the surface of the paper melts, it spreads vertically as well as laterally into the paper. The vertical spreading creates the hydrophobic barrier across the thickness of the paper. However, the lateral spreading decreases the resolution of the printed pattern resulting

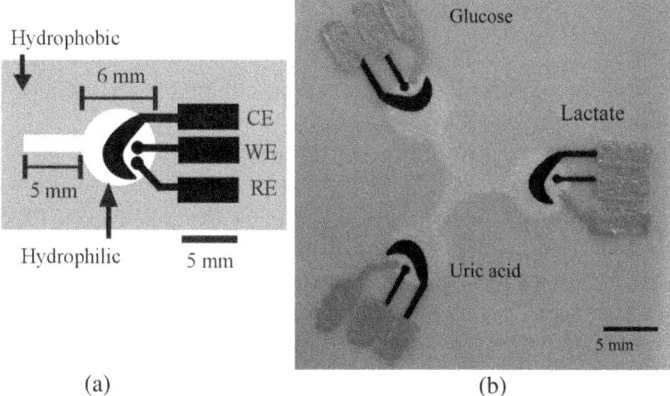

(a) (b)

Figure 1.19 (a) Design of the electrochemical detection cell for paper-based microfluidic device. WE: working electrode; RE: reference electrode; CE: counter electrode. (b) Photography of three electrode paper-based microfluidic device. The hydrophilic area (white colour) at the centre of the device wicks sample into the three separate test zones where independent enzyme reactions occur. [Reprinted with permission from ref. 45. Copyright (2009) American Chemical Society.]

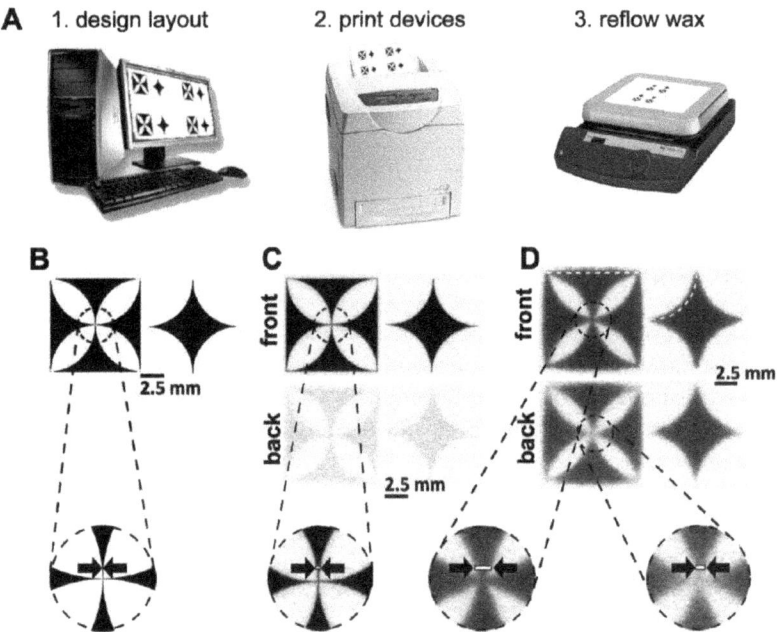

Figure 1.20 The patterning process of hydrophobic barriers in paper using wax printing. (A) Schematic representation of the basic steps required for wax printing. (B) Digital image of a test design. (C) Images of the test design printed on filter paper using the solid ink printer. (D) Images of the test design after heating the paper. [Reprinted with permission from ref. 47. Copyright (2009) American Chemical Society.]

in hydrophobic barriers that are wider than the original printed patterns. Thus, the resolution of the features patterned by wax printing is not as high as those generated by photolithography.

PDMS has been used to pattern paper and is significantly more flexible than photoresist.[49] PDMS dissolved in hexanes is printed on filter paper to define the hydrophobic barriers. After curing, aqueous solutions can only wet the hydrophilic areas in the paper, and no liquid wicks through the PDMS barriers. Figure 1.21 shows a dip-star design that makes it possible to fold the channel system in the paper and dip it into a small volume of sample.

Plasma treatment has been reported to make patterns on paper.[50] The filter paper is first hydrophobised using alkyl ketene dimer (AKD)–heptane solution. The paper is dipped in this solution and immediately removed and placed in a fume cupboard to allow evaporation of the heptane. It is then heated in an oven at 100 °C for 45 min to cure the AKD, causing the treated paper to become strongly hydrophobic. Subsequently, the paper is sandwiched between metal masks having the desired patterns and placed into a vacuum plasma reactor to define the hydrophilic patterns. The plasma-treated areas are strongly wettable by aqueous solutions. The pattern on paper, patterned by plasma treatment, is shown in Figure 1.22.

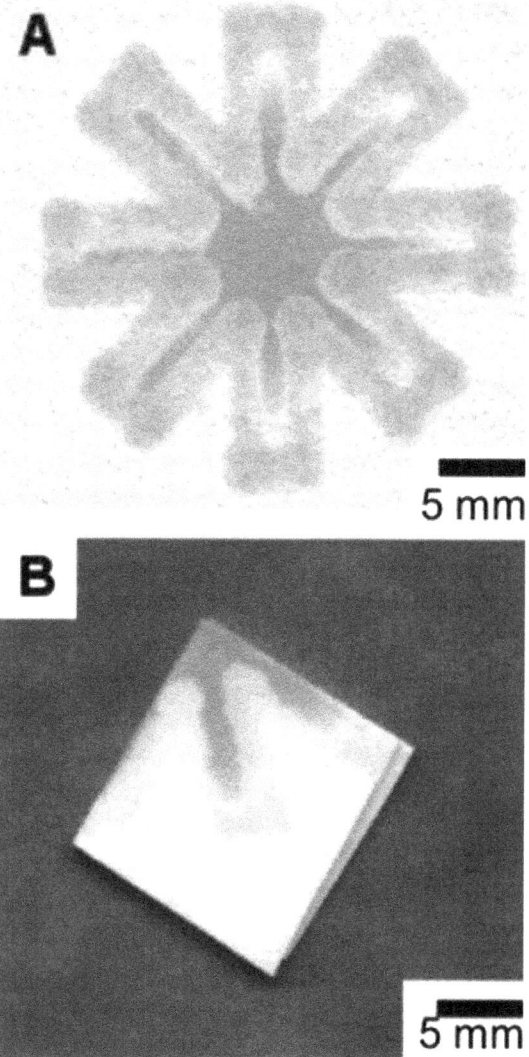

Figure 1.21 Hydrophobic barriers in paper patterned by PDMS. (A) The unfolded printed paper after dipping. (B) The folded printed paper after dipping. [Reprinted with permission from ref. 49. Copyright (2008) American Chemical Society.]

1.4.2 Application Examples

A paper-based microfluidic device has been demonstrated for the determination of glucose and protein simultaneously.[51] A single piece of paper was patterned by photolithography to realise the paper-based assays. The concentration of glucose and protein was determined by matching the colour intensity of the reacted sites and the printed colour on label stock visually.

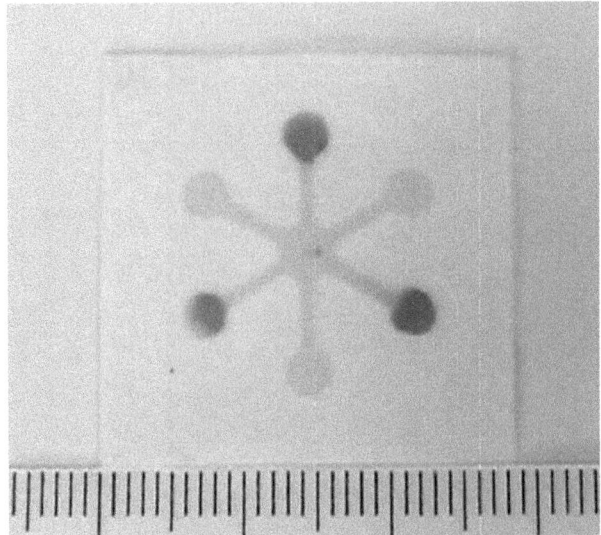

Figure 1.22 A paper-based microfluidic device fabricated by plasma treatment. [Reprinted with permission from ref. 50. Copyright (2008) American Chemical Society.]

However, differentiation of colour intensity by the naked eye was complicated due to the influences of subjective colour perceptions from individuals and lighting. In order to pursue more quantitative analyses based on paper-based microfluidic devices, a camera phone was used for image capturing of the reacted sites and the colour intensity was analysed by the computer software.[52] The strategy is illustrated in Figure 1.23. The idea behind this is to provide real-time and off-site diagnosis, operated by unskilled personnel, in the developing countries, with a higher degree of reliability in results compared to visual inspections. However, the colour intensity of the digital image captured by the camera phone was affected by lighting conditions and the calibration curve of colour from different camera phones. Therefore, electrochemical detection on paper-based microfluidic devices has been proposed for higher analytical sensitivity and selectivity.[45,53] This has been demonstrated in the determination of glucose, lactate, and uric acid in biological samples.[45] The detection electrodes were screen-printed by carbon and Ag/AgCl ink onto a piece of paper. These examples are the most recent developments of the paper-based microfluidics, showing great potential to be good alternatives over traditional microfluidic systems for point-of-care diagnostic applications.

1.5   Summary

The fabrication techniques for nano- and microfluidic devices have matured and excellent detection applications have been demonstrated in the laboratory. From silicon, glass/polymer, to paper-based microfluidic devices, one of

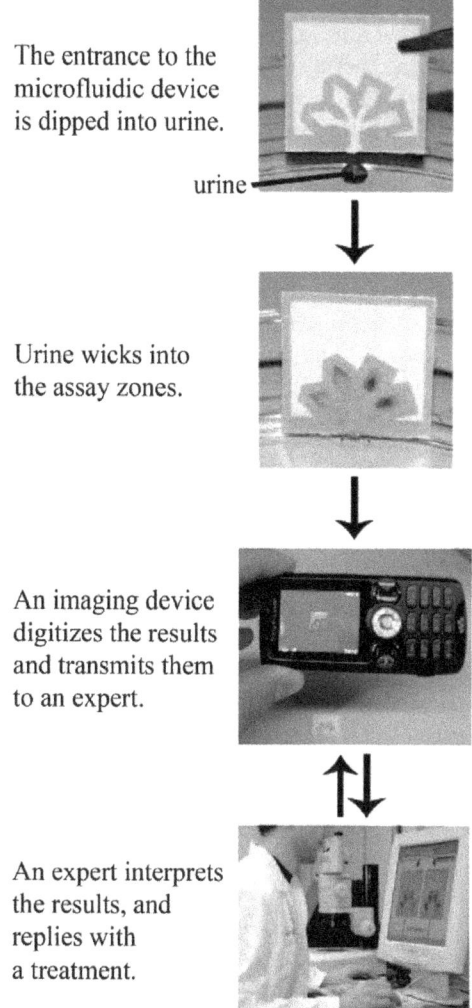

The entrance to the microfluidic device is dipped into urine.

urine

Urine wicks into the assay zones.

An imaging device digitizes the results and transmits them to an expert.

An expert interprets the results, and replies with a treatment.

Figure 1.23 General strategy for performing inexpensive paper-based bioassays in remote locations and for exchanging the results of the tests with off-site technicians. [Reprinted with permission from ref. 52. Copyright (2008) American Chemical Society.]

the main objectives is to provide a portable sample-to-answer diagnostic device, mainly for unskilled personnel. However, microfluidic devices have not yet made a significant impact in the commercial market. A possible reason for this is that these newly developed portable diagnostic devices need time to compete with existing, well-validated, bench-top equipment which have been continually improved on over decades. Moreover, there are still some challenges in the development of microfluidic technology applications, such as purification of raw samples (pretreatment), effective concentration of analytes, reduction of dead volume in microfluidic channels, and the elimination of external bulky and expensive detection systems. In addition, launching of *in vitro* diagnostic devices requires regulatory approval from different countries around the world. These factors can greatly hinder the development of commercial microfluidic diagnostic devices, particularly those arising from smaller research laboratories. However, the range of promising demonstrations of current microfluidic devices, particularly in the biomedical sphere provides a solid foundation for the development of commercial microfluidic diagnostic devices, which are increasingly finding interest by well-established life science and pharmaceutical firms.

References

1. A. Manz, N. Graber and H. M. Widmer, *Sens. Actuators B,* 1990, **1**, 244–248.
2. A. Luque, J. M. Quero, C. Hibert, P. Fluckiger and A. M. Ganan-Calvo, *Sens. Actuators A,* 2005, **118**(1), 144–151.
3. Y. Li, T. Pfohl, J. H. Kim, M. Yasa, Z. Wen, M. W. Kim and C. R. Safinya, *Biomed. Microdevices,* 2001, **3**(3), 239–244.
4. N. R. Harris, M. Hill, S. Beeby, Y. Shen, N. M. White, J. J. Hawkes and W. T. Coakley, *Sens. Actuators B,* 2003, **95**, 425–434.
5. P. Pal and K. Sato, *J. Micromech. Microeng.,* 2009, **19**, 055003–055013.
6. F. Marty, L. Rousseau, B. Saadany, B. Mercier, O. Francais, Y. Mita and T. Bourouina, *Mciroelectron. J.,* 2005, **36**(7), 673–677.
7. A. Daridon, V. Fascio, J. Lichtenberg, R. Wütrich, H. Langen, E. Verpoorte and N. F. de Rooij, *Fresenius' J. Anal. Chem.,* 2001, **371**(2), 261–269.
8. M. A. Unger, H. P. Chou, T. Thorsen, A. Scherer and S. R. Quake, *Science,* 2000, **288**, 113–116.
9. A. W. Martinez, S. T. Philllps and G. M. Whitesides, *Anal. Chem.,* 2010, **82**, 3–10.
10. R. Pelton, *Trends Anal. Chem.,* 2009, **28**(8), 925–942.
11. W. Zhao and A. van den Berg, *Lab Chip,* 2008, **8**, 1988–1991.
12. W. Dungchai, O. Challapakul and C. S. Henry, *Anal. Chem.,* 2009, **81**, 5821–5826.
13. G. E. Moore, *Electronics,* 1965, **38**(8), 114–117.
14. K. E. Petersen, *Proc. IEEE,* 1982, **70**(5), 420–457.
15. G. T. A. Kovacs, *Micromachined Transducers Sourcebook*, McGraw-Hill, 1998, p. 104.
16. W. Kern and D. A. Puotinen, *RCA Rev.,* 1970, **31**, 187–206.
17. V. A. Yunkin, D. Fischer and E. Voges, *Microelectron. Eng.,* 1994, **23**, 373–376.
18. M. Esashi, M. Takinami, Y. Wakabayashi and K. Minami, *J. Micromech. Microeng.,* 1995, **5**, 5–10.
19. R. Zengerle, J. Ulrich, S. Kluge, M. Richter and A. Richter, *Sens. Actuators A,* 1995, **50**(1–2), 81–86.
20. R. Linnermann, P. Woias, C.-D. Senfft and J. A. Ditterich, *Proc. MEMS 98, Heidelberg, Germany*, January 25–29, 1998, pp. 532–537.

21. M. Richter, R. Linnemann and P. Woias, *Sens. Actuators A,* 1998, **68**, 480–486.
22. J. G. Smits, *Sens. Actuators A,* 1990, **21**(1–3), 203–206.
23. M. Bu, T. Melvin, G. Ensell, J. S. Wilkinson and A. G. R. Evans, *J. Micromech. Microeng.,* 2003, **13**, S125–S130.
24. S. S. W. Yeung, T. M. H. Lee and I.-M. Hsing, *Anal. Chem.,* 2008, **80**, 363–368.
25. Y. Xia and G. M. Whitesides, *Ann. Rev. Mater. Sci.,* 1998, **28**, 153–184.
26. A. Gerlach, G. Knebel, A. E. Guber, M. Heckele, D. Herrmann, A. Muslija and Th Schaller, *Microsyst. Technol.,* 2002, **7**, 265–268.
27. A. Bernard, J. P. Renault, B. Michel, H. R. Bosshard and E. Delamarche, *Adv. Mater.,* 2000, **12**, 1067–1070.
28. T. Matsui, K. Komatsu, O. Sugihara and T. Kaino, *Opt. Lett.,* 2005, **30**(9), 970–972.
29. Y. Xia, E. Kim and G. M. Whitesides, *Chem. Mater.,* 1996, **8**(7), 1558–1567.
30. http://www.dowcorning.com/applications/search/default.aspx?r=131en.
31. D. C. Duffy, J. C. McDonald, O. J. A. Schueller and G. M. Whitesides, *Anal. Chem.,* 1998, **70**, 4974–4984.
32. M. K. Chaudhury and G. M. Whitesides, *Langmuir,* 1991, **7**(5), 1013–1025.
33. R. Truckenmuller, R. Ahrens, Y. Cheng, G. Fischer and V. Saile, *Sens. Actuators A,* 2006, **132**(1), 385–392.
34. K. F. Lei, S. Ahsan, N. Budraa, W. J. Li and J. D. Mai, *Sens. Actuators A,* 2004, **114**, 340–346.
35. J. Kim and X. Xu, *J. Laser Appl.,* 2003, **15**, 255.
36. M. A. Unger, H.-P. Chou, T. Thorsen, A. Scherer and S. R. Quake, *Science,* 2000, **288**, 113–116.
37. T. Thorsen, S. J. Maerkl and S. R. Quake, *Science,* 2002, **298**, 580–584.
38. A. Y. Fu, H.-P. Chou, C. Spence, F. H. Arnold and S. R. Quake, *Anal. Chem.,* 2002, **74**, 2451–2457.
39. C.-H. Tai, S.-K. Hsiung, C.-Y. Chen, M.-L. Tsai and G.-B. Lee, *Biomed. Microdevices,* 2007, **9**, 533–543.
40. D. Erickson, X. Liu, U. Krull and D. Li, *Anal. Chem.,* 2004, **76**, 7269–7277.
41. P. J. Hung, P. J. Lee, P. Sabounchi, N. Aghdam, R. Lin and L. P. Lee, *Lab Chip,* 2005, **5**, 44–48.
42. S. Lai, S. Wang, J. Luo, L. J. Lee, S. T. Yang and M. J. Madou, *Anal. Chem.,* 2004, **76**, 1832–1837.
43. N. Honda, U. Lindberg, P. Andersson, S. Hoffmann and H. Takei, *Clinical Chem.,* 2005, **51**, 1955–1961.
44. A. W. Martinez, S. T. Philips and G. M. Whitesides, *Anal. Chem.,* 2010, **76**, 1824–1831.
45. W. Dungchai, O. Challapakul and C. S. Henry, *Anal. Chem.,* 2009, **81**, 5821–5826.
46. A. W. Martinez, S. T. Philips, B. J. Wiley, M. Gupta and G. M. Whitesides, *Lab Chip,* 2008, **8**, 2146–2150.
47. E. Carrilho, A. W. Martinez and G. M. Whitesides, *Anal. Chem.,* 2009, **81**, 7091–7095.

48. Y. Lu, W. Shi, L. Jiang, J. Qin and B. Lin, *Electrophoresis,* 2009, **30,** 1497–1500.
49. D. A. Bruzewicz, M. Reches and G. M. Whitesides, *Anal. Chem.,* 2008, **80,** 3387–3392.
50. X. Li, J. Tian, T. Nguyen and W. Shen, *Anal. Chem.,* 2006, **80,** 9131–9134.
51. A. W. Martinez, S. T. Phillips, M. J. Butte and G. M. Whitesides, *Angew. Chem. Int. Ed.,* 2007, **26,** 1338–1320.
52. A. W. Martinez, S. T. Phillips, E. Carrllho, S. W. Thomas III, H. Sindl and G. M. Whitesides, *Anal. Chem.,* 2008, **80,** 3699–3707.
53. Z. Nie, C. A. Nijhuis, J. Gong, X. Chen, A. Kumachev, A. W. Martinez, M. Narovlyansky and G. M. Whitesides, *Lab Chip,* 2010, **10,** 477–483.

CHAPTER 2

Microfluidics Theory in Practice

MATTHEW J. DAVIES*, MARCO P. C. MARQUES, AND
ANAND N. P. RADHAKRISHNAN

Department of Biochemical Engineering, University College London,
Torrington Place, London, WC1E 7JE, UK
*E-mail: matthew.davies@ucl.ac.uk

2.1 Microfluidics and the Importance of Scale

The requirement for more rapid, sensitive and accurate analyses, for the purposes of biodefense and molecular biology and advances in microelectronics drove the development of miniaturised analytical devices and microfluidics in general.[1,2] Higher sensitivity, throughput and better resolution give microfluidics the edge over other lab-scale analytical techniques, while the potential for reduction in the size of the complete system enabled development of field-deployable analytical microfluidic chips for detecting chemical and biological threats.

Micro-total analysis system, a term created in 1990 by Professor Andreas Manz,[3] was expanded by coining the phrase "lab-on-a-chip". This encompasses not only analytical systems, but also refers to the concatenation and reduction in size of common laboratory processes onto a single microfluidic device (see Figure 2.1).

In general, a fluid can be defined as a substance, "having particles that easily move and change their relative position without a separation of the mass and that easily yield to pressure".[5] However, it can be more technically stated, by comparison to solids, as a substance that is continually deformed

RSC Detection Science Series No. 5
Microfluidics in Detection Science: Lab-on-a-chip Technologies
Edited by Fatima H Labeed and Henry O Fatoyinbo
© The Royal Society of Chemistry 2015
Published by the Royal Society of Chemistry, www.rsc.org

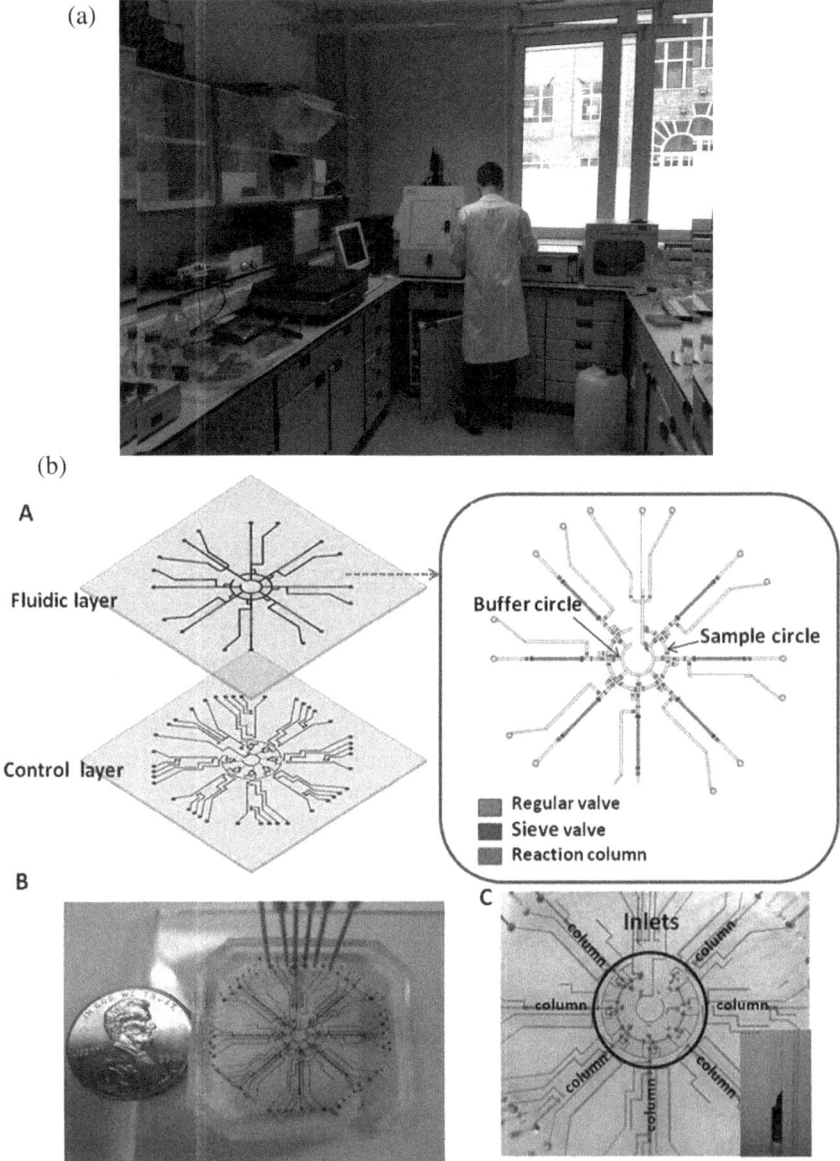

Figure 2.1 (a) Standard analytical laboratory and (b) Lab-on-a-chip device for immunoreaction detection of algal toxins (reproduced from ref. 4).

by a shear stress, while a solid deforms to a stable position. Therefore, both liquids and gases can be defined as fluids, with liquids generally considered incompressible and gases compressible. The study of fluids and the effect of forces upon them, namely fluid mechanics, are subdivided into three main branches, fluid statics, fluid kinematics, and fluid dynamics.

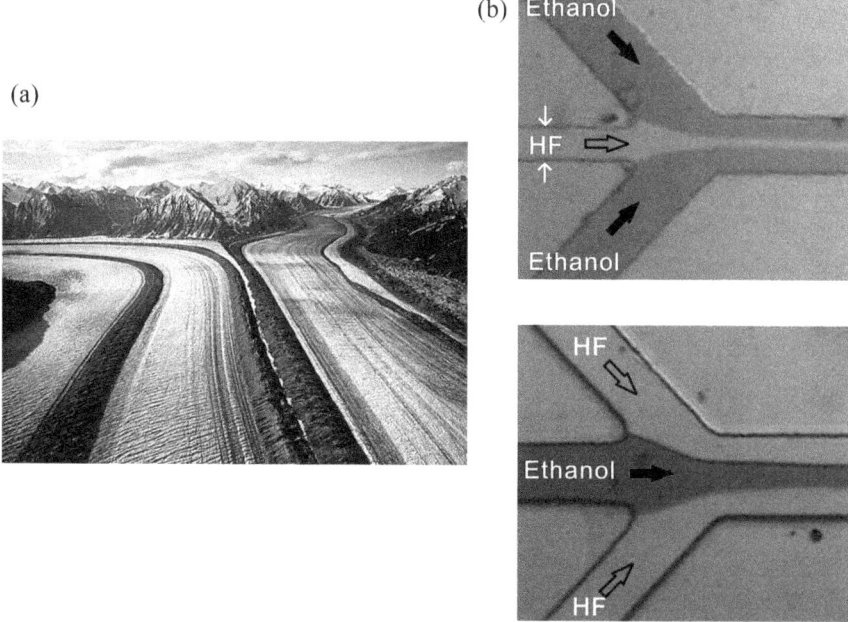

Figure 2.2 Laminar flow on (a) a macroscopic scale (glaciers) and (b) in a microfluidic device in which the two fluids are ethanol and a hydrofluoric acid solution (adapted from ref. 6).

Common experience with fluid behaviour is generally derived from flowing liquids in rivers, pipes or static bodies of water on the kilometre to the centimetre scale. Microfluidic devices are generally manufactured on the sub-mm to sub-100 micrometre range, though sub-1 micrometre dimensions can also be achieved. At these dimensions fluid dynamics phenomena such as laminar flow (see Figure 2.2) dominate and the interaction between the fluid and the surface of the substrate, through which fluid flows, becomes increasingly important. Examining this dimensional scaling on a range of relevant physical scales and fluid-dynamic effects (Table 2.1) reveals how commonly accepted fluid properties are affected with varying dimensional scales, especially at the microfluidic scale. For reasons that will be explained later, these effects may not necessarily be extendable to channels with nanometric dimensions.

While the device size and the fluid effects that dominate at these dimensions may introduce limitations, applying microfluidic technology to detection brings substantial benefits. Consideration must be given to both the difficulty of incorporating some detection methods into lab-on-a-chip devices and the domination of laminar over turbulent flow. This may lead to the requirement of incorporating specific design features to overcome the diffusion-limited mass transfer due to laminar flow, thereby raising

Table 2.1   Scaling laws: dimensional variation at changing length scales

Quantity	Scaling law
Time	l^0
Length	l^1
Area	l^2
Volume	l^3
Velocity	l^1
Acceleration	l^1
Density	l^{-3}
Viscosity	l^{-1}
Diffusion time	l^2
Reynolds number	l^2
Péclet number	l^2
Hydraulic resistance	l^{-4}

fabrication or material issues. However, the internal dimensions inherent to microfluidics enables detection with much smaller volumes of fluids than is usually possible with standard analytical technologies, and the more precise understanding of the fluid flow and physical conditions allows for a more detailed characterisation of the chosen detection method.

The high surface area to volume ratio intrinsic to microfluidics provides numerous advantages in the range of detection technologies, such as surface plasmon resonance (SPR), surface acoustic wave (SAW), microcantilever arrays, *etc.*, that become viable as surface effects are taken into consideration at this scale. However, while the internal dimensions of microfluidic devices are much lower than that of the macroscopic world we experience every day, they are not so small that the molecular nature of the fluid needs to be considered. For the purposes of applying the laws of fluid dynamics we instead consider a fluid "unit cell" that enables the average molecular properties (such as temperature and momentum) to be used.

If we consider a volume of an ideal fluid, of dimensions 100 μm × 100 μm × 100 μm or 1 nl, at standard temperature and pressure, the number of liquid molecules present is of the order of 1.56×10^{16} and the average intermolecular distance is 0.4 nm, while that of a gas at atmospheric pressure is 3.7×10^{13} and 3 nm respectively.[7] Reducing the number of liquid molecules to 100 000 would then be equivalent to a volume of fluid of sides 20 nm × 20 nm × 20 nm. The fluid "unit cell", while not having a defined size, has a minimum size below which molecular properties need to be taken into account and a maximum size over which external forces demonstrate a measurable effect. Hence, as previously indicated, the standard fluid effects may not be extendable to nanofluidic devices.

Differences in intermolecular distances are relevant to the differences not just in compressibility between gases and liquids, but also to the degree of interaction and thus cohesion between the molecules (see Table 2.2).

Table 2.2 Comparison of fluid (liquid and gas) properties (from ref. 7)

Properties	Liquid (water)	Gas (air)
Molecular size (nm)	0.3	0.3
Intermolecular distance (nm)	0.4	3
Mean free path length (nm)	0.2	61
Number density (m^{-3})	3×10^{28}	3×10^{25}
Density (g cm^{-3})	0.998	0.0012
Molecular velocity (m s^{-1})	$\approx 10^{3}$	500
Viscosity (g cm^{-1} s^{-1})	1.002	0.018

As a result of intermolecular distances in liquids being lower than gases, by approximately a factor of 10, they have higher intermolecular forces. Thus, liquids tend to maintain a constant volume and surface under gravity, while gas molecules tend to diffuse until evenly distributed within a container (with some exceptions, such as argon in which the gas will "flow"). However, despite these differences, and the relative lack of interaction between gas molecules, both gases and liquids will stratify under the influence of density differences in stable, multicomponent systems.

The interaction between two molecules can be modelled accurately using the Lennard-Jones potential (eqn (2.1)) in which V_{ij} is the Lennard-Jones potential, ε is the maximum energy of attraction, and σ and r are the collision diameter and intermolecular distance, respectively.[8]

$$V_{ij}(r) = 4\varepsilon \left[\left(\frac{r}{\sigma} \right)^{-12} - \left(\frac{r}{\sigma} \right)^{-6} \right] \qquad (2.1)$$

However, the fluid dynamics effects covered within this chapter do not require an understanding of the intermolecular theory to this depth and therefore the Lennard-Jones potential will not be covered to any greater extent. For a more in-depth discussion the reader is pointed to *Atkins, Physical Chemistry*.[9]

2.2 Flow Characterisation

2.2.1 Navier–Stokes Equations

Calculating fluid flow in streams is non-trivial the moment the channel varies from a straight, cylindrical channel with perfectly smooth walls. The Navier–Stokes equations combine the equations for the conservation of mass, momentum, and energy, using some of the fundamental properties of fluids to form equations from which complex fluid flow can theoretically be calculated. Derived from the conservation of mass, the continuity equation (eqn (2.2)), in which time (t), density (ρ), and velocity ($\mathbf{v}$), applies to incompressible fluids and therefore is relevant to most microfluidic systems. The conservation equations are vector equations and as such can be expanded and expressed in x, y, and z notation. However, using the gradient operator ($\nabla = \partial/\partial x_i$) and vector notation enables the equations to be simplified. When considering that density is constant and applying the gradient operator, the continuity equation simplifies to eqn (2.3).

$$\frac{\partial \rho}{\partial t} + \frac{\partial \rho}{\partial x_i} \cdot (\rho \cdot \mathbf{v}_i) = 0 \qquad (2.2)$$

$$\nabla \cdot \mathbf{v} = 0 \qquad (2.3)$$

Following conservation of mass, conservation of energy is expressed as shown in eqn (2.4), with e being the internal energy, P the pressure, q the heat flux, and τ the total stress tensor.

$$\frac{\partial}{\partial t}(\rho e) + \frac{\partial}{\partial x_i}(\rho \mathbf{v}_i e) = P\frac{\partial \mathbf{v}_i}{\partial x_i} + \tau_{ji}\frac{\partial \mathbf{v}_i}{\partial x_j} + \frac{\partial q_i}{\partial x_i} \qquad (2.4)$$

Conservation of momentum, the final basis for the Navier–Stokes equations applies the law within a fluid "unit cell", as detailed earlier, and extends it by expressing momentum as the product of density and velocity (eqn (2.5)). **F** in this case represents all external body forces (forces acting on a unit volume as supposed to contact forces) (*e.g.*, electrical, centrifugal, and gravitational). Eqn (2.5) can then be simplified to eqn (2.6) and, by considering that the $\mu\nabla\tau_{ji}$ term in an incompressible, homogenous fluid simplifies to $\mu\nabla^2\mathbf{v}$, reduces further to eqn (2.7), where μ is the dynamic viscosity.

$$\frac{\partial}{\partial t}(\rho \mathbf{v}_i) + \frac{\partial}{\partial x_j}(\rho \mathbf{v}_j \mathbf{v}_i) = \rho \mathbf{F}_i - \frac{\partial P}{\partial x_j} + \frac{\partial}{\partial x_j}\tau_{ji} \qquad (2.5)$$

$$\rho\left(\frac{\partial}{\partial t}\mathbf{v} + (\mathbf{v}\cdot\nabla)\mathbf{v}\right) = \rho\mathbf{F} - \nabla P + \nabla\tau_{ji} \qquad (2.6)$$

$$\rho\left(\frac{\partial}{\partial t}\mathbf{v} + (\mathbf{v}\cdot\nabla)\mathbf{v}\right) = \rho\mathbf{F} - \nabla P + \mu\nabla^2\mathbf{v} \qquad (2.7)$$

For eqns (2.5)–(2.7) to apply, several assumptions must be made including fluid continuity and incompressibility, and that the fluid is Newtonian and not moving at relativistic velocities. For almost all microfluidic applications and most likely all detection technologies, these are fairly safe assumptions.

The Navier–Stokes equation can also be used to derive other well-known means of characterising fluid flow. One of these is the Reynolds number (Re), a key dimensionless numbers used to define the degree of laminarity or turbulence of fluid flow (Figure 2.3). The Reynolds number can also be derived by considering it as the ratio between inertial forces and viscous forces. Inertial forces result from the interaction between the density, velocity, and the hydraulic diameter, D_h, of the channel through which the

(a)

(b)

(c)

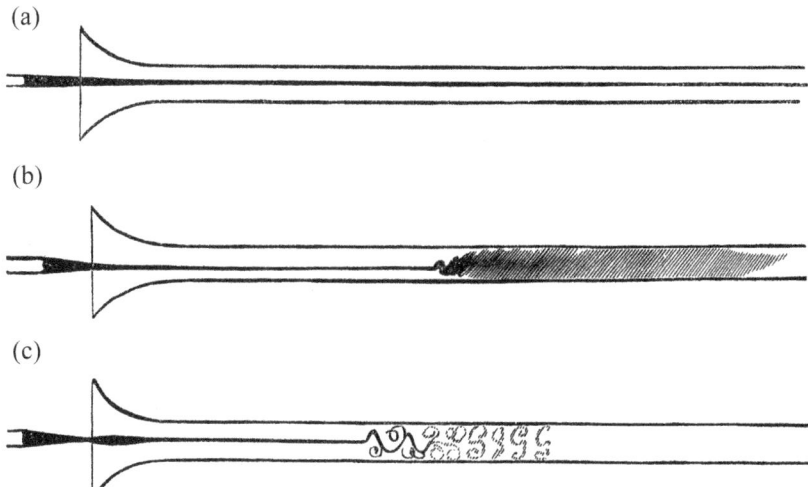

Figure 2.3 Reproductions of the original sketches drawn by Osbourne Reynold's. (a) Dye flowing from a hollow needle into a surrounding flow of water at low Reynolds number, (b) as for (a), but with fluid flowing at higher Reynolds number, and (c) as per (b), but drawn after illuminating the mixing pattern in (b) with a light flash (reproduced from ref. 10).

fluid is flowing (eqn (2.8)). Viscosity is the ratio of the shear stress (*i.e.* the applied force F acting on area A) to the shear rate (*i.e.* the instantaneous rate of change of the fluid velocity) represented by eqn (2.9). Viscous forces, however, are the effects that viscosity has on a fluid as the velocity of the fluid changes (eqn (2.10)).

$$\text{Inertial forces} = \mathbf{v}^2 D_{\text{h}}\rho \tag{2.8}$$

$$\mu = \frac{\text{shear stress}}{\text{shear rate}} = \frac{F/A}{d\mathbf{v}/dy} \tag{2.9}$$

$$\text{Viscous forces} = \mu \cdot \mathbf{v} \tag{2.10}$$

Simplifying the Navier–Stokes equation, by considering that in laminar flow conditions fluid flow becomes more predictable, enables the nonlinear term of the Navier–Stokes equation, $((\mathbf{v} \cdot \nabla)\mathbf{v})$, to be ignored. Determining when the boundary between laminar and turbulent flow occurs requires that the Navier–Stokes equation is expressed in dimensionless form. Redefining all physical variables as dimensionless and using characteristic length and velocity scales to replace the original variable dimensions, the Navier–Stokes equation is itself defined as dimensionless (eqn (2.11)). The Reynolds number (eqn (2.12)), from eqn (2.8) and (2.10), combines the characteristic length and velocity scales with density and viscosity to simplify the Navier–Stokes equation.

$$Re\left(\frac{\partial}{\partial t}\mathbf{v}^* + (\mathbf{v}^* \cdot \nabla^*)\mathbf{v}^*\right) = -\nabla^* P^* + \nabla^{*2}\mathbf{v}^* \tag{2.11}$$

$$Re = \frac{\text{inertial forces}}{\text{viscous forces}} = \frac{\rho v D_h}{\mu} \tag{2.12}$$

Following from eqn (2.8), the length scale in the equation for Reynolds number relates to the hydraulic diameter of the microchannel. In a channel with a circular cross section the hydraulic diameter is the channel diameter. However, the equation for hydraulic diameter (eqn (2.13)) enables the Reynolds number to be calculated for more complex microchannel cross-sectional shapes by inputting the cross-sectional area, A, and the wetted perimeter, P_w.

$$D_h = \frac{4A}{P_w} \tag{2.13}$$

A general solution to the Navier–Stokes equation has not been found, despite a US \$1 million prize being offered for a successful solution.[11] Several solutions have been calculated that apply to specific flow conditions. Three of these, Couette (see Figure 2.4), Poiseuille (see Figure 2.5(a)), and Stokes flow (see Figure 2.5(b)) are well characterised. Couette flow is the flow imparted to a fluid contained between two infinitely long plates by the movement of one of the plates relative to the other. Poiseuille flow is the steady-state flow generated in a fluid in a channel by pressure driven flow. Finally, Stokes flow, also known as creeping flow, is fluid flow occurring at very low Reynolds number, due to very low inertial forces compared to the viscous forces at small length scales.

As Couette flow is not deemed to be directly relevant to the theme of detection in microfluidics devices and there are multiple reference texts in which this is covered in greater detail,[12,13] no further discussion will be given. Poiseuille flow and particles moving in Stokes flow, however, are covered in more detail in later sections.

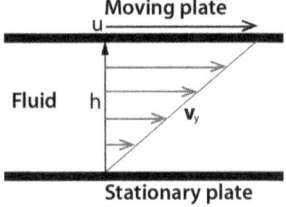

Figure 2.4   Couette flow schematic.

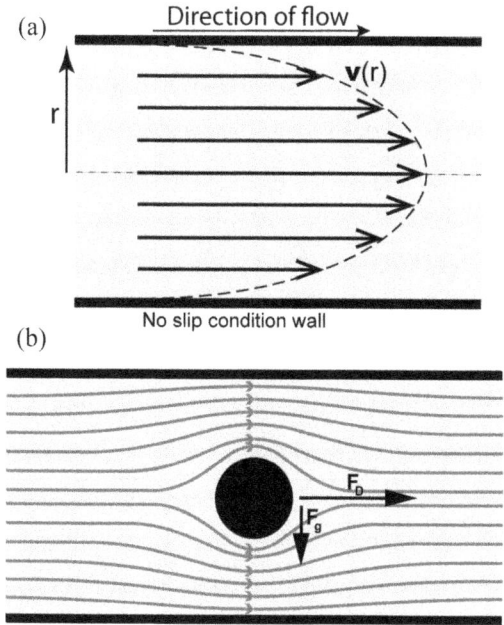

Figure 2.5 Schematic of (a) Poiseuille and (b) Stokes flow.

2.2.2 The Poiseuille Flow

Poiseuille flow is the standard pattern observed in pressure driven fluid flow in a microchannel, at Reynolds numbers for which laminar flow would be demonstrated. As shown in Figure 2.5(a), Poiseuille flow generates a characteristic parabolic fluid velocity flow profile for a Newtonian, incompressible fluid.

Fluid in the centre of a microchannel has the highest velocity with the velocity decreasing as the distance to the wall decreases. Finally, a fluid is generally considered to have zero velocity, known as the no-slip condition, at the walls of a microchannel (this is a simplification, however, for most conditions the idea holds). For a cylindrical microchannel the velocity profile can be represented as a series of ring-shaped lamellae (see Figure 2.6).

The velocity distribution under pressure driven flow is obviously dependent upon the cross-sectional shape of the microchannel. As a result, the Hagen–Poiseuille equation that enables the volumetric flow rate, Q, to be calculated as a function of the specified cross-sectional dimensions, viscosity μ, channel length L and applied pressure P, varies as well. Equations for a variety of channel cross sections are shown below with the associated fabrication methods (Table 2.3). Of course, the fabrication methods may not give exactly the cross section shown as there are always limitations in every method's resolution.

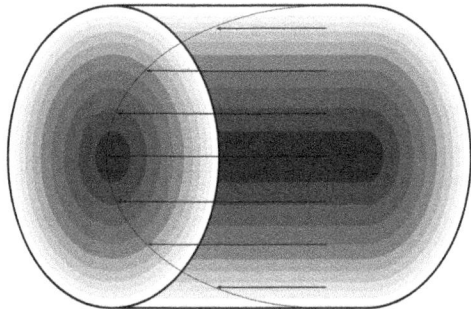

Figure 2.6 Cylindrical microchannel velocity profile.

Table 2.3 Poiseuille flow equations for a variety of channel cross sections with fabrication methods that produce those cross sections

Shape of cross section	Cylindrical	Rectangular	Triangular
Volumetric flow rate	$Q = \dfrac{\pi r^4 \Delta P}{8\mu L}$	$Q = \dfrac{h^3 w \Delta P}{12\mu L}\left(1 - 0.630\dfrac{h}{w}\right)$	$Q = \dfrac{\sqrt{3 \cdot a^4} \cdot \Delta P}{320\mu L}$
Common fabrication methods	Isotropic wet etching Ball-end mill Most tubing	Casting on a photolithographically fabricated mould Milling	Anisotropic wet etching

Determining the fluid velocity at a specific point within a microchannel's cross-section is possible by changing the appropriate Hagen–Poiseuille equation to the form demonstrated in eqn (2.14) for a circular microchannel.

$$\mathbf{v}(r) = \frac{\pi\left(r^2 - r_1^2\right)\Delta P}{4\mu L} \tag{2.14}$$

The reciprocal of the volumetric flow rate within a pressure-driven flow being pressure dependent is that a set flow rate requires a pressure differential between the applied pressure and the outlet pressure. As this "pressure drop" along a section of the channel increases, the corollary of this is that the resistance to fluid flow increases (eqn (2.15)), with the flow resistance, R, being a function of the flow rate, Q, and the pressure drop, ΔP.

$$R = \frac{\Delta P}{Q} \tag{2.15}$$

It can be observed that the pressure drop along the length of the channel will decrease as the measurements are made closer to the end of the channel, leading to the practical effect of the increasing hydraulic diameter of channels fabricated in flexible materials as the flow resistance increases. This is important when attempting to determine the relative fluid velocity at different distances along a long microchannel. Establishing stable flow rates also becomes an exercise in patience if a proportionally large stepwise reduction in flow rate is made, as the relaxation of the channel and any flexible tubing, have a secondary pumping effect.

The application of eqn (2.15) depends on the type of actuation driving the fluid in a system. Both syringe and peristaltic pumps operate by driving fluid at a particular flow rate. With Q and R constant, if the fluid system does not behave as intended it can be difficult to observe. For instance, if a poor connection has been made, leading to a pinch point and localised higher fluid resistance, a syringe pump doesn't change the flow rate as a result unless the force required to pump liquid through at the programmed flow rate is higher than the pump can apply. If hydraulic pumping is used then the applied pressure is controlled, with Q varying as a result. Unexpected behaviour, resulting from poor system assembly or device fabrication, becomes immediately apparent as the flow rate changes.

Fluid circuits can, to a first approximation, be compared to electrical circuits by considering certain analogies.[14] Comparing fluid and electrical resistance, pressure drop and voltage, current and flow rate, leads to equivalent equations for flow resistance (eqn (2.15)) and Ohm's law (eqn (2.16)), in which R is the electrical resistance. The previously mentioned secondary pumping effect, resulting from relaxation of a flexible substrate as the fluid pressure is released, can even be considered as similar to the effect of a capacitor.

$$R = \frac{V}{i} \tag{2.16}$$

This analogy can be further extended to the serial addition of resistances (eqn (2.17)) or the sum of the reciprocals of resistance (eqn (2.18)) that equally apply to electrical and hydraulic circuits (Figure 2.7).

$$R = R_1 + R_2 + \cdots + R_n \tag{2.17}$$

$$\frac{1}{R} = \frac{1}{R_1} + \frac{1}{R_2} + \cdots + \frac{1}{R_n} \tag{2.18}$$

Eqn (2.17) and (2.18) theoretically allow calculation of the total fluid resistance of a fluid system. However, realistically it only enables an approximate fluid resistance to be determined. Previous discussion on fluid flow was centred around a number of assumptions. One of these, that the flow is well developed and stable, is made due to "end effects" and the cyclical variability in many pumping systems' rates.

End effects, also known as entrance effects, are the observation that fluid entering or leaving a channel does not have a completely developed parabolic flow profile. As the majority of microfabrication methods lend themselves to single channel-layer devices, much of the functionality within devices is arrived at by changing the channel dimensions or by splitting a single channel into multiple channels. Fluid entering or leaving the channel is thus assumed to enter with a particular flow profile. As the fluid enters the new channel the

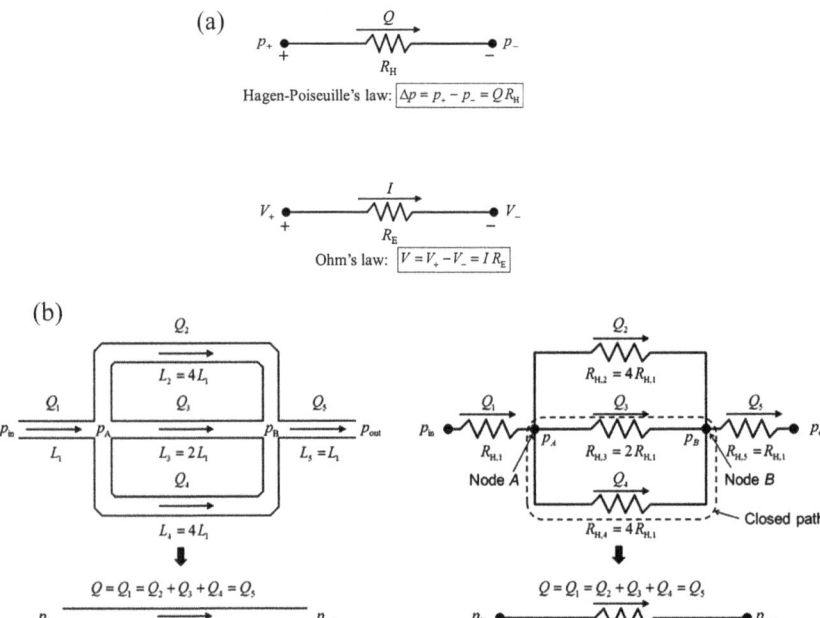

Figure 2.7   Schematic representations of: (a) fluidic and electrical resistance; and (b) the equivalent serial and parallel circuits (reproduced from ref. 14).

flow profile changes to suit the new channel dimensions. This change does not happen instantaneously, but rather develops over a certain distance.

Development of the fluid flow profile from the entrance is initially rapid, then tapers off very quickly as it approaches full development, but is never fully achieved. However, the flow profile is defined as fully developed when an arbitrary value of the fluid velocity at the centre of the channel reaches 99% of its maximum value. At low Reynolds numbers generally found in microfluidic devices, eqn (2.19)[15] provides the best agreement with experiments and relates the entrance length to Reynolds number and the hydraulic diameter.

$$\frac{L_c}{D_h} \approx \frac{0.6}{1 + 0.035 \cdot Re_{D_h}} + 0.056 \cdot Re_{D_h} \tag{2.19}$$

2.3   Two-Phase Systems Flow

Despite the first microfluidic device being a gas chromatograph, developed by Terry *et al.* at Stanford in 1979,[16] the majority of microfluidic devices have been developed solely for liquids to flow through the primary functional channel. While several devices have been designed specifically to enable both gas and liquids to be used as the operating fluid, in most microfluidic devices, the presence of gas bubbles would lead to device failure. Rather than physically breaking the device, bubbles instead prevent the intended application from being fulfilled by interfering with the device operation: by introducing recirculating flows; or reducing or even preventing fluid flow at the intended pressures.

In the field of detection the primary reason for including gas flow within a microfluidic device is to enable the detection of a component of the gas mixture,[17] normally a vapour. Several different microfluidic implementations of gas or vapour transfer into liquid have been developed. Broadly, they can generally be categorised into one of two formats: coflowing,[18] in which gas and liquid flow side by side; and segmented flow,[19] in which gas and liquid are either sequentially injected or gas is injected into a liquid flow.

While liquids are generally treated as incompressible, macroscopic experience with gases generally leads to them being considered compressible. The dimensionless Mach number defines the compressibility of a gas by determining the ratio between the local fluid velocity $\mathbf{v}$, and the speed of sound in air c_s (eqn (2.20)). A fluid can generally be treated as incompressible if the Mach number is less than about 0.2–0.3. While there are applications in microfluidics in which very high gas flow rates, relative to the liquid, are required, this would rarely be the case in detection applications. It is highly unlikely that gas velocities would approach the approximately 70 m s^{-1} necessary for a gas to be treated as compressible.

$$\mathrm{Ma} = \frac{\mathbf{v}}{c_s} \tag{2.20}$$

Gases, having an intermolecular distance approximately 10 times greater than that of liquids, require a larger "unit cell" (see Section 2.1) in order for the average molecular properties to be considered. While the device and "unit cell" dimensions may be different by several orders of magnitude, a larger "unit cell" may influence the flow conditions close to the channel wall, namely the no-slip condition.[8] The Knudsen number, Kn, the ratio of the mean free path length, λ, to a characteristic length, L_0, (eqn (2.21)) performs a similar function to the Reynolds number by providing approximate ranges within which flow conditions can be characterised. Likewise the Knudsen number, Mach number and Reynolds number can all be related by eqn (2.22), in which k is the ratio of the specific heats (ratio of heat capacity at constant pressure to heat capacity at constant volume). While the Knudsen number, as with the Mach number, is unlikely to be relevant for gas flow in detection applications, in rare cases where the gas density is low relative to the characteristic length scale (generally the hydraulic diameter) the Knudsen number may become important (Table 2.4).

$$\mathrm{Kn} = \frac{\lambda}{L_0} \tag{2.21}$$

$$\mathrm{Kn} = \sqrt{\frac{k\pi}{2}} \frac{\mathrm{Ma}}{\mathrm{Re}} \tag{2.22}$$

Table 2.4   Relationship between Knudsen number in different ranges and the Navier–Stokes equations

$Kn < 10^{-3}$	Navier–Stokes equations and no-slip condition valid.
$10^{-3} < Kn < 10^{-1}$	Navier–Stokes equations valid, slip flow regime.
$10^{-1} < Kn < 10$	Transitional flow regime in which Navier–Stokes equations must be modified.
$10 < Kn$	Navier–Stokes equations invalid, Boltzmann equations used instead.

Capillary flow is a common means of pumping fluid in lab-on-a-chip devices. Capillary flow results from the pressure differential that occurs between the liquid on one side of a meniscus and gas on the other and the interaction of the channel surface with the liquid. Surface tension is the tension created by the minimisation of the energy difference between molecules in the bulk of a liquid and those at the surface. The energy difference is caused by the energy of interaction between liquid molecules being lower than that between liquid molecules and gas at the liquid surface. This results in the liquid molecules at the surface having a higher energy than those in the bulk. The equations describing the surface tension, γ, can be expressed in a number of ways, but it can simply be considered as the force, F, necessary to stretch the meniscus by a certain distance, ΔL, (eqn (2.23)). While this does not describe the shape of the meniscus or the pressure difference between the liquid and gas sides of the surface, the Young–Laplace equation (eqn (2.24)) relates the 2-dimensional curvature of the meniscus, R_1 and R_2, and the surface tension (see Figure 2.8), to the pressure drop, ΔP, or for a circular channel (eqn (2.25)) solely, R, the radius of curvature of the meniscus.

$$\gamma = \frac{F}{\Delta L} \tag{2.23}$$

$$\Delta P = \gamma \left(\frac{1}{R_1} + \frac{1}{R_2} \right) \tag{2.24}$$

$$\Delta P = \gamma \frac{2}{R} \tag{2.25}$$

One extension of the Young–Laplace equation is the capillary rise of liquid in narrow tubes placed vertically over a liquid source. The liquid level is raised above the bulk level of the liquid (Figure 2.9) until the potential energy due to gravity is balanced by the energy change at the meniscus (eqn (2.26)). As the pressure drop is dependent on the channel dimensions, any change in channel geometry will affect the priming or

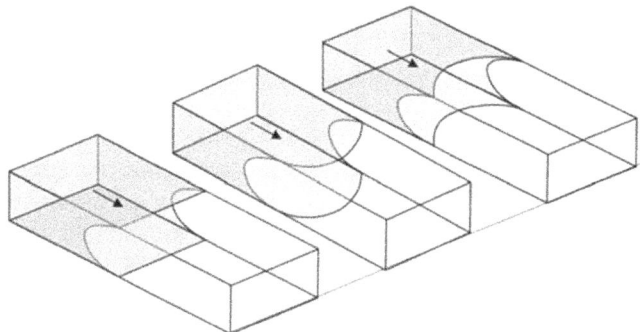

Figure 2.8 Meniscus shape and thus surface tension change due to the contact angle change resulting from changing hydrophilicity of the device substrate (reprinted with kind permission from Springer Science + Business Media from ref. 20).

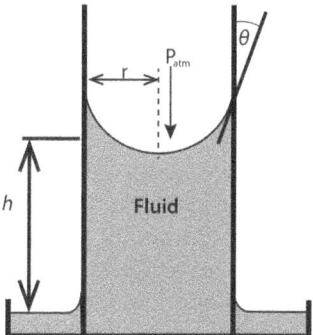

Figure 2.9 Schematic illustration of capillary rise in a narrow tube resulting from the pressure differential of the meniscus within the tube.

pumping of a fluid in a microchannel *via* capillary forces, as shown in Figure 2.9.

$$h = \frac{2\gamma \cdot \cos\theta}{\rho g r} \tag{2.26}$$

The Young–Laplace equation not only describes the meniscus shape at a gas–liquid interface, but also at interfaces between immiscible liquids. This has implications for digital microfluidics, and this will be discussed in more detail in Chapter 4 dealing with the effect of electrowetting on dielectrics (EWOD), and the interaction between surface tension and the changing solid–liquid contact angle as a charge is applied. However, several dimensionless numbers will be discussed here that relate the surface tension force to either external forces or the fluid's properties and are applicable to both gas–liquid and liquid–liquid interfaces.

The shapes of bubbles moving through a surrounding fluid are not only defined by surface tension,[21-23] but also by external forces, such as those resulting from changes to velocity or gravity. To characterise the effect quantitatively, the Bond number, Bo, is considered, which is the ratio of the external force to surface tension forces. While the Bond number applies for any external force,[24] generally it is used to determine the effect of gravity relative to surface tension (eqn (2.27)), in which g represents the force due to gravity. In flow within microchannels surface tension forces tend to dominate and thus Bo $\ll 1$.

$$\text{Bo} = \frac{\text{gravitational force}}{\text{surface tension force}} = \frac{\rho g l^2}{\gamma} \tag{2.27}$$

Likewise the Weber number, We, is an indicator of the relative importance of a fluid's inertia to its surface tension (eqn (2.28)). Therefore, at higher Weber numbers breaking of the meniscus and the formation of small droplets or bubbles becomes more likely.[25] While in most microchannels We $\ll 1$, in specific circumstances such as when a large chamber is included[26] or high fluid velocities[27] are used, We > 1 are observed (Figure 2.10).

$$\text{We} = \frac{\text{Inertia}}{\text{surface tension}} = \frac{\rho \mathbf{v}^2 l}{\gamma} \tag{2.28}$$

Finally, the capillary number, Ca, relates the effect of viscous forces to surface tension forces (eqn (2.29)). Similar to the Bond number, the capillary number provides a rough quantification of the size and degree of distortion of a bubble or droplet in gas–liquid or liquid–liquid systems, respectively.[28-30] Viscous forces will tend to elongate the interface between two fluids, while surface tension forces tend to minimise the surface area.[31,32]

$$\text{Ca} = \frac{\text{viscous force}}{\text{surface tension force}} = \frac{\mu \mathbf{v}}{\gamma} \tag{2.29}$$

2.4   Transport Phenomena

A characteristic of fluid flow in microchannels is that it is laminar, in contrast with fluid flow encountered at the macroscale. While laminar flow can be advantageous in many biomicrofluidic applications (*e.g.* liquid–liquid extraction in two-phase flows or electrophoretic separations) it can hinder many applications that rely on mixing of components or phases. The laminar regime implies that mixing depends mainly on molecular diffusion.

Fick's first law (eqn (2.30)) summarises the fact that molecules travel from higher concentration zones to lower concentration zones as

$$J = -D\frac{\partial C}{\partial z} \tag{2.30}$$

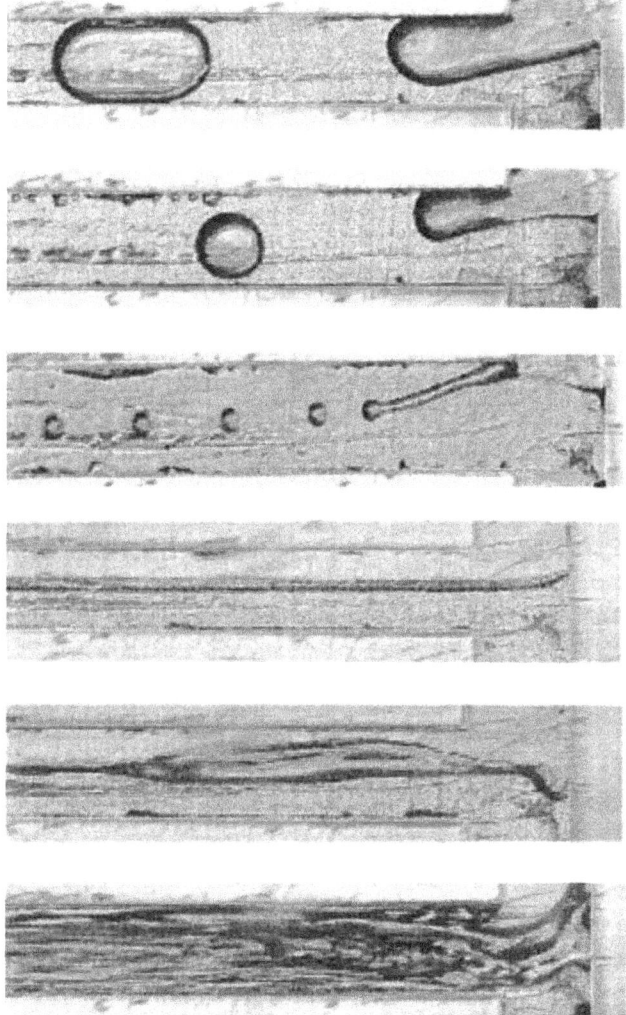

Figure 2.10 Development of two-phase flow pattern as the Weber number increases from 5.94×10^{-6} to 53.50: (reproduced from ref. 25 with permission. Copyright John Wiley and Sons).

where, D is the diffusion coefficient, C is the concentration, $\partial C/\partial z$ is the change in concentration of the diffusing molecule per unit length and J the flux of its concentration gradient. Typically in liquids, the values of D are in the range of 10^{-5} cm^2 s^{-1}, except for high molecular weight compounds where diffusion can be significantly slower.

In dilute solutions, the penetration distance (eqn (2.31)) is

$$z = \sqrt{4Dt} \qquad (2.31)$$

while the diffusion time is (eqn (2.32))

$$\tau_D \sim \frac{w^2}{D} \tag{2.32}$$

where w is the width of the channel and D the diffusion coefficient. Typically, for microfluidic distance $w = 100$ μm and molecules with values of $D = 2 \times 10^{-5}$ cm^2 s^{-1}, the diffusion time is roughly 5 s.

The diffusion coefficient in liquids can be estimated by several correlations,[33] but the most commonly encountered method is the Stokes–Einstein's method (for solute molecules having higher molecular weight than the solvent) and the Wilke–Chang method (for solvents and solutes with similar molecular weights). Nonetheless, all empirical correlations share the same viscosity and temperature dependence.

By taking advantage of the knowledge on diffusion time and residence time (the relationship between channel dimensions and fluid velocity) we can computationally design and simulate different types of systems to suit their applications prior to fabrication. This is interesting if we want to simulate the performance of different liquid–liquid extraction systems and characterise different solvent polarities and, the use of cocurrent (parallel fluid stream) or stratified systems, among others.

The convection–diffusion equation describes the molecular transport at the microscale (eqn (2.33)):

$$\frac{\partial C}{\partial t} + \nabla \cdot (C\mathbf{v}) = D\nabla^2 C \tag{2.33}$$

where, $\mathbf{v}$ denotes the velocity field. The dimensionless number that expresses the relative importance of convection to diffusion is the Péclet number (eqn (2.34)):

$$Pe = \frac{\text{advection rate}}{\text{diffusion rate}} = \frac{w\mathbf{v}}{D} \tag{2.34}$$

For high Péclet numbers, where the diffusion time is higher than the convection time, convection is much faster than diffusion. Conversely, for low Péclet numbers (below 1), diffusion happens much faster than advection, resulting in a diffusion-dominated regime and convection has secondary importance. In microfluidic environments, the critical variable for the determination of Pe is the channel width, w. Typically, due to the small dimensions, diffusion is dominant. However, Pe $\gg$ 1 can be encountered where there is a dominance of advection.[34–36] A typical application of Pe is to establish if Taylor dispersion is of relevance.[37] This phenomena corresponds to the enhancement of the rate of axial dispersion when compared to the expected rate of molecular diffusion alone in the absence of flow.[38]

Mixing is essential in many applications of microfluidic systems, especially in (bio)chemical analysis. The overall process of mixing based solely on molecular diffusion is relatively slow, but can be enhanced by decreasing the diffusion length using, for example, a high aspect ratio channel. This is a critical aspect if very fast reactions are considered or unstable compounds are used. Micromixers have been developed to tackle this by increasing the contact surface area and depending on the principle to induce mixing they can be classified as passive or active.[35,39]

Passive mixing relies entirely on diffusion or chaotic advection, not requiring an external force (except those for fluid delivery) to induce perturbation in the flow. This is accomplished by parallel lamination, serial lamination, injection, chaotic advection or droplet formation. The most commonly found microstructure designs for passive mixing are zig-zag microchannels, the incorporation of flow obstacles within the channels (grooves, rivets or posts), T-, ψ- and Y-flow inlet structures,[35,39] *etc.* On the other hand, active mixers rely on an external field to induce flow disturbance and can be categorised by the external source, like acoustic, electrical, thermal, pressure disturbance or integrated microvalves/pumps.[35,39]

Choosing an appropriate micromixer sometimes requires a compromise between mixing efficiency and the ease of fabrication and cost. As a rule of thumb, mixing efficiency is enhanced in active mixers compared to passive mixers but fabrication of the former is more complex and expensive than the latter (additionally, the integration of external devices has to be accounted for as well). Moreover, high temperature gradients may occur in some approaches used for active mixing, which may prove deleterious to biological particles.

The fluid properties and flow hydrodynamics are key issues in mass transport and reaction applications (including detection). The importance of forces, energy and timescale can be depicted from several dimensionless numbers.

When working with microfluidic devices, both the Bond and Reynolds numbers, and thereby, the interfacial tension and viscosity respectively, are of considerable importance. The Capillary number (eqn (2.29)) expresses this relationship between viscosity and the interfacial tension. Günther and Jensen[40] proposed a relationship between the Re, We and Bo numbers with respect to the channel hydraulic diameter (D_h) and the fluid velocity (v) in different types of fluid systems, namely: organic–gaseous, organic–aqueous and aqueous–gaseous systems. Based on the conditions at which interfacial forces dominate over gravitational, inertial and viscous ones, different types of multiphase flows can be observed.

Other important dimensionless relationships can be encountered based on fluid properties and operating conditions. Within these, are the density ratio (α), viscosity ratio (β) and the flow rate ratio (φ).[41] Marques *et al.*[42] combined the viscosity and flow ratio in order to achieve a parallel laminar flow in an aqueous–organic two-phase system in a Y-shaped microreactor. This defined interphase along the entire channel length allowed clear phase separation in the microreactor end. The application was applied to perform simultaneously the biotransformation of steroids and their extraction towards the organic solvent.

When considering biochemical reactions alongside fluid-flow properties and phenomena, a more complex approach is necessary. The Damköhler number (eqn (2.35)) takes this into consideration and verifies if an overall process is limited either by the reaction time or by the transport time:

$$Da = \frac{\text{transport time}}{\text{reaction time}} = \frac{\tau_t}{\tau_r} \tag{2.35}$$

In certain annotations, reaction rate over the convective transport rate is considered the first Damköhler number (Da_I), whereas the second Damköhler number (Da_{II}), considers the relationship between the diffusive transport rate and the reaction rate.[41] Conventionally, when $Da \ll 1$ the overall process is reaction limited, whereas for higher values of Da the process is transport limited.

Besides mass transport, heat transport and its control is also a critical parameter in several applications of microfluidic systems. The control of heat transfer is more important when applications such as polymerase chain reaction (PCR), temperature-gradient focusing for electrophoresis (TGF), digital microfluidics, mixing and protein crystallisation are considered (ref. 43 and references therein). The heating or cooling of fluids in microchannels can be achieved either by external fields (*e.g.* preheated liquids that flow through the microfluidic setup) or by integrating heating systems (*e.g.* Peltier components, microwaves or lasers among others [ref. 43 and references therein]). The former is easier to implement but lack, to some extent, the necessary space–temporal control.

Despite recent reports on the difference between heat transfer at the nanoscale (observed by experimental techniques and computational fluid dynamics) and its prediction by classical heat-transfer laws,[44] we will focus only on the latter, since the former is out of the scope of this chapter.

Related to mass transport, heat transfer by conduction can be expressed by the Fourier equation (eqn (2.36)).

$$q = -K\nabla T \tag{2.36}$$

where, q is the heat flux, K is the thermal conductivity and T is the local temperature. The heat flux can also be expressed by the following scale law (eqn (2.37)).

$$q \sim KL_0\Delta T \tag{2.37}$$

where L_0 is the characteristic length, and ΔT is the temperature difference. Dynamically, when a temporal variation of the temperature is considered the heat flux can be expressed as (eqn (2.38)).

$$q = \rho C_p \frac{dT}{dt} \tag{2.38}$$

where ρ is the density and C_p is the specific heat at constant pressure. Like mass, heat can also be described by the convection–diffusion equation (eqn (2.39)) (considering incompressible Newtonian fluids and neglecting variations in thermal conductivity).

$$\rho C_p \frac{dT}{dt} = \rho C_p \left(\frac{\partial T}{\partial t} + \mathbf{v}\nabla T \right) = K\Delta T + \varepsilon + s(x,t) \tag{2.39}$$

where ε is the viscous dissipation and $s(x,t)$ is the heat source. Analogous to the diffusion time, the thermalisation time, the time taken to achieve thermal equilibrium, can be expressed as,

$$\tau_T \sim \frac{L_0^2}{k} \tag{2.40}$$

where, k is the thermal diffusivity. This shows that it is possible to achieve a strict control over temperature in a microfluidic device with a small characteristic length. This control is of considerable importance when, *e.g.*, biochemical applications are envisaged.

Dimensionless numbers that are commonly used in heat transfer, which depict the importance and relationship of the several forces present, are the Grashof, Nusselt, Prandtl and the thermal Péclet numbers.

The Nusselt number (eqn (2.41)) expresses the ratio of convective heat transfer to conductive transfer.

$$\mathrm{Nu} = \frac{\text{convective heat transfer}}{\text{conductive heat transfer}} = \frac{L_0 h}{K} \tag{2.41}$$

In a microfluidic system, the Nusselt number can be increased by using large aspect ratio channels or micromixers. The Prandtl number (eqn (2.42)) gives an understanding of the thickness of the hydrodynamic boundary layer and the thermal boundary layer, and is expressed as the ratio of the molecular diffusivity to thermal diffusivity

$$\mathrm{Pr} = \frac{\text{molecular diffusivity}}{\text{Thermal diffusivity}} = \frac{\mu C_p}{K} \tag{2.42}$$

The Prandtl number does not have a length scale unlike Re or Pe. The Grashof number (eqn (2.43)) relates the inertial to viscous forces in a buoyancy-driven flow and can additionally be used to study the advective mixing in fluids,

$$\mathrm{Gr} = \frac{\text{inertial forces}}{\text{viscous forces}} = \frac{L_0^3 \rho^2 \beta g \Delta T}{\mu^2} \tag{2.43}$$

where, β is the fluid expansion coefficient. Low Grashof numbers are generally encountered in microfluidic chips.

Finally, the thermal Péclet number (eqn (2.44)) characterises the ratio between advection and diffusive heat transfer.

$$\text{Pe}_{\text{T}} = \frac{\text{thermal advection}}{\text{thermal diffusion}} = \frac{L_0 \mathbf{v} \rho C_{\text{p}}}{K} \qquad (2.44)$$

Generally, in microfluidic systems rapid heat transfer and thermal homogeneity are encountered.

2.5 Application in Detection Systems

Finally, we present three secondary sources of fluid and particle motion that are indirectly related, but may be practically useful, for detection systems. The first, Stokes drag on spheres moving through a liquid, describes the interaction between a particle's radius, velocity and the viscosity of a fluid; directly leading to the second, a discussion of Dean flow for particle focusing. Thirdly, droplet and digital microfluidics are increasingly important detection tools in which Taylor flows occur as a result of the influence of friction on the interface between two immiscible phases.

Many microfluidic devices incorporate particles of different forms in the fluids pumped through the device. Devices for the separation and detection of biological particles from cells[45–47] through to proteins,[48,49] are relatively common and utilise a variety of active and passive technologies. Artificial particles, such as magnetic microparticles or polymer beads, are also commonly used in microfluidics.[50] For example, detector technology such as surface plasmon resonance (SPR), in which surface-modified gold nano-particles can be used as in-flow, as supposed to surface-bound, detectors,[51] or separation methods utilising magnetic microbeads.[52,53] The passive forces acting on the particles are the same, with variations occurring as a result of the geometry and size of the particles.[54]

Particles in a viscous fluid flowing at a low Reynolds number are subject to friction between the particle and the fluid (Figure 2.5(b)). The resulting drag pulls the particle through the fluid.[55] Stokes drag (eqn (2.45)), named after George Gabriel Stokes who initially derived the equation, quantifies the drag force, F_{D}, exerted on a spherical particle of radius, R, in a fluid of dynamic viscosity, μ, moving at a velocity of, V_0 as:

$$F_{\text{D}} = 6\pi\mu R V_0 \qquad (2.45)$$

Eqn (2.45) is a reduction of the Navier–Stokes equation (eqn (2.7)) to the linear Stokes equation (eqn (2.46)) resulting from operating at Re $\ll$ 1, and deriving the equation for drag forces in polar coordinates for fluid close to the particle surface. Several corrections to the Stokes equation have been made. That derived by Proudman and Pearson,[56] combines a solution to the Navier–Stokes equations, for distances at which the Stokes equation is valid, and an improvement of a correction for infinite distances from the sphere, into an equation that is valid at distances from the particle at which both equations are applicable (eqn (2.47)).

$$\nabla^2 \mathbf{v} - \frac{1}{\mu}\nabla P = 0 \qquad (2.46)$$

$$F_D = 6\pi\mu R\mathbf{v}\left(1 + \frac{3}{8}\text{Re} + \frac{9}{40}\text{Re}^2\text{ln Re} + O(\text{Re}^2)\right) \qquad (2.47)$$

where O is the error in the ratio between viscous stress and inertia, in the Stokes approximation, at distances far from the particle ($r \gg a$).

A significant problem with detection and analysis in microfluidics is its applicability to real-world samples, such as blood, urine, or drinking-water samples.[57] Many aspects of standard "sample clean-up" methods have yet to be implemented *via* microfluidics or integrated into a complete collection-to-detection system. One area in which significant advances has been made is particle separation. Several methods have been demonstrated to be effective including deterministic lateral displacement,[58] multiorifice flow fractionation,[59] and Dean flow.[57] Separation *via* Dean flow clearly demonstrates how multiple forces acting on a particle can result in a simple solution to a potential detection problem.

In a straight channel a particle experiences Stokes drag and two sources of inertial lift forces: shear-induced; and wall inertial lift forces.[60,61] As previously discussed, Stokes drag is responsible for particle movement along the main axis of fluid flow. The two inertial lift forces act orthogonally to this primary axis. Shear-induced lift forces are a result of the parabolic flow profile of Poiseuille flow and direct the particle away from the centre of the microchannel. The asymmetry of the wake around the particle as it nears the wall creates wall inertial lift forces that push the particle away from the wall. These two inertial lift forces therefore act in opposition. However, the shear-induced lift force decays as the distance from the channel centre increases and, likewise, the wall inertial lift force decays as the distance from the wall increases. Therefore, a point exists between each wall and the channel centre, where these forces cancel and particles are more likely to coalesce (see Figure 2.11).

A curved channel adds Stoke drag forces from Dean flows to these forces.[62] Dean flows, secondary flows caused by the curvature of a channel, induce recirculation patterns as the fluid on the outside curve of the channel is forced to travel faster than fluid on the inside curve (see Figure 2.12(a)). Adding the Dean flow drag forces to the inertial lift forces, results in the previous four-force equilibrium points being reduced to a single point (see Figure 2.12(b) and (c)). The magnitude of the secondary flows is quantified by the dimensionless Dean number, De (eqn (2.48)). Likewise, the average particle velocity resulting from Dean flows, is characterised using eqn (2.49), in which the average Dean flow velocity, $\mathbf{v}_D$, is used alongside the equation for Stokes drag. One caveat for almost all separation methods is that they are size dependent. In the case of separation *via* Dean flow, the cut-off point for

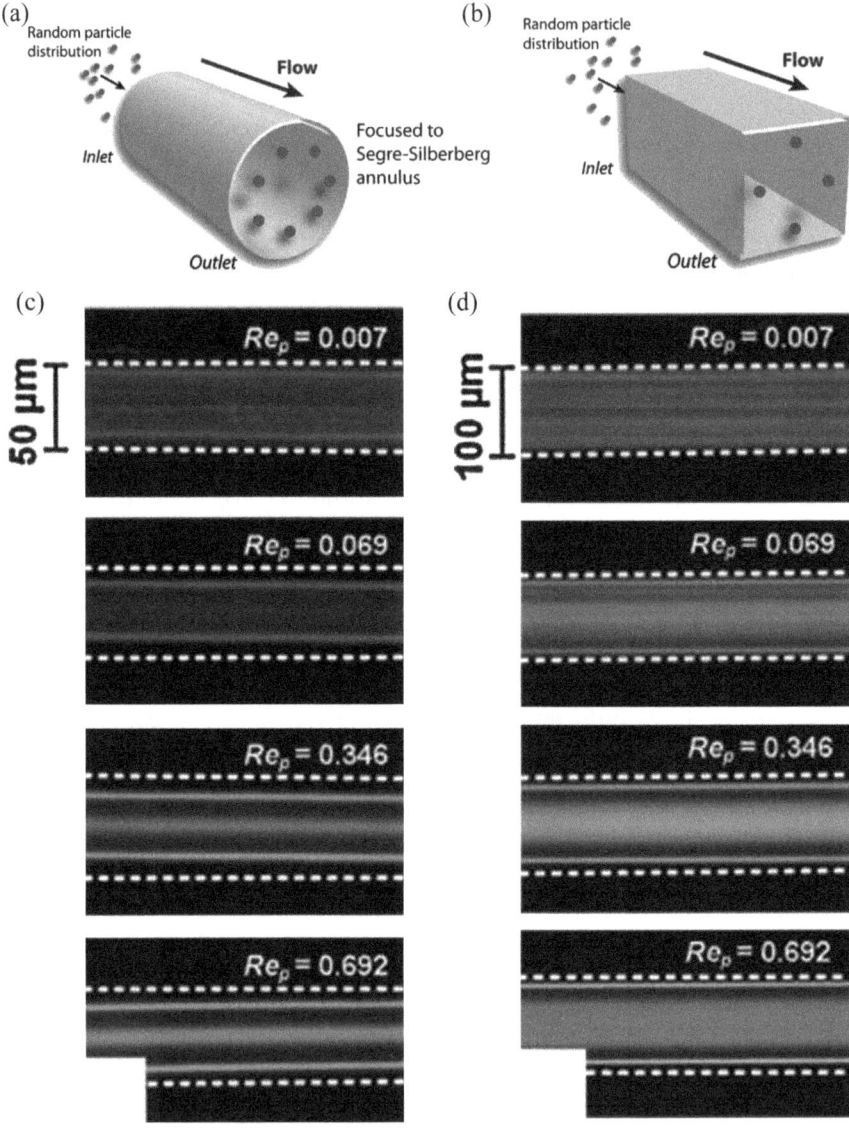

Figure 2.11 Wall and shear-induced inertial lift forces balancing to produce stable particle localisation points within (a) a cylindrical and (b) a square channel (reproduced from ref. 60), and images of inertial lift force focusing of fluorescent (c) 4.16 μm and (d) 1.9-μm particles in 50-μm and 100-μm width channels respectively, at a variety of Reynolds number (reprinted with kind permission from Springer Science + Business Media from ref. 61).

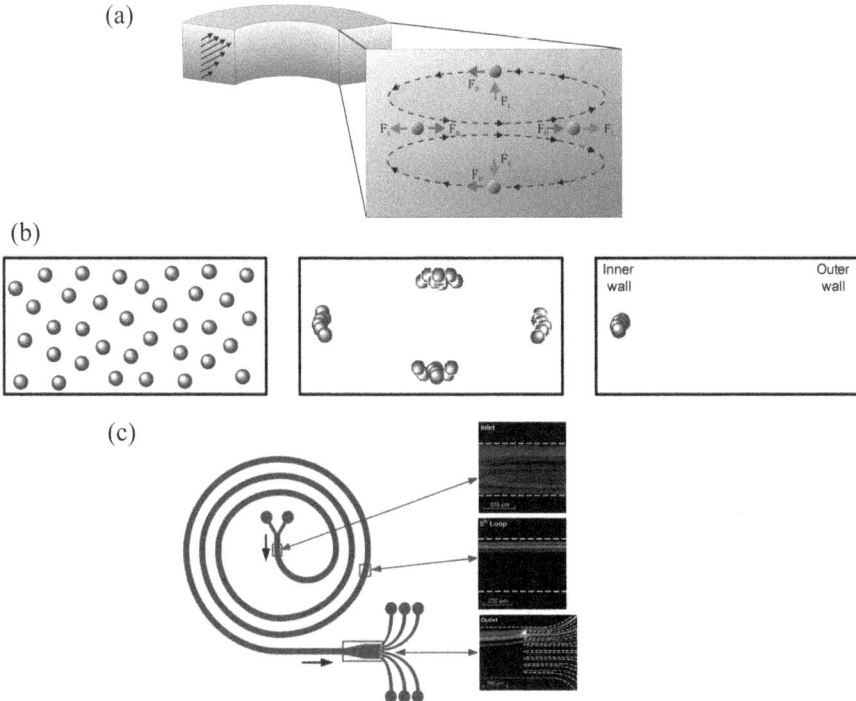

Figure 2.12 (a) Illustration of the recirculation flow and the lift and Dean forces as applied to particles at the four equilibrium positions found in a straight channel (reproduced from ref. 62), (b) particles distributed randomly, at the lift force equilibrium points, and after focusing by Dean forces (reproduced from ref. 63), and (c) focusing of 10-μm (purple), 15-μm (green), and 20-μm (orange) particles at the inlet to, during, and outlet of a spiral separation device (adapted from ref. 64).

particle focusing has been demonstrated to be approximately 0.05 for the ratio of particle size to hydraulic diameter (a_p/D_h).[64]

$$De = Re \sqrt{\frac{D_h}{2 \cdot R}} \qquad (2.48)$$

$$v_D = 1.8 \times 10^{-4} \times De^{1.63} \qquad (2.49)$$

Finally, Taylor flows[66] (Figure 2.13) combine the concept of secondary and multiphase flows with the previous discussion into the difficulty of mixing in microfluidics. The secondary flows, counter-rotating vortices caused by friction at the interface between two immiscible fluids or the interface with the channel wall, can be used for efficient convective mixing or heat transfer.[67–69] As discussed previously, mixing in microfluidics tends to be diffusion limited at the low fluid velocities normally used, with most passive mixing methods operating by increasing the surface area of the boundaries between

$\lambda \ll 1$

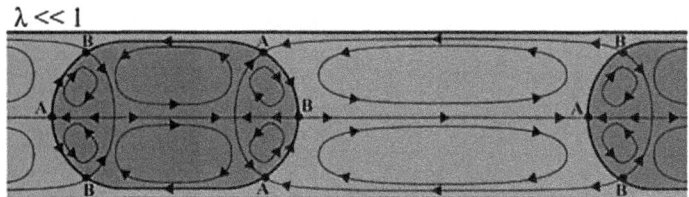

Figure 2.13 Model of the complex Taylor flow patterns found in liquid–liquid flow when the droplet liquid has a low viscosity (reproduced from ref. 65).

the fluid layers. The Péclet number (eqn (2.34)) normally enables the contribution to mass transfer, or mixing, resulting from convection, to be calculated. However, due to the complexity of Taylor flows this simple version of the Péclet equation is no longer valid.

Two-phase flow can take many forms,[70,72,73] however, both gas–liquid (see Figure 2.14(a)) and liquid–liquid (see Figure 2.14(b)) systems exhibit Taylor flow.[67] The pattern of secondary flow and the extent to which they propagate into the fluids is dependent upon the relative viscosities of the two fluids and the size of the droplets in a liquid–liquid system or bubbles in a gas–liquid flow.[65] Secondary flows within digital microfluidics (used in this context to refer to devices using EWOD to manoeuvre droplets) are complicated by electrohydrodynamically induced flows.[74] Likewise, droplet microfluidics operating in the plug-flow regime may exhibit multiple recirculation patterns depending upon whether the plug is in contact with the wall or surrounded by liquid, in which case the recirculation patterns along the length of the plug are reversed.[71]

2.6 Summary

At the outset, the readers are briefly introduced to the different physical laws that govern and affect fluid flow, heat and mass transfer at the micrometre scale. Since microfluidics deals with a different physical scale from conventional analytical systems, a scaling law that depicts the importance of physical quantities in this scale is presented. To design a robust lab-on-a-chip (LoC) detection system, consideration must be given to the presence of a dominant laminar-flow regime and an inherent diffusion-based transport mechanism. Describing laminar flow, the readers are directed towards the Navier–Stokes equation, further developing into the Poiseuille flow, the Couette flow and the Stokes flow. As regards the transport phenomena, various physical quantities and dimensionless numbers have been mentioned to better understand the behaviour of the analyte in a system. In an effort to help the readers develop a better understanding of LoC detection systems through this chapter, the authors assert that even though numerous systems have been fully developed and characterised in the past, there still lies a lot of potential that is yet to be tapped in this field.

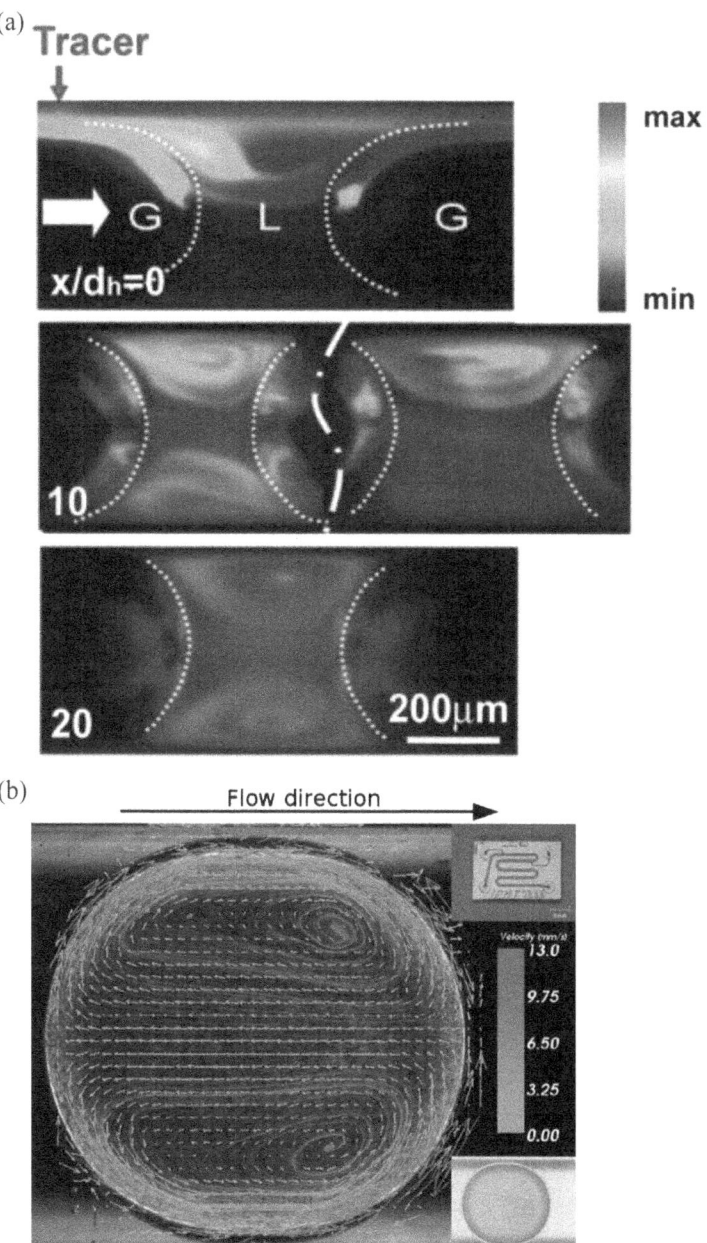

Figure 2.14 Images of Taylor flow in (a) gas–liquid and (b) liquid–liquid flow taken using microparticle image velocimetry (µPIV). (a) (Reproduced from ref. 70) and (b) (reproduced from ref. 71).

Symbols

F	Applied force (N)
A	Area (m^2)
h	Convective heat transfer coefficient
A	Channel cross-sectional area (m^3)
H	Channel height (m)
L	Channel length (m)
r	Channel radius (m)
R	Channel radius of curvature
w	Channel width (m)
L_0	Characteristic length (m)
σ	Collision diameter (m)
C	Concentration (mol l^{-1})
J	Concentration gradient flux
i	Current (A)
Da	Dahmköhler number
De	Dean number
ρ	Density (g l^{-1})
D	Diffusion coefficient
τ_{D}	Diffusion time (s)
ΔL	Distance a meniscus is stretched by (m)
μ	Dynamic viscosity
R	Electrical resistance (Ω)
L_{c}	Entrance length (m)
$\mathbf{F}$	External body forces (N)
R	Flow resistance (kg s^{-1} m^{-4})
β	Fluid expansion coefficient
$\mathbf{v}$	Fluid velocity (m s^{-1})
∇	Gradient operator
Gr	Grashof number
g	Gravity (m s^{-2})
q	Heat flux
$s(x,t)$	Heat source
D_{h}	Hydraulic diameter (m)
r	Intermolecular distance (m)
e	Internal energy
Kn	Knudsen number
$\mathbf{v}_{\mathrm{D}}$	Lateral particle velocity from Dean flow
V_{ij}	Lennard-Jones constant
Ma	Mach
λ	Mean free path length (m)
h	Meniscus driven capillary rise height (m)
Nu	Nusselt number
F_{D}	Particle drag force (N)
R	Particle radius (m)

V_0	Particle velocity (m s^{-1})
Pe	Péclet numbers
z	Penetration distance (m)
ε	Potential energy well
Pr	Prandtl number
P	Pressure (Pa)
ΔP	Pressure difference across a meniscus (Pa)
r_1	Radius of specific point within channel (m)
k	Ratio of the specific heats
Re	Reynolds number
Re$_{D_h}$	Reynolds number
C_p	Specific heat
c_s	Speed of sound (m s^{-1})
γ	Surface tension
T	Temperature (°C)
K	Thermal conductivity
k	Thermal diffusivity
Pe$_T$	Thermal Péclet number
t	Time (s)
τ	Total stress tensor
ε	Viscous dissipation
V	Voltage (V)
Q	Volumetric flow rate (l s^{-1})
P_w	Wetted perimeter (m)

References

1. G. M. Whitesides, *Nature,* 2006, **442**, 368–373.
2. D. J. Harrison, K. Fluri, K. Seiler, Z. Fan, C. S. Effenhauser and A. Manz, *Science,* 1993, **261**, 895–897.
3. A. Manz, N. Graber and H. M. Widmer, *Sens. Actuators B,* 1990, **1**, 244–248.
4. J. Zhang, S. Liu, P. Yang and G. Sui, *Lab Chip,* 2011, **11**, 3516–3522.
5. http://www.merriam-webster.com.
6. X. Mu, Q. Liang, P. Hu, K. Ren, Y. Wang and G. Luo, *Lab Chip,* 2009, **9**, 1994–1996.
7. W. M. Deen, *Analysis of Transport Phenomena (Topics in Chemical Engineering)*, Oxford University Press, New York, 1998, vol. 3.
8. M. Gad-el-Hak, *J. Fluid Eng.,* 1999, **121**, 5–33.
9. P. Atkins and J. de Paula, *Atkins' Physical Chemistry*, Oxford University Press, Oxford, 8th edn, 2006.
10. O. Reynolds, *Proc. Roy. Soc. London,* 1883, **35**, 84–99.
11. http://claymath.org.
12. F. M. White, *Fluid Mechanics*, McGraw-Hill, 4th edn, 1998.

13. H. Bruus, *Theoretical Microfluidics*, Oxford University Press, Oxford, 1st edn, 2008.
14. K. W. Oh, K. Lee, B. Ahn and E. P. Furlani, *Lab Chip*, 2012, **12**, 515–545.
15. R. K. Shah and A. L. London, *Laminar Flow Forced Convection in Ducts: A Source Book for Compact Heat Exchanger Analytical Data*, Academic Press, 1st edn, 1978.
16. S. C. Terry, J. H. Jerman and J. B. Angell, *IEEE Trans. Electron Devices*, 1979, **26**, 1880–1886.
17. M. Da Silva and C. Pasquini, *Anal. Chim. Acta*, 1997, **349**, 377–384.
18. K. Sakamoto, H. Nakanishi, M. Tokeshi, Y. Yoshida and T. Kitamori, in *Micro Total Analysis Systems*, 2004, Malmo, 213–215.
19. A. Günther, M. Jhunjhunwala, M. Thalmann, *et al.*, *Langmuir*, 2004, **21**, 1547–1555.
20. Y. Zhu and K. Petkovic-Duran, *Microfluid. Nanofluid.*, 2009, **8**, 275–282.
21. V. Ajaev and G. Homsy, *Annu. Rev. Fluid Mech.*, 2006, **38**, 277–307.
22. J. Coleman and S. Garimella, *Int. J. Heat Mass Trans.*, 1999, **42**, 2869–2881.
23. C. G. Cooney, C.-Y. Chen, M. R. Emerling, A. Nadim and J. D. Sterling, *Microfluid. Nanofluid.*, 2006, **2**, 435–446.
24. K. Curran, S. Colin, L. Baldas and M. Davies, *Microfluid. Nanofluid.*, 2005, **1**, 336–345.
25. Y. Zhao, G. Chen and Q. Yuan, *AIChE J.*, 2006, **52**, 4052–4060.
26. J. J. Chen, W. Z. Liu, J. D. Lin and J. W. Wu, *Sens. Actuators A*, 2006, **132**, 597–606.
27. A. S. Utada, A. Fernandez-Nieves, H. A. Stone and D. A. Weitz, *Phys. Rev. Lett.*, 2007, **99**, 094502.
28. P. Garstecki, M. Fuerstman, H. Stone and G. M. Whitesides, *Lab Chip*, 2006, **6**, 437–446.
29. A. Gupta and R. Kumar, *Microfluid. Nanofluid.*, 2010, **8**, 799–812.
30. P. A. Romero and A. R. Abate, *Lab Chip*, 2012, **12**, 5130–5132.
31. T. Squires and S. R. Quake, *Rev. Mod. Phys.*, 2005, **77**, 977–1026.
32. B. Ghidersa, M. Worner and D. Cacuci, *Chem. Eng. J.*, 2004, **101**, 285–294.
33. E. L. Cussler, *Diffusion*, Cambridge University Press, Cambridge, 3rd edn, 2009.
34. S.-E. Ong, S. Zhang, H. Du and Y. Fu, *Front. Biosci.*, 2008, **13**, 2757–2773.
35. N.-T. Nguyen and Z. Wu, *J. Micromech. Microeng.*, 2004, **15**, R1–R16.
36. T. M. Squires, R. J. Messinger and S. R. Manalis, *Nature Biotechnol.*, 2008, **26**, 417–426.
37. O. Geschke, H. Klank and P. Telleman, *Microsystem Engineering of Lab-on-a-Chip Devices*, Wiley-VCH, Weinheim, 2008.
38. S. Datta and S. Ghosal, *Lab Chip*, 2009, **9**, 2537–2550.
39. L. Capretto, W. Cheng, M. Hill and X. Zhang, *Microfluidics: Technologies and Applications*, Topics in Current Chemistry, Heidelberg, 2011, vol. 304.
40. A. Günther and K. Jensen, *Lab Chip*, 2006, **6**, 1487–1503.
41. M. P. C. Marques and P. Fernandes, *Molecules*, 2011, **16**, 8368–8401.

42. M. P. C. Marques, J. M. S. Cabral and P. Fernandes, *J. Chem. Technol. Biotechnol.,* 2010, **85**, 1184–1198.
43. V. Miralles, A. Huerre, F. Malloggi and M.-C. Jullien, *Diagnostics,* 2013, **3**, 33–67.
44. T. Luo and G. Chen, *Phys. Chem. Chem. Phys.,* 2013, **15**, 3389–3412.
45. T. F. Didar and M. Tabrizian, *Lab Chip,* 2010, **10**, 3043–3053.
46. S. Gabriele, M. Versaevel, P. Preira and O. Théodoly, *Lab Chip,* 2010, **10**, 1459–1467.
47. Y. Chen, P. Li, P.-H. Huang, Y. Xie, J. D. Mai, L. Wang, N.-T. Nguyen and T. J. Huang, *Lab Chip,* 2014, **14**, 626–645.
48. E. Eteshola and D. Leckband, *Sens. Actuators B,* 2001, **72**, 129–133.
49. J. Fruetel, R. Renzi, V. VanderNoot, J. Stamps, B. Horn, J. West, S. Ferko, R. Crocker, C. Bailey, D. Arnold, B. Wiedenman, W. Choi, D. Yee, I. Shokair, E. Hasselbrink, P. Paul, D. Rakestraw and D. Padgen, *Electrophoresis,* 2005, **26**, 1144–1154.
50. M. A. M. Gijs, F. Lacharme and U. Lehmann, *Chem. Rev.,* 2010, **110**, 1518–1563.
51. H.-I. Peng and B. L. Miller, *Analyst,* 2011, **136**, 436–447.
52. K.-Y. Lien, Y.-H. Chuang, L.-Y. Hung, K.-F. Hsu, W.-W. Lai, C.-L. Ho, C.-Y. Chou and G.-B. Lee, *Lab Chip,* 2010, **10**, 2875–2886.
53. J. H. Kang, S. Krause, H. Tobin, A. Mammoto, M. Kanapathipillai and D. E. Ingber, *Lab Chip,* 2012, **12**, 2175–2181.
54. X. Xuan, J. Zhu and C. Church, *Microfluid. Nanofluid.,* 2010, **9**, 1–16.
55. D. Yoon, J. Ha, Y. Bahk, T. Arakawa, S. Shoji and J. Go, *Lab Chip,* 2009, **9**, 87–90.
56. I. Proudman and J. Pearson, *J. Fluid Mech.,* 1957, **2**, 237–262.
57. D. J. Caruana, *Analyst,* 2011, **136**, 4641–4652.
58. L. Huang, E. Cox, R. Austin and J. Sturm, *Science,* 2004, **304**, 987–990.
59. T. S. Sim, K. Kwon, J. C. Park, J.-G. Lee and H.-I. Jung, *Lab Chip,* 2011, **11**, 93–99.
60. D. Di Carlo, *Lab Chip,* 2009, **9**, 3038–3046.
61. A. A. S. Bhagat, S. S. Kuntaegowdanahalli and I. Papautsky, *Microfluid. Nanofluid.,* 2009, **7**, 217–226.
62. E. W. M. Kemna, R. M. Schoeman, F. Wolbers, I. Vermes, D. A. Weitz and A. van den Berg, *Lab Chip,* 2012, **12**, 2881–2887.
63. A. A. S. Bhagat, S. S. Kuntaegowdanahalli and I. Papautsky, *Lab Chip,* 2008, **8**, 1906–1914.
64. S. S. Kuntaegowdanahalli, A. A. S. Bhagat, G. Kumar and I. Papautsky, *Lab Chip,* 2009, **9**, 2973–2980.
65. C. N. Baroud, F. Gallaire and R. Dangla, *Lab Chip,* 2010, **10**, 2032–2045.
66. G. I. Taylor, *J. Fluid Mech.,* 1961, **10**, 161–165.
67. R. Gupta, S. S. Y. Leung, R. Manica, D. F. Fletcher and B. S. Haynes, *Chem. Eng. Sci.,* 2013, **92**, 180–189.
68. P. Garstecki, M. Fuerstman, M. Fischbach, S. K. Sia and G. M. Whitesides, *Lab Chip,* 2006, **6**, 207–212.

69. O. K. Castell, C. J. Allender and D. A. Barrow, *Lab Chip,* 2008, **8**, 1031–1033.
70. A. Günther, S. Khan, M. Thalmann, F. Trachsel and K. F. Jensen, *Lab Chip,* 2004, **4**, 278–286.
71. D. Malsch, M. Kielpinski, R. Merthan, J. Albert, G. Mayer, J. Kohler, H. Suse, M. Stahl and T. Henkel, *Chem. Eng. J.,* 2008, **135**, S166–S172.
72. M. J. Jebrail, M. S. Bartsch and K. D. Patel, *Lab Chip,* 2012, **12**, 2452–2463.
73. S.-Y. Teh, R. Lin, L.-H. Hung and A. P. Lee, *Lab Chip,* 2008, **8**, 198–220.
74. R. Malk, Y. Fouillet and L. Davoust, *Sens. Actuators B,* 2011, **154**, 191–198.

CHAPTER 3

Electrokinetics and Rare-Cell Detection

ALIREZA SALAMANZADEH AND RAFAEL V. DAVALOS*

Virginia Tech-Wake Forest University, Blacksburg, VA 24061, USA
*E-mail: davalos@vt.edu

3.1   Introduction

The isolation and enrichment of rare cells from background cells is imperative for disease detection, especially in presymptomatic stages of the disease as well as for disease treatment and individualised medicine.[1] Isolating rare cells is challenging, primarily due to their rarity compared to background cells. For example, in the case of detecting circulating tumour cells (CTCs), it is estimated that there are fewer than 5 CTCs detectable per 7.5 ml of blood using immunomagnetic labelling in patients diagnosed with metastatic breast cancer.[2] In some applications, such as isolation of cancer stem cells (CSCs) from normal cancer cells, the cells of interest are similar in size to the background cells, which make rare-cell isolation difficult.

Applications for lab-on-a-chip devices have been expanding rapidly in the last decade due to advantages including lower required sample volume, faster analysis, smaller tools, more control, the possibility of portability, and ease of parallelisation compared to their macroscale counterparts. Microfluidic devices have been used for isolation and enrichment of target cells through various methods including antibody-,[1] size-,[3,4] inertia-,[5] streamline-,[6] and electrokinetic-based[7–12] techniques. Although powerful and commonly used antibody-based methods such as FACS[13] and magnetic cell sorting rely

RSC Detection Science Series No. 5
Microfluidics in Detection Science: Lab-on-a-chip Technologies
Edited by Fatima H Labeed and Henry O Fatoyinbo
© The Royal Society of Chemistry 2015
Published by the Royal Society of Chemistry, www.rsc.org

upon surface marker expression labelling which is time consuming and requires special training. The use of antibodies also requires *a priori* knowledge, which is not yet available for all cell types. Moreover, irreversible binding of markers to the cell exterior can permanently disturb the functionality of the cell during isolation, making post-studies difficult to perform.[14]

In this chapter, we focus on the isolation and enrichment (*i.e.* manipulation) of bioparticles from cell mixtures using electrokinetic-based lab-on-a-chip devices (see Pratt *et al.*[15] for a review on other microfluidic sorting techniques). Electrokinetic-based particle manipulation techniques can be divided into subcategories including electrowetting,[16] electrothermal,[17] electro-osmosis,[18] electrostatics,[19,20] electrophoresis,[21–23] and dielectrophoresis (DEP).[24] Each of these electrokinetic techniques has its own advantages and limitations. Electro-osmosis and electrothermal techniques affect the bulk fluid, while electrophoresis and DEP are forces that only affect particles and therefore apply best to particle sorting. Different phenomena can exist simultaneously; for example DEP, AC-electro-osmosis, and AC-electrothermal can occur in the same system. The dominant phenomena depends on the applied voltage and frequency, microelectrode geometry and aspect ratio within the microdevice, conductivity and permittivity of the sample, and the size of the suspended particles as will be explained in greater detail later in this chapter. Among electrokinetic techniques, DEP is the arguably the most commonly used technique for particle sorting. Other electrokinetic techniques are less applicable for particle sorting, such as electro-osmosis and electrothermal that influence the fluid not the particles, and electrophoresis that is a well-established technique that can separate charged particles from the background flow but cannot induce motion of neutral particles.

3.2 Electrohydrodynamics

In almost all electrokinetic applications, particles must be suspended in a fluid. Manipulating particles in a suspending fluid is especially important for biological systems such as in the manipulation of blood cells in a blood sample;[25,26] chemical systems;[27] medicine, such as drug delivery;[28,29] and electronics, with single-electron transistors and nanoparticle memory.[30] Electrohydrodynamics is a branch of fluid mechanics that focuses on the mutual effects of an electrical force and a fluid.[31] Having a thorough understanding of the interaction between fluid flow and electrokinetics can lead to more precisely controlled, fast, and affordable particle manipulation, specifically particle sorting[32] and micropumping.[33]

3.2.1 Navier–Stokes Equations

The velocity of flow in microfluidic systems is usually less than 1 mm s^{-1} and the flow can be assumed to be incompressible.[34] The continuity and the Navier–Stokes equations for an incompressible fluid can be written as (for more details about Navier–Stokes equations see Section 2.2.1 or *Viscous Fluid Flow* by F. M. White):

$$\vec{\nabla} \cdot \vec{v} = 0 \tag{3.1}$$

$$\rho_m \left(\frac{\partial \mathbf{v}}{\partial t} + (\vec{v} \cdot \vec{\nabla})\vec{v} \right) = -\vec{\nabla}p + \eta \nabla^2 \vec{v} + \vec{F}_{ext} + \rho_m \vec{g} \tag{3.2}$$

In this study $\vec{F}_{ext}$ can be assumed to bean electrokinetic force. Reynolds number (eqn (3.3)), defined as the ratio of convection (inertial term) to viscous effects, is usually very low in microfluidic applications

$$Re \equiv \frac{\rho_m v l}{\eta} \tag{3.3}$$

where ρ_m and η are the density and viscosity of the medium, respectively, $\mathbf{v}$ is the velocity, and l is the length scale. Considering $\mathbf{v}$ on the order of 100 μm s^{-1} and l on the order of 100 μm, the Reynolds number will be of the order of 10^{-2}, where the flow is well within the laminar region. Thus, the convective term in the Navier–Stokes equation is negligible.

Also, comparing the time dependent and viscous terms, the time required to reach steady-state condition can be estimated as

$$t = \frac{\rho_m l^2}{\eta} \tag{3.4}$$

which is of the order of 10^{-2}. Thus, the time-dependent term in the Navier–Stokes equation is also negligible.

3.2.2 Electrical Reynolds Number

Studying Maxwell's electromagnetic equations is outside the scope of this chapter. However, an important note for electrohydrodynamic systems is that the electrical equations are decoupled from the mechanical equations. The electrical Reynolds number is defined as the ratio of a charge-relaxation time to a time for the fluid to move a characteristic length l at a characteristic velocity, $\mathbf{v}$:[31]

$$Re_{electrical} = \frac{|\nabla \cdot (\varepsilon \vec{E})\vec{v}|}{|\sigma \vec{E}|} \approx \frac{\varepsilon/\sigma}{l/v} \tag{3.5}$$

The electrical Reynolds number can also be interpreted as the ratio of convection current to conduction current.[31,35] Considering $\mathbf{v}$ on the order of 100 μm s^{-1}, l on the order of 100 μm, and conductivity (σ) about 0.01 S m^{-1}, the electrical Reynolds number will be on the order of 10^{-7}. This means that the velocity of the charge in response to an electric field ($\mathbf{E}$) is much larger than the fluid velocity,[33] which suggests that the electrical equations are decoupled from the mechanical equations.[31]

3.2.3 Hydrodynamic Force

As particles move under the influence of an electrokinetic force, such as DEP, relative to the suspending fluid, they interact with the surrounding fluid and experience a hydrodynamic force. Assuming the medium viscosity, η, is relatively high (such as with water), and that the particles are small, spherical, and are moving with a relatively low velocity, the hydrodynamic force can be approximated using Stokes' drag, which is given by

$$\vec{\mathbf{F}}_{\text{Drag}} = 6\eta r\pi\left(\vec{\mathbf{u}}_{\text{p}} - \vec{\mathbf{u}}_{\text{f}}\right) \tag{3.6}$$

where r is the particle radius, $\vec{\mathbf{u}}_{\text{p}}$ is the velocity of the particle, and $\vec{\mathbf{u}}_{\text{f}}$ is the medium velocity. To study the movement of a particle in a fluid, we can write

$$m_{\text{p}}\frac{d\vec{\mathbf{u}}_{\text{p}}}{dt} = -6\eta r\pi\left(\vec{\mathbf{u}}_{\text{p}} - \vec{\mathbf{u}}_{\text{f}}\right) + \vec{\mathbf{F}}_{\text{ext}} \tag{3.7}$$

By solving this equation, $\vec{\mathbf{u}}_{\text{p}}$ can be found as

$$\vec{\mathbf{u}}_{\text{p}} = \left(\vec{\mathbf{u}}_{\text{p}} - \vec{\mathbf{u}}_{\text{f}} - \frac{\vec{\mathbf{F}}_{\text{ext}}}{6\eta r\pi}\right)e^{-\left(\frac{6\eta r\pi}{m}\right)t} + \vec{\mathbf{u}}_{\text{f}} + \frac{\vec{\mathbf{F}}_{\text{ext}}}{6\eta r\pi} \tag{3.8}$$

For a spherical particle, the characteristic time of acceleration can be defined as

$$\tau_{\text{a}} = \frac{m}{6\eta r\pi} = \frac{\rho 4/3\pi r^3}{6\eta r\pi} = \frac{2}{9}\frac{\rho r^2}{\eta} \tag{3.9}$$

For instance, for a 10-μm diameter cell, assuming that the density of the cells is close to the density of water and the viscosity of the suspending medium is equal to that of water, τ_{a} will be on the order of 10^{-5}. Thus, it can be assumed that the particle is moving at terminal velocity:

$$\vec{\mathbf{u}}_{\text{p}} = \vec{\mathbf{u}}_{\text{f}} + \frac{\vec{\mathbf{F}}_{\text{ext}}}{6\eta r\pi} \tag{3.10}$$

This means that the terminal velocity of a particle is the velocity of the fluid plus the velocity induced by the external force acting on the particle, such as the DEP force. To consider the acceleration of fluid, refer to Clift *et al.*[36] If the fluid is at rest, then the velocity of the particle will be proportional to the applied external force. Also, if the particle is moving and the force is stopped, the particle stops moving almost immediately because the deceleration time is so short that it cannot be observed. It will be seen in the rest of the chapter that $\vec{\mathbf{u}}_{\text{f}}$ can be replaced by the electro-osmotic and/or pressure-driven flow velocity, and $\vec{\mathbf{F}}_{\text{ext}}$ can be replaced by any external force, including the DEP force.

3.2.4 Electrical Double Layer

In most microfluidic applications, bioparticles are suspended in an aqueous solution of ions, known as an electrolyte. Ion concentration and type changes the conductivity and even permittivity of the solution.[33] Ions also gather at interfaces between the particle, electrodes and the electrolyte to maintain electroneutrality. When an ion is suspended in an electrolyte, polarised water molecules gather around the ion to maintain electro-neutrality of the solution. This is known as the electrical double layer. Study of the electrical double layer is important as it influences the forces applied on the particle or the fluid, such as in AC-electro-osmosis. The length scale of this layer of polarised molecules is defined by the Debye length,

$$\kappa \equiv \sqrt{\frac{\sigma}{D\varepsilon}} \qquad (3.11)$$

where D is the diffusion constant. When an electric field is applied to an electrode that is in contact with an electrolyte, a couple of layers of oppositely charged ions attach to the electrode and form a plane also known as the electrical double layer. The potential at this slip plane is defined as the zeta potential, ζ.[33] The zeta potential is a function of viscosity and permittivity of the electrolyte as well as the electrolyte pH and concentration. As pH increases, the magnitude of zeta potential also increases (see Section 7.1 for more details on the electrical double layer).

3.3 Dielectrophoresis for Cellular Manipulation

Dielectrophoresis (DEP), the motion of particles due to their polarisation in a nonuniform electric field, is a nondestructive, noninvasive particle manipulation technique. One of the advantages of DEP is that it does not rely solely upon particle size, but also upon the intrinsic electrical properties of the particle. This makes it possible to separately manipulate similarly sized particles that have different intrinsic electrical properties. The magnitude and direction of the DEP force depends upon the size, conductivity, and permittivity of the particle, the conductivity and permittivity of the media, and the applied voltage and frequency. It is also a popular technique due to the ease of integration with other microfluidic systems and the ability to generate both positive and negative forces. DEP has been shown to be a successful particle manipulation technique useful for mixing,[37] separation,[7–9] enrichment,[10,11] detection,[12] and characterisation.[38,39]

When a polarisable particle is placed in a nonuniform electric field, positive and negative charges move to form the interfacial charge layer[40] (see Figure 3.1(a)). This polarisation acts as an effective dipole moment that is dependent upon the applied electric-field frequency and the properties of the particle as well as the suspending fluid. Thus, the particle can be assumed as a dipole consisting of two opposite charges, q, separated by a distance

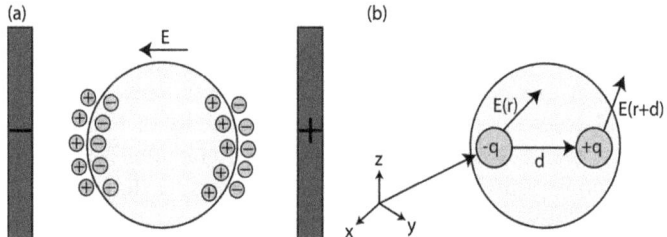

Figure 3.1 (a) Schematic of a polarised particle in an electric field. (b) Schematic of a dipole. The direction of the dipole depends on the applied frequency (see Figure 3.2).

$\vec{\mathbf{d}}$ (see Figure 3.1(b)). The dipole moment, $\vec{\mathbf{p}}$, can be defined as a vector pointing from the negative to the positive charge:

$$\vec{\mathbf{p}} = q\vec{\mathbf{d}} \tag{3.12}$$

The force, $\vec{\mathbf{F}}$, a dipole experiences in an electric field, $\vec{\mathbf{E}}$, can be approximated, by applying the Taylor series, as:

$$\vec{\mathbf{F}} = (\vec{\mathbf{p}} \cdot \nabla)\vec{\mathbf{E}}(\vec{\mathbf{r}}) \tag{3.13}$$

and the torque that works to align the dipole with the electric field is

$$\vec{\mathbf{\Gamma}} = \vec{\mathbf{p}} \times \vec{\mathbf{E}} \tag{3.14}$$

This torque causes electrorotation, defined as the rotation of polarised particles suspended in a liquid due to an induced torque in a rotating electric field (for more details see *AC Electrokinetic: Colloids and Nanoparticles* by H. Morgan and N. G. Green).

When a neutral spherical particle is suspended in a medium and an electric field, $\vec{\mathbf{E}}$, is applied, the Laplace equation must be satisfied. By assuming the particle is a dielectric, meaning that it is polarisable in an electric field, the effective dipole moment on a circular particle becomes:

$$\vec{\mathbf{p}} = 4\pi\varepsilon_1 \mathrm{Re}\{f_{\mathrm{CM}}(\omega)\}r^3\vec{\mathbf{E}} \tag{3.15}$$

where ω is the radial frequency of the electric field. The Clausius–Mossotti (CM) factor, $f_{\mathrm{CM}}(\omega)$, is defined as,[41]

$$f_{\mathrm{CM}}(\omega) = \frac{\varepsilon_{\mathrm{p}}^* - \varepsilon_{\mathrm{m}}^*}{\varepsilon_{\mathrm{p}}^* + 2\varepsilon_{\mathrm{m}}^*} = \frac{\sigma_{\mathrm{p}}^* - \sigma_{\mathrm{m}}^*}{\sigma_{\mathrm{p}}^* + 2\sigma_{\mathrm{m}}^*} \tag{3.16}$$

where $\varepsilon*$ and $\sigma*$ are the complex permittivity and conductivity of the particle, with subscript p, or suspending medium, subscript m, respectively. The complex permittivity is defined as $\varepsilon* = \varepsilon - j\sigma/\omega$, where ε and σ are the real permittivity and conductivity, respectively, and $j^2 = -1$. The real part of $f_{\mathrm{CM}}(\omega)$ is constrained between $-\frac{1}{2}$ and 1, and the imaginary part between $-\frac{3}{4}$ and $\frac{3}{4}$.

By substituting the effective dipole moment into eqn (3.13), the time-averaged DEP force acting on a spherical particle can be found by,[41]

$$\vec{F}_{DEP} = 2\pi\varepsilon_0\varepsilon_m r^3 \text{Re}\{f_{CM}(\omega)\}\nabla(\vec{E}\cdot\vec{E}) \tag{3.17}$$

Depending on the sign of $\text{Re}\{f_{CM}(\omega)\}$, given by eqn (3.16), the DEP force can either be positive, in which case it is directed towards the strongest regions of the electric-field gradient, or negative, in which case it is directed towards the weakest regions of the electric-field gradient. This dependence is related to the polarisability of the particle with respect to the surrounding medium and enables sorting of particles based on their physical and dielectric properties.

The translational velocity of the particle can thus be estimated from eqn (3.10) by a balance between the dielectrophoretic force and Stokes' drag force on a particle, giving

$$\vec{u}_p = \mu_{DEP}\nabla(\vec{E}\cdot\vec{E}) + \vec{u}_f \tag{3.18}$$

where μ_{DEP} is the dielectrophoretic mobility, defined as

$$\mu_{DEP} = \frac{r^2\varepsilon_m\text{Re}\{f_{CM}\}}{6\eta} \tag{3.19}$$

It is important to note that in the above relations for DEP force, it is assumed that the sample is diluted enough that interactions between particles (*e.g.*, giving rise to pearl chaining effects) are negligible. However, for a dense suspension of spherical particle, the DEP force can be found from,[42]

$$\vec{F}_{DEP} = \frac{2}{3}\pi r^3 \left(\frac{\partial\text{Re}\{\varepsilon_m^*\}}{\partial c}\right)_{\omega t_c} \nabla(\vec{E}\cdot\vec{E}) \tag{3.20}$$

where t_c is the relaxation time and c is the particle volume fraction. In this relation, the electrical permittivity of the suspending medium depends on the particle volume fraction. t_c is defined as

$$t_c = \left(\frac{\varepsilon_p + 2\varepsilon_m}{\sigma_p + 2\sigma_m}\right) \tag{3.21}$$

3.3.1 Multishell Models

Multishell models are a group of models to predict the dielectric properties of cells.[43] In this model the cell membrane, the nucleus, and even the nucleus membrane can be considered a shell. For instance, in the single-shell model, the bioparticles' thin lipid bilayer and the internal cytoplasm are considered. Then, ε_p^* will be written as

$$\varepsilon_p^* = \varepsilon_2^* \frac{\gamma_{12}^3 + 2\left(\dfrac{\varepsilon_3^* - \varepsilon_2^*}{\varepsilon_3^* + 2\varepsilon_2^*}\right)}{\gamma_{12}^3 - \left(\dfrac{\varepsilon_3^* - \varepsilon_2^*}{\varepsilon_3^* + 2\varepsilon_2^*}\right)} \tag{3.22}$$

where $\gamma_{12}^3 = r_1/r_2$ and $r_1 = r_2 + d$, d is the thickness of the membrane, and $r_1 \gg d$. In the two-shell model, the nucleus is added as an additional layer and a similar relation as eqn (3.22) can be written for the complex permittivity of the new shell.

The multishell model neglects the complicated internal structure of a biological particle and does not take into account the membrane inhomogeneity, cytoplasm, and nuclear structural features. This limits its ability such that it cannot correlate the membrane morphological complexity with the specific membrane capacitance. Despite this, it has been shown by several studies[44] that this model is still accurate enough to demonstrate differences in the dielectrophoretic properties of cells and the possibility of sorting them based on these differences.

3.3.2   Crossover Frequency

For each cell type, within a specific type of medium, there typically exist two unique frequencies at which the cells and the media have an equivalent complex permittivity. These are known as the crossover frequency, f_{xo}. At these frequencies the real part of the f_{CM} equals zero, $\text{Re}\{f_{CM}(\omega)\} = 0$, thus there is no net DEP force acting on the cell; the cell is practically invisible to the electric field (see Figure 3.2(b)).

The first crossover frequency of mammalian cells typically occurs between 10 and 100 kHz, and the second crossover frequency is typically on the order of 10 MHz for a sample with a conductivity of 100 μS cm^{-1}.[45] A subtle change in the biophysical properties of the cell will also change the crossover frequency of the cell. Thus, the crossover frequency can be used as a tool to monitor changes that occur in cells such as after treating cells with drugs or toxicants. Cell size, shape, cytoskeleton, and membrane morphology affect the first crossover frequency and cytoplasm conductivity, nuclear envelope permittivity, nucleus/cytoplasm (N/C) volume ratio, and endoplasmic

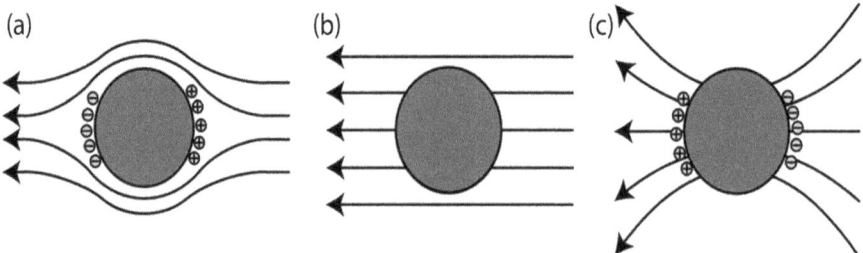

Figure 3.2   Polarity of mammalian cells changes by changing the frequency. (a) At low frequencies cells behave as an electrically insulating particle. (b) At the crossover frequency the induced dipole moment of the cell is zero and the cell behaves as if it is transparent to the electric field. (c) At high frequencies cells are more polarisable than the suspending medium.

reticulum affect the second crossover frequency.[46] The biophysical cell parameters affecting the two crossover frequencies and $\text{Re}\{f_{\text{CM}}(\omega)\}$ are shown in Figure 3.3 for a typical mammalian cell using the two-shell model and will be discussed further in this chapter.

The material properties of cells and biological tissues change with the applied frequency. This disparity in properties is known as dispersion. At frequencies less than 10 kHz, counterion polarisation happens along the cell membrane (α dispersion). In the higher frequency range, interfacial polarisation of the cell membrane occurs, which develops due to the polarisation of proteins and other macromolecules (β dispersion).[47] For frequencies well below β dispersion (<1 MHz), the cells dielectric properties are related to membrane properties.[48] Then, the specific capacitance of the cell membrane, C_{mem}, can be defined as

$$C_{\text{mem}} = \frac{\varepsilon_{\text{mem}}}{d} \qquad (3.23)$$

and the first crossover frequency can be written then as[49]

$$f_{\text{xo1}} = \frac{\sqrt{2}\sigma_{\text{m}}}{2\pi r C_{\text{mem}}} \qquad (3.24)$$

At frequencies higher than 10 MHz, the permittivity of a cell's internal organelles will have a more dominant effect on the permittivity of the cell. Changing $\text{Re}\{f_{\text{CM}}(\omega)\}$ from positive to negative at the second crossover frequency shows that the permittivity of the cells decreases to less than the buffer permittivity, $\sim$80. This change is mostly due to changes in the nucleus. Due to the counterion fluctuations, the DNA shows dispersion and the proteins show relaxation of polar groups.[14] Since the second crossover

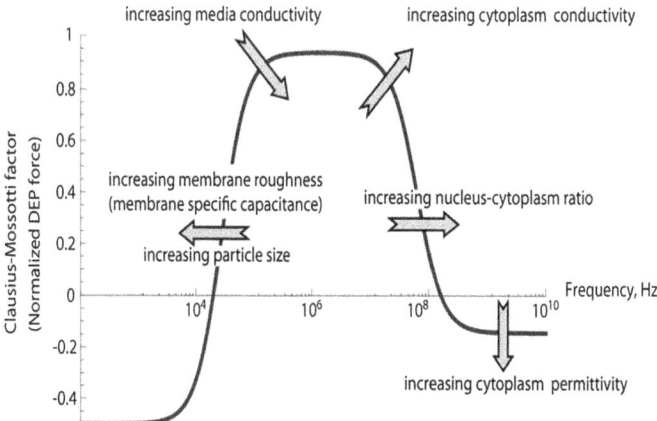

Figure 3.3 Clausius–Mossotti factor, $\text{Re}\{f_{\text{CM}}(\omega)\}$, for a typical mammalian cell. Arrows show how $\text{Re}\{f_{\text{CM}}(\omega)\}$ of cells changes by changing the cells characteristics. Adapted from Pethig.[53]

frequency occurs at frequencies higher than 10 MHz where suitable and safe electronics are more difficult to build, it has not been studied as in depth as the first crossover frequency.[14]

A membrane-specific area parameter, φ, can be defined to show the ratio of the actual membrane area to the membrane area that would be required to cover a smooth cell with the same radius.[50] The parameter, φ, quantifies the amount of surface foldings and protrusions due to morphological features such as microvilli, villi, ruffles, ridges, and blebs.[50] These complexities on the membrane increase the surface area and consequently the membrane capacitance. The parameter, φ, can also be defined as the ratio of the membrane capacitance of a cell to the membrane capacitance of a completely smooth cell, 9 mF m^{-2}.[51,52] Thus, φ will be unity for a perfectly smooth cell and have higher values for cells with uneven surfaces.

Differences in cell viability, morphology, and structural architecture can also affect the polarisability of the cells, which consequently changes the cells' dielectric properties. For instance, a nonviable cell has a more permeable membrane and ions can pass through it. This increases the conductivity of a cell from about 10^{-7} to around 10^{-3} that, in turn, changes the polarisability of the cell. Also, different cell types and cells at different stages of differentiation have different morphological and cytoskeleton structure, which makes their polarisability different, and consequently changes the DEP force they experience under the same conditions.

3.4 Separation of Biological Cells

3.4.1 Separation of Apoptotic and Necrotic Cells

DEP can also distinguish the changes in cells under apoptosis.[45,54,55] Since apoptosis causes loss of structural features such as microvilli and formation of blebs[55] it changes the dielectric properties of cells.[45] This suggests that apoptosis may cause different morphological changes in different cell types.[54] It is also suggested that DEP might be able to distinguish between apoptosis and necrosis of cells and to sort these from normal cells.[55]

3.4.2 Separation of Stem Cells from Somatic Cells

One of the important applications of DEP can be studying and selectively isolating stem cells from somatic cells. There are several certain and possible differences in biophysical properties of stem cells and nonstem cells. These biophysical differences include protein expression, size, morphology, nucleus to cytoplasm (N/C) ratio, nucleic acid content, presence and conductivity of internal membrane-bound vesicles, surface charge, and charged cytoplasmic molecules. The N/C ratio decreases by cell differentiation and maturity[14] and changes the effective conductivity and permittivity of cells. For instance, human embryonic stem (hES) cells have much higher N/C ratio than ordinary somatic cells. Thus, the cells with different N/C ratios will have different dielectric properties and can be sorted based on that. Specifically, differences in the N/C ratio change the second crossover frequency of cells (Figure 3.4(a)). There is also a network of cytoskeleton structures around the nucleus in somatic cells. Possible lack of this network due to smaller cytoplasmic volume in hES cells can be another reason for difference in their dielectric properties when compared to somatic cells.[56]

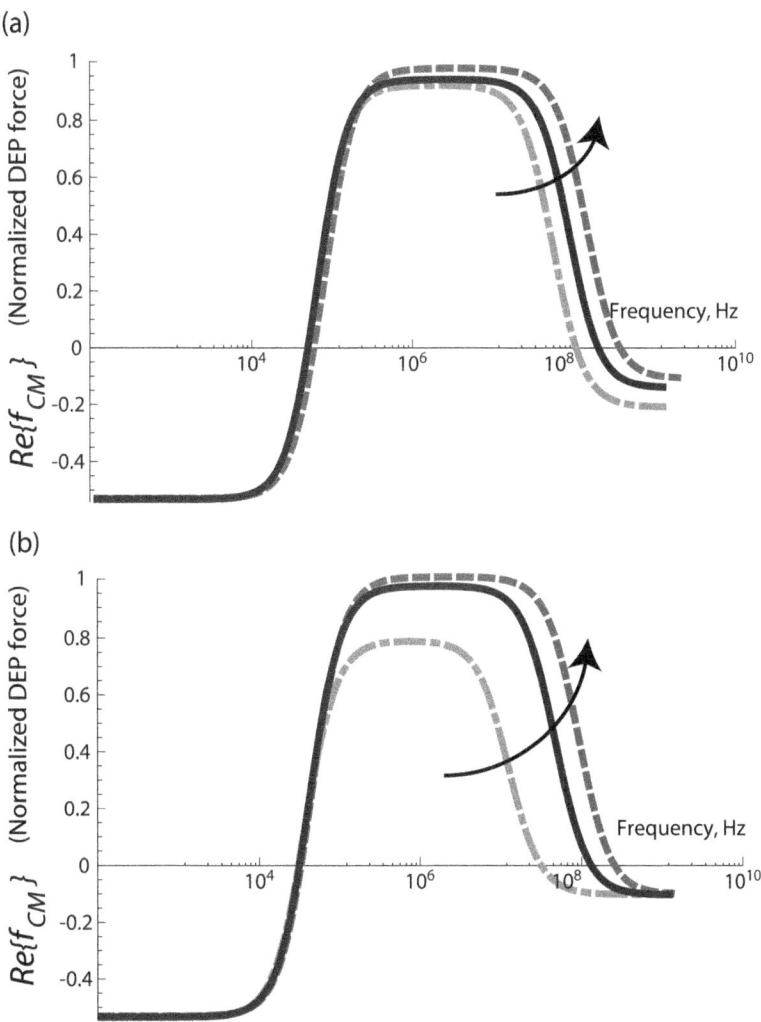

Figure 3.4   Clausius–Mossotti factor changes by increasing (a) N/C ratio of cells and (b) cytoplasm conductivity.

3.4.3 Separation of Cancer Stem Cells from Normal Cancer Cells

Cancer stem cells (CSCs) are a rare population of cancer cells that have the ability to transplant a new tumour from an existing one. CSCs are thought to be responsible for the metastatic properties of tumours. In the case of pancreatic cancer, less than 1% of cancer cells expressed positive to specific biomarkers and showed significantly high tumourigenic potential, indicating they are CSCs.[57]

The polarisation of endoplasmic reticulum, nucleus, and mitochondria affects cell dielectric properties at frequencies less than 1 MHz and increases the relative permittivity of the cells due to interfacial polarisations of these organelles.[53] On the other hand, it must also be taken into account that cells' organelles change during differentiation. For instance, Figure 3.4(a) demonstrates how $Re\{f_{CM}(\omega)\}$ changes by changing the N/C ratio of cells. This is consistent with the presented results that CSCs (assuming they have a higher N/C ratio than non-CSCs) need lower DEP forces to be trapped than non-CSCs at the same frequency in the range of 200–600 kHz.[58]

As another example, mitochondrial membrane potential heterogeneity is different in CSCs and normal cancer cells and therefore can be used as a tool for isolating CSCs.[59] Differentiation, transformation, and tumourigenicity of cells cause a significant change in mitochondrial membrane potential.[59] There is a 60-mV difference between mitochondrial membrane potential of normal cells and cancer cells.[60] Carcinoma cells have a higher mitochondrial membrane potential than normal cells.[60–62] There is also a relation between tumourigenicity of stem cells and their mitochondrial membrane potential; stem cells with higher mitochondrial membrane potential have higher tendency to form tumours[63] Ye *et al.*[59] sorted cells by decreasing mitochondrial membrane potential in the order of CSC > cancer cells > normal epithelial cells. This relation could be due to the difference in the metabolism of CSCs and non-CSCs,[59] as mitochondria play a key role in metabolic functions. Distinct differences in the morphology of human pancreatic cancer stem cells and non-CSCs[64] can also be used to distinguish CSCs and non-CSCs. CSCs are more spindle-shaped and have greater metastatic and ascites formation capability. CSCs also can grow without anchorage, a property of malignant cells.[64] For instance, Kabashima *et al.*[65] reported that a CSC-enriched population of pancreatic cancer cells has a more mesenchymal-like shape and less adhesion proteins, which enables them to migrate.

3.4.4 Separation of Differentiated Cells

Cells' dielectric properties can reveal the developmental progression of progenitor cell populations and can reflect their differentiated fate. Populations of mouse neural stem/precursor cells (NSPCs) can be isolated from differentiated neurons and differentiated astrocytes using DEP.[66]

By comparing these results with the results of ref. 67 for glioma and neuroblastoma, they showed that the differences in cell dielectric properties remains in the cancerous states of these cells.[66] Thus, DEP can solve the challenges of isolating stem cells from more differentiated progeny.

3.4.5 Separation of Progenitor Cells from the Stromal Cells

Progenitor cells have been separated from the stromal vascular fraction that includes cell debris, erythrocytes, and nucleated cells based on their differences between the density, morphology, and size of putative stem cells and other cells.[68] $CD34^+$ stem cells can also be isolated from peripheral blood stem cell harvests using differences in the cell's dielectrophoretic properties.[11] DEP also has been used to isolate human breast cancer cells from hematopoietic $CD34^+$ stem cells due to differences in cell polarisation, morphological characteristics, and density.[69] The specific membrane capacitance of breast cancer cells and hematopoietic $CD34^+$ stem cells are reported as 23.0 ± 7.1 and 10.2 ± 1.5 mF m^{-2}, respectively.[69] These studies showed that differences in cell density can play a role in isolations of some rare-cell subpopulations.

3.4.6 Separation of Different Stages of Cancer Cells

It has also been shown that cancerous and normal cells have different electrical properties. Oral squamous cell carcinomas have distinctly different electrical properties than more normal keratinocyte populations[70] and noncancer-derived oral epithelial cells.[71] Additionally, transformed and nontransformed rat kidney cells,[72] malignant human breast cancer epithelial cells and benign breast epithelial cells,[73,74] and healthy and infected erythrocytes have been shown to have different electrical properties.[75] Similarly, normal, precancerous and cancerous oral keratinocytes have distinct electrical properties.[76] By progression of the disease to more malignant stages, C_{mem} increases from 6.9 ± 0.6 to 10.9 ± 2, 15.1 ± 2.6, and 14.3 ± 4.5 mF m^{-2}, and the cytoplasm conductivity decreases from 0.71 ± 0.08 to 0.42 ± 0.26, 0.26 ± 0.06, and 0.25 ± 0.10 S m^{-1} for primary normal oral keratinocytes, dysplastic, and malignant oral keratinocyte cell lines, respectively.[76] Change in the membrane capacitance is associated with the complexity of the surface morphology, specifically microvilli and the density of complex features. Thus, the membrane capacitance reflects membrane morphology.

Based on these discussions, DEP can be used to investigate the bioelectrical properties of cells during treatment and differentiation as well as changes in cell-membrane morphology because the cells' physiological activities and induction of pathologic state are related to the dielectric properties of cells.[48,77–84] For instance, DEP can be used to distinguish cell-membrane properties, such as membrane protrusions,[85] membrane conductivity,[79] as well as the membrane skeleton.[85,86]

3.4.7   Separation of Cancer Cells from Blood Cells

Isolating cancer cells from blood cells is essential in early cancer detection and monitoring therapeutic outcomes.[87] There are several differences in surface morphology and size of cancer cells and blood cells which makes it possible to isolate them utilising DEP.[50,88–94] A membrane-specific area parameter, φ, can quantify subtle variances between cancer cells and normal blood cells due to differences in membrane morphology.[50] Different stages of cultured breast cancer cells and leukemia cells have been isolated from blood cells using DEP.[50,90,95] For normal cells, such as T-lymphocytes, φ has been reported to be close to 1, however, for cancerous cells having more surface protrusions, φ can be up to around 4 times higher.[96] DEP can also be combined with other cell-sorting techniques, such as multiorifice flow fractionation (MOFF),[88] to isolate cancer cells from blood cells by taking advantage of the higher throughput of MOFF and higher sensitivity of DEP.[88]

3.5   Other Electrokinetic-Based Microseparation Techniques

3.5.1   Insulator-Based Dielectrophoresis (iDEP)

Although DEP has been a very successful technique for manipulating microparticles,[97] it does have drawbacks such as electrolysis (bubble formation), electrode delamination, and sample contamination.[98,99] Traditionally, the nonuniform electric fields necessary to induce dielectrophoresis are generated by patterning metal electrodes onto the bottom of a microfluidic channel or by using insulating structures to distort a uniform electric field.[100] The alternative to using insulators to create nonuniformity in an electric field is known as insulator-based dielectrophoresis (iDEP). iDEP arose to overcome the drawbacks of electrode-based dielectrophoresis (eDEP) such as complex and expensive fabrication as well as loss of functionality due to fouling.[101,102]

iDEP is one of the practical methods used to overcome traditional DEP difficulties. The patterned electrodes required by traditional DEP can be complex and expensive to fabricate. In contrast, iDEP uses insulating obstacles rather than the geometry of the electrodes to generate nonuniformities in the electric field. iDEP microdevices can thus be fabricated using simple etching techniques from a single master, allowing for more economical large-scale systems and mass production.[103–106] iDEP microdevices have been successfully tested for different applications, ranging from macromolecules such as DNA,[107,108] to cells such as bacteria,[7,109] yeast,[110] and microalgae.[111] However, it is still difficult to use iDEP with highly conductive biological samples because Joule heating and bubble formation still occur.

3.5.2 Contactless Dielectrophoresis (cDEP)

Contactless dielectrophoresis (cDEP) is an extension of these techniques in which metal electrodes are exchanged for conductive fluid electrode channels. In cDEP, an electric field is created in a microchannel by inserting electrodes into two side channels filled with conductive solution. These side channels are separated from the main channel by thin insulating barriers that exhibit a capacitive behaviour. The application of a high-frequency electric field to the electrode reservoirs causes their capacitive coupling to the main channel and an electric field is induced across the sample fluid. Not having direct contact between the sample and the metallic electrodes eliminates many of the challenges with conventional DEP such as electrolysis in the sample channel and electrode delamination. Furthermore, the fabrication process is relatively simple because it is not necessary to pattern microelectrodes in the main channels. This method is well suited for mass-fabrication techniques such as hot embossing and injection moulding.

cDEP has recently been used in various applications. Cancer stem cells (CSCs) have been separated from nontumour-initiating cells[112] as well as cancer cells from erythrocytes.[23] The technique has also been used to isolate THP-1 human leukemia monocytes from a heterogeneous mixture of live and dead cells[113] and to segregate breast cancer cells from different cell lines of varying metastatic potential.[114]

3.5.3 Electrophoresis

Electrophoresis is the movement of charged particles in an electric field. The electrophoretic mobility of a charged particle moving in an electric field is defined as the ratio of its velocity to the electric field

$$\mu_{EP} \equiv \frac{u}{\mathbf{E}} \qquad (3.25)$$

The electrical double layer, and in particular its thickness, plays an important role in μ_{EP} of particles. The ratio of the radius of the particle to the thickness of the double layer, Debye length κ^{-1}, is shown as κr. Then, the mobility can be written in a general form as,

$$\mu_{EP} = \frac{\varepsilon \zeta}{\eta} f(\kappa r) \qquad (3.26)$$

where ε and η are the permittivity and viscosity of the medium, respectively, and ζ is the zeta potential of the particle.

For a thin double layer, $f(\kappa r) = 1$, is known as the Helmholtz–Smolunchowski limit, and for a thick double layer $f(\kappa r) = 2/3$, is known as the Huckel–Onsager limit (for very dilute electrolyte or very small particles).[33]

Electrophoresis plays an important role in separation of charged particles, including small molecules,[115,116] DNA,[117,118] and proteins[119] within a sample.

Electrophoresis separation has been used in the food industry, antibioterrorism, and environmental safety through uses of portable microdevices.[120]

3.5.4 Electro-osmosis

Electro-osmosis occurs when a tangential electric field is applied to a surface that is in contact with electrolytes. The tangential electric field applies a force to the double-layer charges on the surface and consequently pulls the fluid and creates a flow. The electro-osmotic velocity in a channel can be calculated by considering the balance between the electrical force (Coulomb force) and the friction force:

$$\vec{\mathbf{u}}_{EO} = \frac{\varepsilon \zeta}{\eta} \vec{\mathbf{E}} = \mu_{EO} \vec{\mathbf{E}} \tag{3.27}$$

To find this relation it is assumed that the channel depth is infinite and a no-slip boundary condition is satisfied on the electrode surface. The velocity of the fluid is proportional to the zeta potential of the surface, ζ, permittivity, ε, and viscosity, η, of the fluid, and the applied electric field. The electro-osmotic mobility is defined as, μ_{EO}. The electrokinetic mobility can be defined as the summation of electro-osmotic and electrophoretic mobilities:

$$\mu_{EK} = \mu_{EO} + \mu_{EP} \tag{3.28}$$

The oscillatory displacement of a particle due to EO is equal to $\vec{\mathbf{u}}_{EO}/\omega$.[35] If the electric field is about 10^5 V m^{-1}, the frequency on the order of kHz, and zeta potential about 10 mV, then the velocity will be on the order of 1 mm s^{-1} and the oscillatory displacement of the particle will be on the order of 1 μm. The oscillatory displacement can be negligible compared to the displacement of the particles due to DEP forces. It is important to note that there is a difference in the flow velocity profile between AC-EO and pressure-driven flow. The induced flow due to AC-EO is plug-like, while in pressure-driven flow, the velocity profile takes a parabolic shape (zero close to walls and has its maximum value at the centre).

AC-EO has the advantage over DC-EO in that it requires less voltage for operation. Two of the most frequent applications of AC-EO are fluid micropumping[121,122] and micromixing.[17,123] The fluid flow rate in AC-EO pumping depends on the applied voltage and frequency, electrode shape and arrangement, the properties of the fluid, and the zeta potential. It is important to mention that in AC-electro-osmosis, the electric field should be nonuniform to have a nonzero time-averaged flow. Flow will be zero in a uniform AC field.[33] This is similar to dielectrophoresis which is also zero in a uniform electric field.

The magnitude of AC-EO velocity depends on the charge density and charge density depends on the applied frequency; thus, AC-EO depends on the frequency. At low frequencies (<10 kHz) AC-EO is almost zero since there is no field in the bulk of the fluid. At high frequencies (>100 kHz) AC-EO is again zero since there is not enough time to induce charges to form the double layer. At frequencies higher than 100 kHz usually electrothermal effects are dominant, depending on the conductivity of the sample, as is explained in the next section.

In an insulating, impermeable channel for the flux of particles, j, on the channel boundaries, we can write

$$\vec{j} \cdot \vec{n} = 0. \tag{3.29}$$

And for the inside of the channel, ignoring the diffusion, we can write[124]

$$\vec{j} = C \sum \vec{v}_i, \tag{3.30}$$

where C is the concentration of particles and $\vec{v}_i$ is the velocity of particles due to different phenomena such as pressure-driven flow, dielectrophoresis, and electrokinetics (electro-osmosis and electrophoresis),

$$\vec{j} = C\left(\vec{v}_{\text{pressure}} + \vec{v}_{\text{DEP}} + \vec{v}_{\text{EK}}\right), \tag{3.31}$$

where $\vec{v}_{\text{pressure}}$, $\vec{v}_{\text{DEP}}$, and $\vec{v}_{\text{EK}}$ are the velocities due to pressure-driven flow, DEP and electrokinetics.

In the case of capturing targeted cells, the DEP force experienced by target cells should overcome electro-osmotic and electrophoretic effects. This can be achieved when the flux along the electric field lines is zero

$$\vec{j} \cdot \vec{E} = 0. \tag{3.32}$$

Then,

$$\left(\mu_{\text{EK}}\vec{E} + \mu_{\text{DEP}}\nabla(\vec{E} \cdot \vec{E})\right) \cdot \vec{E} = 0 \tag{3.33}$$

For trapping a particle due to DEP force, we can write[125]

$$\frac{\mu_{\text{DEP}}\nabla(\vec{E} \cdot \vec{E})}{\mu_{\text{EK}}\vec{E}} \cdot \vec{E} > 1 \tag{3.34}$$

3.5.5 Electrothermal Effects in the Fluid

Due to the application of a nonuniform AC field, a nonuniformity in the temperature of the fluid occurs which can generate gradients in the conductivity, permittivity, density, and/or viscosity within the fluid locally. This imbalance in the properties of the fluid creates electrothermal (ET) forces, which then leads to fluid motion.[17]

There are some differences between AC-EO and ET fluid flow. AC-EO usually happens at frequencies less than 100 kHz, but the frequency to observe ET is usually greater than 100 kHz. The conductivity of the sample in AC-EO is usually very low (<0.01 mS m^{-1}) and most of the time DI water is used, but the conductivity of the sample in ET pumping is usually high (>0.01 S m^{-1}).[126,127] AC-EO happens because of the charges on the surface of the electrodes or walls due to Faradaic or capacitive charging.[126] However, ET is a volumetric force. Since ET fluid movement does not play a role in particle separation, it is not discussed any further, but should be borne in mind when designing cell-separation processes, particularly where physiological media of high conductivities can play and influential role in its occurrence within electrokinetic-based microsystems.

3.6 Summary

The movement of particles in a microfluidic device subjected to an electric field and their interaction with the suspending fluid was studied in this chapter. Emphasis was placed on AC nonuniform electric fields for rare cell-sorting applications. Many electrokinetic phenomena discussed above may act in concert upon a particle within a microfluidic system under a nonuniform electric field. Nonuniformity in the electric field can create gradients in permittivity and conductivity. This can cause electrothermal forces on the fluidor forces on the electrical double-layer charges therefore driving electroosmotic fluid flow. Among electrokinetic processes, DEP plays an important role in separating particles, specifically in isolating subpopulations of cells. DEP can detect and separate cells based on the differences in their size, morphology, nucleus/cytoplasm ratio, nucleic acid content, surface charge, and charged cytoplasmic molecules. The lack of labelling cells through their membrane proteins makes DEP-based isolation a promising technique for crucial applications, such as early cancer detection, performance monitoring of cancer therapy, regenerative medicine, and cell characterisation for personalised medicine.

References

1. S. Nagrath, L. V. Sequist, S. Maheswaran, D. W. Bell, D. Irimia, L. Ulkus, M. R. Smith, E. L. Kwak, S. Digumarthy, A. Muzikansky, P. Ryan, U. J. Balis, R. G. Tompkins, D. A. Haber and M. Toner, *Nature*, 2007, **450**, 1235–1239.
2. D. F. Hayes, M. Cristofanilli, G. T. Budd, M. J. Ellis, A. Stopeck, M. C. Miller, J. Matera, W. J. Allard, G. V. Doyle and L. W. Terstappen, *Clin. Cancer Res.*, 2006, **12**, 4218–4224.
3. V. VanDelinder and A. Groisman, *Anal. Chem.*, 2006, **78**, 3765–3771.
4. V. VanDelinder and A. Groisman, *Anal. Chem.*, 2007, **79**, 2023–2030.
5. D. Di Carlo, *Lab Chip*, 2009, **9**, 3038–3046.
6. S. Choi, S. Song, C. Choi and J. K. Park, *Anal. Chem.*, 2009, **81**, 1964–1968.
7. B. H. Lapizco-Encinas, B. A. Simmons, E. B. Cummings and Y. Fintschenko, *Anal. Chem.*, 2004, **76**, 1571–1579.
8. G. H. Markx, M. S. Talary and R. Pethig, *J. Biotechnol.*, 1994, **32**, 29–37.

9. K. H. Kang, Y. Kang, X. Xuan and D. Li, *Electrophoresis,* 2006, **27**, 694–702.
10. H. Shafiee, J. L. Caldwell and R. V. Davalos, *JALA,* 2010, **15**, 224–232.
11. M. Stephens, M. S. Talary, R. Pethig, A. K. Burnett and K. I. Mills, *Bone Marrow Transplant.,* 1996, **18**, 777–782.
12. Z. Gagnon and H. C. Chang, *Electrophoresis,* 2005, **26**, 3725–3737.
13. A. Lostumbo, D. Mehta, S. Setty and R. Nunez, *Exp. Mol. Pathol.,* 2006, **80**, 46–53.
14. R. Pethig, A. Menachery, S. Pells and P. De Sousa, *J. Biomed. Biotechnol.,* 2010, **2010**, 182581.
15. E. D. Pratt, C. Huang, B. G. Hawkins, J. P. Gleghorn and B. J. Kirby, *Chem. Eng. Sci.,* 2011, **66**, 1508–1522.
16. S. K. Fan, P. W. Huang, T. T. Wang and Y. H. Peng, *Lab Chip,* 2008, **8**, 1325–1331.
17. W. Y. Ng, S. Goh, Y. C. Lam, C. Yang and I. Rodriguez, *Lab Chip,* 2009, **9**, 802–809.
18. R. Johann and P. Renaud, *Electrophoresis,* 2004, **25**, 3720–3729.
19. H. Kawamoto and K. Tsuji, *Adv. Powder Technol.,* 2011, **22**, 602–607.
20. S. Saito, H. Himeno, K. Takahashi and M. Urago, *Appl. Phys. Lett.,* 2003, **83**, 2076–2078.
21. T. Yasukawa, K. Nagamine, Y. Horiguchi, H. Shiku, M. Koide, T. Itayama, F. Shiraishi and T. Matsue, *Anal. Chem.,* 2008, **80**, 3722–3727.
22. A. H. Harakuwe and P. R. Haddad, *TrAc, Trends Anal. Chem.,* 2001, **20**, 375–385.
23. L. C. Campbell, M. J. Wilkinson, A. Manz, P. Camilleri and C. J. Humphreys, *Lab Chip,* 2004, **4**, 225–229.
24. C. Zhang, K. Khoshmanesh, A. Mitchell and K. Kalantar-zadeh, *Anal. Bioanal. Chem.,* 2010, **396**, 401–420.
25. M. Toner and D. Irimia, *Annu. Rev. Biomed. Eng.,* 2005, **7**, 77–103.
26. R. Carlson, C. V. Gabel, S. S. Chan, R. H. Austin, J. P. Brody and J. Winkleman, *Mol. Biol. Cell,* 1997, **8**, 2407.
27. S. Ostergaard, G. Blankenstein, H. Dirac and O. Leistiko, *J. Magn. Magn. Mater.,* 1999, **194**, 156–162.
28. L. Pasqua, S. Cundari, C. Ceresa and G. Cavaletti, *Curr. Med. Chem.,* 2009, **16**, 3054–3063.
29. N. F. Kushchevskaya, *Powder Metall. Met. Ceram.,* 1997, **36**, 668–672.
30. W. Q. Ding, *J. Adhes. Sci. Technol.,* 2008, **22**, 457–480.
31. J. R. Melcher and G. I. Taylor, *Annu. Rev. Fluid Mech.,* 1969, **1**, 111–146.
32. Z. Wu, B. Willing, J. Bjerketorp, J. K. Jansson and K. Hjort, *Lab Chip,* 2009, **9**, 1193–1199.
33. H. Morgan, N. G. Green, *AC Electrokinetics: Colloids and Nanoparticles*, Research Studies Press Ltd, 2003.
34. A. Castellanos, *Electrohydrodynamics*, Springer, 1998.
35. A. Castellanos, A. Ramos, A. Gonzalez, N. G. Green and H. Morgan, *J. Phys. D: Appl. Phys.,* 2003, **36**, 2584–2597.
36. R. Clift, J. R. Grace and M. E. Weber, *Bubbles, Drops and Particles*, Academic, New York, 1978.

37. A. Salmanzadeh, H. Shafiee, R. V. Davalos and M. A. Stremler, *Electrophoresis,* 2011, **32**, 2569–2578.
38. M. P. Hughes and H. Morgan, *Anal. Chem.,* 1999, **71**, 3441–3445.
39. A. R. Minerick, R. Zhou, P. Takhistov and H. C. Chang, *Electrophoresis,* 2003, **24**, 3703–3717.
40. A. D. Goater and R. Pethig, *Parasitology,* 1998, **117**, S177–S189.
41. H. A. Pohl, *Dielectrophoresis,* Cambridge University Press, 1978.
42. A. Kumar, Z. Qiu, A. Acrivos, B. Khusid and D. Jacqmin, *Phys. Rev. E: Statist., Nonlinear, Soft Matter Phys.,* 2004, **69**, 021402.
43. A. Irimajiri, T. Hanai and A. Inouye, *J. Theor. Biol.,* 1979, **78**, 251–269.
44. A. Salmanzadeh, M. B. Sano, R. C. Gallo-Villanueva, P. C. Roberts, E. M. Schmelz and R. V. Davalos, *Biomicrofluidics,* 2013, **7**, 011809.
45. R. Pethig and M. S. Talary, *IET Nanobiotechnol.,* 2007, **1**, 2–9.
46. R. Pethig, *Biomicrofluidics,* 2010, **4**, 022811.
47. D. Miklavčič, N. Pavšelj and F. X. Hart, in *Wiley Encyclopedia of Biomedical Engineering*, Wiley, 2006.
48. Y. Huang, X. B. Wang, F. F. Becker and P. R. C. Gascoyne, *Biochem. Biophys. Acta, Biomembr.,* 1996, **1282**, 76–84.
49. R. Pethig, V. Bressler, C. Carswell-Crumpton, Y. Chen, L. Foster-Haje, M. E. Garcia-Ojeda, R. S. Lee, G. M. Lock, M. S. Talary and K. M. Tate, *Electrophoresis,* 2002, **23**, 2057–2063.
50. P. R. Gascoyne, X. B. Wang, Y. Huang and F. F. Becker, *IEEE Trans. Ind. Appl.,* 1997, **33**, 670–678.
51. R. Pethig and D. B. Kell, *Phys. Med. Biol.,* 1987, **32**, 933–970.
52. W. M. Arnold and U. Zimmermann, *Naturwissenschaften,* 1982, **69**, 297–298.
53. R. Pethig, in *BioMEMS and Biomedical Nanotechnology*, SpringerLink, 2007, pp. 103–126.
54. F. H. Labeed, H. M. Coley and M. P. Hughes, *Biochem. Biophys. Acta, Gen. Subj.,* 2006, **1760**, 922–929.
55. X. J. Wang, F. F. Becker and P. R. C. Gascoyne, *Biochem. Biophys. Acta, Biomembr.,* 2002, **1564**, 412–420.
56. B. C. Heng, H. Liu and T. Cao, *Med. Hypotheses,* 2005, **64**, 1242–1243.
57. C. W. Li, D. G. Heidt, P. Dalerba, C. F. Burant, L. J. Zhang, V. Adsay, M. Wicha, M. F. Clarke and D. M. Simeone, *Cancer Res.,* 2007, **67**, 1030–1037.
58. A. Salmanzadeh, L. Romero, H. Shafiee, R. C. Gallo-Villanueva, M. A. Stremler, S. D. Cramer and R. V. Davalos, *Lab Chip,* 2012, **12**, 182–189.
59. X. Q. Ye, G. H. Wang, G. J. Huang, X. W. Bian, G. S. Qian and S. C. Yu, *Stem Cell Rev.,* 2011, **7**, 153–160.
60. L. B. Chen, *Annu. Rev. Cell Biol.,* 1988, **4**, 155–181.
61. G. Kroemer and J. Pouyssegur, *Cancer Cell,* 2008, **13**, 472–482.
62. D. M. Hockenbery, *Cancer Cell,* 2002, **2**, 1–2.
63. S. M. Schieke, M. Ma, L. Cao, J. P. McCoy, Jr, C. Liu, N. F. Hensel, A. J. Barrett, M. Boehm and T. Finkel, *J. Biol. Chem.,* 2008, **283**, 28506–28512.

64. M. K. Hassanein, A. Suetsugu, S. Saji, H. Moriwaki, M. Bouvet, A. R. Moossa and R. M. Hoffman, *J. Cell. Biochem.*, 2011, **112**, 3549–3554.
65. A. Kabashima, H. Higuchi, H. Takaishi, Y. Matsuzaki, S. Suzuki, M. Izumiya, H. Iizuka, G. Sakai, S. Hozawa, T. Azuma and T. Hibi, *Int. J. Cancer*, 2009, **124**, 2771–2779.
66. L. A. Flanagan, J. Lu, L. Wang, S. A. Marchenko, N. L. Jeon, A. P. Lee and E. S. Monuki, *Stem Cells*, 2008, **26**, 656–665.
67. Y. Huang, S. Joo, M. Duhon, M. Heller, B. Wallace and X. Xu, *Anal. Chem.*, 2002, **74**, 3362–3371.
68. J. Vykoukal, D. M. Vykoukal, S. Freyberg, E. U. Alt and P. R. C. Gascoyne, *Lab Chip*, 2008, **8**, 1386–1393.
69. Y. Huang, J. Yang, X. B. Wang, F. F. Becker and P. R. Gascoyne, *J. Hematother. Stem Cell Res.*, 1999, **8**, 481–490.
70. L. M. Broche, N. Bhadal, M. P. Lewis, S. Porter, M. P. Hughes and F. H. Labeed, *Oral Oncol.*, 2007, **43**, 199–203.
71. L. Yang, L. R. Arias, T. S. Lane, M. D. Yancey and J. Mamouni, *Anal. Bioanal. Chem.*, 2011, **399**, 1823–1833.
72. Y. Huang, X. B. Wang, F. F. Becker and P. R. Gascoyne, *Biochim. Biophys. Acta*, 1996, **1282**, 76–84.
73. J. An, J. Lee, S. H. Lee, J. Park and B. Kim, *Anal. Bioanal. Chem.*, 2009, **394**, 801–809.
74. E. A. Henslee, M. B. Sano, A. D. Rojas, E. M. Schmelz and R. V. Davalos, *Electrophoresis*, 2011, **32**, 2523–2529.
75. T. Braschler, N. Demierre, E. Nascimento, T. Silva, A. G. Oliva and P. Renaud, *Lab Chip*, 2008, **8**, 280–286.
76. H. J. Mulhall, F. H. Labeed, B. Kazmi, D. E. Costea, M. P. Hughes and M. P. Lewis, *Anal. Bioanal. Chem.*, 2011, **401**, 2455–2463.
77. X. B. Wang, Y. Huang, P. R. C. Gascoyne, F. F. Becker, R. Holzel and R. Pethig, *Biochem. Biophys. Acta, Biomembr.*, 1994, **1193**, 330–344.
78. P. R. C. Gascoyne, R. Pethig, J. P. H. Burt and F. F. Becker, *Biochim. Biophys. Acta*, 1993, **1149**, 119–126.
79. J. P. H. Burt, R. Pethig, P. R. C. Gascoyne and F. F. Becker, *Biochim. Biophys. Acta*, 1990, **1034**, 93–101.
80. P. R. C. Gascoyne, J. Noshari, F. F. Becker and R. Pethig, *IEEE Trans. Ind. Appl.*, 1994, **30**, 829–834.
81. J. Gimsa, P. Marszalek, U. Loewe and T. Y. Tsong, *Biophys. J.*, 1991, **60**, 749–760.
82. Y. Huang, X. B. Wang, P. R. C. Gascoyne and F. F. Becker, *Biochem. Biophys. Acta, Biomembr.*, 1999, **1417**, 51–62.
83. X. J. Wang, J. Yang, Y. Huang, M. Andreeff, F. F. Becker and P. R. C. Gascoyne, *Blood*, 1998, **92**, 167b.
84. J. Yang, Y. Huang, X. J. Wang, X. B. Wang, F. F. Becker and P. R. C. Gascoyne, *Biophys. J.*, 1999, **76**, 3307–3314.
85. X. B. Wang, Y. Huang, P. R. Gascoyne, F. F. Becker, R. Holzel and R. Pethig, *Biochim. Biophys. Acta*, 1994, **1193**, 330–344.

86. P. R. Gascoyne, R. Pethig, J. P. Burt and F. F. Becker, *Biochim. Biophys. Acta,* 1993, **1149**, 119–126.

87. M. Yu, S. Stott, M. Toner, S. Maheswaran and D. A. Haber, *J. Cell. Biol.,* 2011, **192**, 373–382.

88. H.-S. Moon, K. Kwon, S.-I. Kim, H. Han, J. Sohn, S. Lee and H.-I. Jung, *Lab Chip,* 2011, **11**, 1118–1125.

89. X. Wang, J. Yang and P. R. Gascoyne, *Biochim. Biophys. Acta,* 1999, **1426**, 53–68.

90. F. F. Becker, X. B. Wang, Y. Huang, R. Pethig, J. Vykoukal and P. R. Gascoyne, *Proc. Natl. Acad. Sci. USA,* 1995, **92**, 860–864.

91. P. R. C. Gascoyne, J. Noshari, T. J. Anderson and F. F. Becker, *Electrophoresis,* 2009, **30**, 1388–1398.

92. A. Salmanzadeh, M. B. Sano, H. Shafiee, M. A. Stremler and R. V. Davalos, presented in part at the EMBC, San Diego, CA, 2012.

93. J. Cheng, E. L. Sheldon, L. Wu, M. J. Heller and J. P. O'Connell, *Anal. Chem.,* 1998, **70**, 2321–2326.

94. F. Becker, X. Wang, Y. Huang, R. Pethig, J. Vykoukal and P. Gascoyne, *J. Phys. D: Appl. Phys.,* 1994, **27**, 2659–2662.

95. P. R. Gascoyne, J. Noshari, T. J. Anderson and F. F. Becker, *Electrophoresis,* 2009, **30**, 1388–1398.

96. P. R. Gascoyne, S. Shim, J. Noshari, F. F. Becker and K. Stemke-Hale, *Electrophoresis,* 2012, **34**, 1042–1050.

97. K. Khoshmanesh, S. Nahavandi, S. Baratchi, A. Mitchell and K. Kalantar-zadeh, *Biosens. Bioelectron.,* 2011, **26**, 1800–1814.

98. E. B. Cummings and A. K. Singh, *Anal. Chem.,* 2003, **75**, 4724–4731.

99. C. Chou, J. Tegenfeldt, O. Bakajin, S. Chan, E. Cox, N. Darnton, T. Duke and R. Austin, *Biophys. J.,* 2002, **83**, 2170–2179.

100. C. F. Gonzalez and V. T. Remcho, *J. Chromatogr. A,* 2005, **1079**, 59–68.

101. H. B. Li and R. Bashir, *Sens. Actuators B,* 2002, **86**, 215–221.

102. G. H. Markx, Y. Huang, X. F. Zhou and R. Pethig, *Microbiology,* 1994, **140**, 585–591.

103. B. H. Lapizco-Encinas, S. Ozuna-Chacón and M. Rito-Palomares, *J. Chromatogr. A,* 2008, **1206**, 45–51.

104. B. A. Simmons, G. J. Mcgraw, R. V. Davalos, G. J. Fiechtner, Y. Fintschenko and E. B. Cummings, *MRS Bull.,* 2006, **31**, 120–124.

105. L. M. Barrett, A. J. Skulan, A. K. Singh, E. B. Cummings and G. J. Fiechtner, *Anal. Chem.,* 2005, **77**, 6798–6804.

106. P. Sabounchi, A. M. Morales, P. Ponce, L. P. Lee, B. a. Simmons and R. V. Davalos, *Biomed. Microdevices,* 2008, **10**, 661–670.

107. C. F. Chou, J. O. Tegenfeldt, O. Bakajin, S. S. Chan, E. C. Cox, N. Darnton, T. Duke and R. H. Austin, *Biophys. J.,* 2002, **83**, 2170–2179.

108. R. C. Gallo-Villanueva, C. E. Rodriguez-Lopez, R. I. Diaz-De-La-Garza, C. Reyes-Betanzo and B. H. Lapizco-Encinas, *Electrophoresis,* 2009, **30**, 4195–4205.

109. M. D. Pysher and M. A. Hayes, *Anal. Chem.,* 2007, **79**, 4552–4557.

110. H. Moncada-Hernandez and B. H. Lapizco-Encinas, *Anal. Bioanal. Chem.,* 2010, **396**, 1805–1816.
111. R. C. Gallo-Villanueva, N. M. Jesus-Perez, J. I. Martinez-Lopez, A. Pacheco and B. H. Lapizco-Encinas, *Microfluid. Nanofluid.,* 2011, **10**, 1305–1315.
112. A. Salmanzadeh, L. Romero, H. Shafiee, R. C. Gallo-Villanueva, M. A. Stremler, S. D. Cramer and R. V. Davalos, *Lab Chip,* 2012, **12**, 182–189.
113. H. Shafiee, M. B. Sano, E. A. Henslee, J. L. Caldwell and R. V. Davalos, *Lab Chip,* 2010, **10**, 438–445.
114. E. Henslee, M. B. Sano, A. Rojas, E. Schmelz and R. V. Davalos, *Electrophoresis,* 2011, **32**, 2523–2529.
115. L. D. Hutt, D. P. Glavin, J. L. Bada and R. A. Mathies, *Anal. Chem.,* 1999, **71**, 4000–4006.
116. N. P. Beard, J. B. Edel and A. J. deMello, *Electrophoresis,* 2004, **25**, 2363–2373.
117. E. T. Lagally, P. C. Simpson and R. A. Mathies, *Sens. Actuators B,* 2000, **63**, 138–146.
118. A. Wainright, U. T. Nguyen, T. Bjornson and T. D. Boone, *Electrophoresis,* 2003, **24**, 3784–3792.
119. J. Vieillard, R. Mazurczyk, C. Morin, B. Hannes, Y. Chevolot, P. L. Desbene and S. Krawczyk, *J. Chromatogr. B,* 2007, **845**, 218–225.
120. D. P. Wu, J. H. Qin and B. C. Lin, *J. Chromatogr. A,* 2008, **1184**, 542–559.
121. C. T. Culbertson, R. S. Ramsey and J. M. Ramsey, *Anal. Chem.,* 2000, **72**, 2285–2291.
122. I. M. Lazar and B. L. Karger, *Anal. Chem.,* 2002, **74**, 6259–6268.
123. I. Glasgow, J. Batton and N. Aubry, *Lab Chip,* 2004, **4**, 558–562.
124. B. H. Lapizco-Encinas, B. A. Simmons, E. B. Cummings and Y. Fintschenko, *Electrophoresis,* 2004, **25**, 1695–1704.
125. R. V. Davalos, G. J. McGraw, T. I. Wallow, A. M. Morales, K. L. Krafcik, Y. Fintschenko, E. B. Cummings and B. A. Simmons, *Anal. Bioanal. Chem.,* 2008, **390**, 847–855.
126. E. Du and S. Manoochehri, *J. Appl. Phys.,* 2008, **104**, 064902.
127. N. Sasaki, *Anal. Sci.,* 2012, **28**, 3–8.

CHAPTER 4

Digital Microfluidics

KAILIANG WANG[a] AND HENRY O. FATOYINBO[*b]

[a]GenMark Diagnostics, 5964 La Place Court, Carlsbad, CA 92008, USA;
[b]University of Surrey, Faculty of Engineering and Physical Sciences,
Department of Mechanical and Engineering Sciences, Centre for Biomedical
Engineering, Guildford, Surrey, GU2 7XH, UK
*E-mail: henry.fatoyinbo@uclmail.net

4.1 Introduction

Transport of fluids within microsystems occurs primarily through the actuation of bulk aqueous solutions within enclosed microfluidic channel networks. This continuous-flow regime is commonly accomplished by pressure-driven or electrokinetic-driven methods.[1] On the other hand, discrete fluid volumes or droplets can also be manipulated within microsystems, overcoming certain perceived challenges once associated with earlier microfluidic systems. Although advancements in microfabrication techniques have significantly improved the operational reliability of newer microfluidic devices where bulk flow is concerned, actuation of droplets has a few distinct advantages over bulk volume actuation. For instance, complex microchannel geometries or micromechanical components are not essential, meaning dead-volume spaces are dramatically reduced for volume-critical samples. Secondly, the controlled dispensation from reservoirs, merging, mixing and splitting of picolitre- to microlitre-sized droplets provides an environment suitable for rapid multiplexing of a range of microscale processes in different phases, including biomolecular assays, sequential chemical reactions, cell-based assays, microbioreactors and tissue engineering.

RSC Detection Science Series No. 5
Microfluidics in Detection Science: Lab-on-a-chip Technologies
Edited by Fatima H Labeed and Henry O Fatoyinbo
Published by the Royal Society of Chemistry, www.rsc.org

4.1.1 Electrostatic Droplet Actuation

Microactuation of droplets[†] has been demonstrated using thermocapillary forces,[2-4] electrostatic forces,[5,6] optical forces,[7] and acoustic forces.[8,9] Within the electrostatics field is a mechanism known as *electrowetting* (*EW*). This mechanism, which involves the electrical control of interfacial tension between media phases, is a popular route for fluid microactuation due to low power requirements, rapid reversal of contact angles, and stability with wide-ranging applicability in areas such as microdevice thermal management,[10,11] optical displays,[12-14] variable-focus microlenses,[15-17] microconveyor systems,[18] microbubble manipulation,[19] and microviscometry.[20] A common configuration for the technology widely referred to as droplet based EW is *electrowetting-on-dielectrics* (*EWOD*). It has gained significant traction since the early 2000s as a contender in lab-on-a-chip system development and integration for processes from sample preparation to combinatorial chemistry and analytical detection, in fields as varied as life sciences to environmental monitoring.[21] Whilst the generation and bulk transportation of discrete droplets in two-phase microfluidic systems (*e.g.*, water-in-oil emulsions) has been demonstrated in traditional microfluidic channels,[22,23] this chapter focuses on the physics underpinning individual droplet manipulation through electrostatic forces on microchannel-less microelectrode arrays and their evolving applications in lab-on-a-chip systems.

4.2 Principles of Electrowetting

4.2.1 Lippmann's Equation

Since the early investigations into electrocapillarity by Gabriel Lippmann (see ref. 24 for translation of original works), who observed a mercury drop covered in dilute sulfuric acid contracted when touched by an iron wire then returned to its original shape upon removal was influenced by electrical polarisation of its surface, there has been much work into the effects of electric fields at phase boundaries as a means for small-scale fluid actuation within microfluidic systems.[12,25,26] In the case of electrocapillary, surface tension is a function of the electric potential across the interface of a liquid metal (*e.g.*, mercury) and a liquid electrolyte, with motion occurring due to pressure differences caused by induced surface tension imbalances $(\gamma_A < \gamma_R)$ at the advancing and receding menisci by an applied voltage as shown in Figure 4.1(a). Based on various determinants, such as temperature, concentrations and pairing of liquid metal and liquid electrolyte an electrical double layer (EDL) of variable thickness is formed at the interface, with electrical isolation of the liquid metal from the electrolyte.[27] It should be noted that the

---

[†]Actuation of dispersed droplets in pressure-driven microfluidic channels is primarily based on immiscible two-phase encapsulating systems. They are a distinct paradigm from individual droplet manipulation techniques, where three-phase systems are taken into account, which is the focus of digital microfluidics in this chapter.

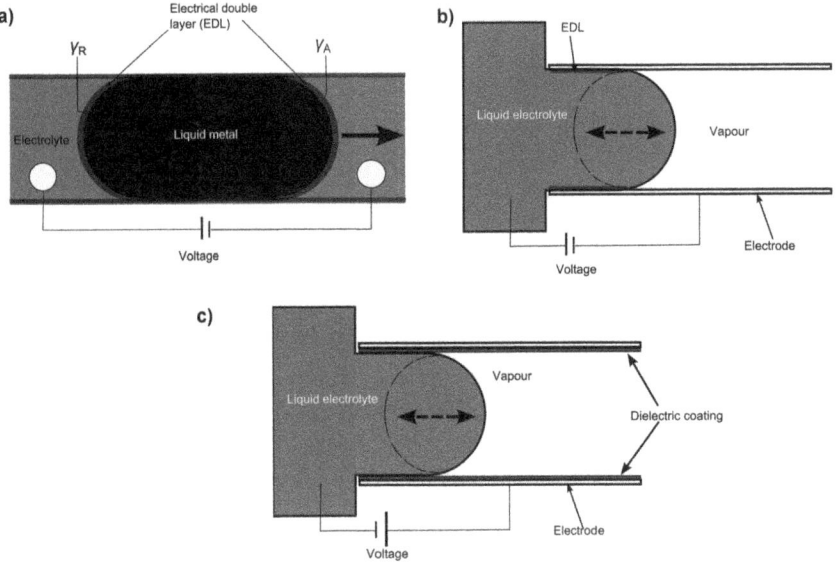

Figure 4.1   Types of voltage-induced fluid-motion phenomena; (a) electrocapillarity, (b) electrowetting and (c) electrowetting on dielectric.

EDL presented in this system differs from that occurring in AC electro-osmotic flow regimes. The relationship between surface tension (γ) and voltage difference (V) across the electrical double layer is expressed through Lippmann's equation (eqn (4.1)), where γ_0 is the initial surface tension with no voltage applied and C_{DL} is the capacitance per unit area ($F \cdot m^{-2}$) of the electrical double layer.[25]

$$\gamma = \gamma_0 - \frac{1}{2}C_{DL}V^2 \tag{4.1}$$

Electrowetting (EW), (see Figure 4.1(b)), differs from electrocapillary in that instead of an electrically induced curvature change in meniscus between liquid metal and electrolyte, EW is typically used to describe the electrically induced *spreading* of a liquid on to a solid and the observed contact-angle reduction which may result (*i.e.* wetting).[12,28,29]

4.2.2   Surface and Interfacial Tensions

Surface tension is a property caused by the cohesive forces of similar molecules at the surface of a material, such as a liquid, to minimise surface energy, thus enabling the material to withstand external pressures. Molecules distributed within the bulk liquid form intermolecular attractions (*e.g.*, van der Waals forces, hydrogen bonds, ionic bonds and London dispersion forces) with neighbouring molecules giving a net force of zero, while in the interfacial surface region an imbalance of intermolecular forces exists with

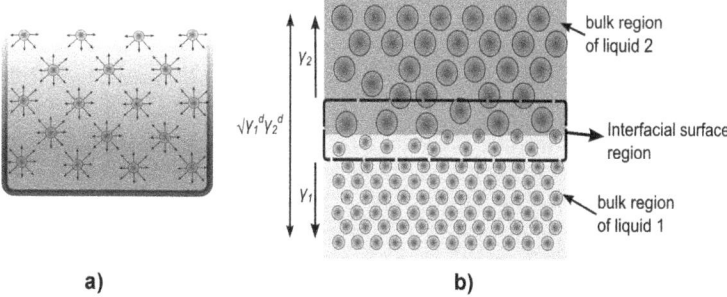

Figure 4.2 (a) Net intermolecular forces at the surface is directed towards the bulk of the fluid, while a net force of zero in the bulk fluid exists due to intermolecular attraction on all sides of a molecule; (b) Interfacial surface tension of two dissimilar liquid materials is made up of each material's surface tension and the London dispersion forces between the two materials at the interfacial region.

a net attractive force normal to the surface and towards the bulk of the liquid (Figure 4.2(a)). A reduction in molecule numbers in the interfacial surface region results in an increase in intermolecular distance. The work (W) returned to the system, to reduce the intermolecular distance, causes the surface molecules to contract leading to the existence of a surface tension or *surface free energy*.[30] Hence, surface tension can be defined as either,[31]

(i) The energy needed to increase the fluid's surface area (A) by one unit (*i.e.* $dW = \gamma \, dA$) with SI units of joules per square metre $(J \cdot m^{-2})$, or,
(ii) As a capillary force per unit length trying to minimise the fluid's surface area (*i.e.* $F \, dx = 2\gamma l \, dx$).

Consider at the interface of a liquid and dissimilar material (see Figure 4.2(b)), an interfacial tension exists composed of two adjacent interfacial surface regions, in which the London dispersion forces, *i.e.* attractive forces arising from fluctuating electron clouds of atoms that appear as oscillating dipoles,[32] are the principle interfacial interaction forces. In the interfacial region of material (1), molecules are attracted to the bulk of the material by intermolecular forces, producing a tension equal to a surface tension of the material (γ_1). Simultaneously, at the interface, attraction by London dispersion forces of material (1) molecules at the interfacial region by material (2) surface molecules occur. As a result, the interfacial tension between the two materials is a function of the surface tension of both materials and the London dispersion force (γ^d) interaction between both materials. This can be approximated using Fowkes equation (eqn (4.2)),[30,33]

$$\gamma_{12} = \gamma_1 + \gamma_2 - 2\sqrt{\gamma_1^d \gamma_2^d} \qquad (4.2)$$

4.2.3   Contact Angles: Young's Equation

Following on from Thomas Young's 1805 essay on the cohesion of fluids,[34] in which he proposed a liquid in contact with both a solid and a gas must have an mechanical equilibrium angle of contact, which vanishes as the solid becomes perfectly wetted, Zisman and coworkers conducted extensive investigations into equilibrium contact angles of pure liquids on low- and high-energy solid surfaces with a range of monolayer surface modifications, causes of nonspreading on high-energy surfaces and the effects of homology on spreading by pure liquids, to elucidate constitutive relationships of critical surface tension of wetting.[35] A fluid drop (L) immobilised on a solid surface (S), also in contact with a gas phase (V), exhibits an equilibrium contact angle (θ_Y) between 0–180° (see Figure 4.3(a)), hence from Young's equation:

$$\cos \theta_Y = \frac{\gamma_{SV} - \gamma_{SL}}{\gamma_{LV}} \tag{4.3}$$

The contact angle is a measure of a fluid's wettability (or adhesion capabilities) on a solid, which all fluids are able to achieve to some extent, *i.e.* $\theta_Y \neq 180°$. A fluid on a solid surface is said to have a high *spreadability* or *wettability* as $\theta_Y \rightarrow 0°$ (also termed the critical surface tension for wetting). While fluids with $\theta_Y > 0°$ are described as nonspreading.[36] Due to the difficulties in obtaining measurements of interfacial surface tensions of solid–vapour (γ_{SV}) and solid–liquid (γ_{SL}), a thermodynamic approach for the reversible work (W_{12}) of a fluid adhered to another immiscible fluid was introduced by Athanase Dupré (eqn (4.4)).[37,38]

$$W_{SL} = \gamma_{SV} + \gamma_{LV} - \gamma_{SL} \tag{4.4}$$

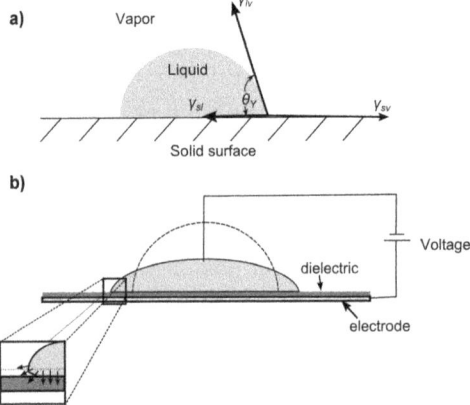

Figure 4.3   (a) Contact angle (θ_Y) of a sessile droplet on a solid surface; (b) Voltage-induced contact-angle reduction *via* EWOD. The box region shows the electrostatic force distribution at the 3-phase contact line.

This expression shows that the work to separate a fluid from a solid must be equal to the change in the free energy of the system. Substituting Young's equation into eqn (4.4), we obtain the Young–Dupré equation (eqn (4.5)), with θ_Y the equilibrium contact angle of the fluid on the solid or the wetting angle.

$$W_{SL} = \gamma_{LV}(1 + \cos \theta_Y) \tag{4.5}$$

Derivation of the Young–Dupré equation can also be obtained through a thermodynamic treatment of the system when in thermal and mechanical equilibrium, with the assumption that contact angles are dependent on Helmholtz free energy, the phase contact areas, the chemical potentials of each phase component and temperature. In addition, the effects of vapour adsorption on the solid phase has been indicated as playing a major role in the determination of equilibrium relations concerning contact angles and spreading effects on solids.[39,40] Accordingly, the free surface energy at the solid-vacuum interface (γ_{S^o}) is lowered when vapour adsorption saturates at the solid interface (γ_{SV^o}), such that,

$$\gamma_{S^o} - \gamma_{SL} = (\gamma_{S^o} - \gamma_{SV^o}) + \gamma_{LV^o}\cos \theta_Y \tag{4.6}$$

and,

$$W_{SL} = \gamma_{S^o} + \gamma_{LV^o} - \gamma_{SL} \tag{4.7}$$

thus giving the expression for the reversible work of adhesion (W_{SL}) of the liquid to a solid when coated with an adsorbed film of saturated vapour and the free energy decrease on immersion of the solid in the saturated vapour phase (eqn (4.9)); where R is the gas constant, T is absolute temperature, p_o is the pressure of the saturated vapour and Γ is the Gibb's surface excess adsorption per unit area squared.

$$d\gamma_{S^o} = -\Gamma RT d \ln p \tag{4.8}$$

and,

$$W_{SL} = RT \int_0^{p_o} \Gamma d(\ln p) + \gamma_{LV^o}\cos \theta_Y \tag{4.9}$$

4.2.4 Young–Laplace Equation

Electrowetting (EW) is generally used to describe the wetting capabilities of a conductive liquid solution, influenced by an electric field, on a hydrophobic solid substrate (see Figure 4.3(b)). Excellent reviews on the physics of electrowetting and droplet actuation have been given by Mugele and Baret, Nelson and Kim, and Zeng and Korsmeyer that provide further depth and background in this dynamic field.[24,39,41]

In the absence of an electric field (*i.e.* static condition), droplet wetting is a function of surface tension only. For droplets of the order of less ~1 mm height (h), in ambient conditions (*e.g.*, air), the influence of gravitational forces with respect to capillary forces are negligible if the Bond number $(\mathrm{Bo} = \rho_L h^2 g/\gamma_{LV})$ is less than unity. The shape of the droplet is assumed to be spherical if the gravitational force is negligible. On a horizontal surface, a droplet will spread until it reaches its minimum potential energy as determined by the fluid's cohesion forces and the adhesion forces to the solid surface. This is characterised by the equilibrium contact angle made between the liquid-vapour phases at the contact line as described by Young's equation. Furthermore, deviations in observed θ_Y values at the nanoscopic interface level (*i.e.* 3-phase contact line),where disjoining pressures exist between vapour and liquid, influence the local curvature of the droplet due to effects such as surface irregularities (*e.g.*, contact-angle hysteresis develops), drop evaporation, molecular orientation, electrical double-layer formation and van der Waal's interactions.[42–44] At the liquid/vapour interface the associated pressure difference (Δp) between the phases is described by the Young–Laplace equation, such that R_1 and R_2 are the principal radii of curvature of the interface that gives a constant curvature for a spherical-shaped droplet in mechanical equilibrium, on a homogeneous substrate.

$$\Delta p = \gamma_{12}\left(\frac{1}{R_1} + \frac{1}{R_2}\right) \qquad (4.10)$$

As long as the radii of curvature are significantly larger than the length scale of the long-range surface forces (*i.e.* ~0.01 µm) these effects on contact angle, and thus the droplet's surface energies, are usually ignored, though computer simulations using techniques such as molecular-kinetic modelling at the solid/liquid interface or kinetic and diffusive models of evaporation/condensation have shown their influences on droplet spreading and on the characteristic size of the droplet.[45–47]

4.3 Theories of Electrowetting-on-Dielectrics

When a sessile droplet partially wets a homogeneous solid surface, the 3-phase contact line delimits the three phase system, as seen in Figure 4.3(a). At equilibrium (*i.e.* constant chemical potential (μ) for each component (i), pressure (p) and temperature (T)) the thermodynamic parameters defining θ_Y must be stationary according to eqn (4.11) for the surface free energy per unit area. A shift in the contact line position (dx) should not have any effect on either the core energy from a simple translational movement of the droplet body or the bulk energies as pressures on either side of the liquid/vapour interface remains constant.[36,42]

$$\gamma_{12} = \left(\frac{\partial F}{\partial A_{12}}\right)_{p,T,\mu_i} \qquad (4.11)$$

In the presence of an electric field with appropriate electrode configurations, two effects on the hydrostatic equilibrium of a droplet can be seen; (i) a voltage dependency on the apparent contact angle, (ii) a net electrostatic force acting on the free surface of the liquid volume causing droplet motion.[48] Moving away from earlier EW configurations where fluid actuation was *via* direct contact with energising electrodes,[10] the most common configurations for electrically induced droplet actuation nowadays are where a thin layer of dielectric material is deposited over the actuating electrodes, sandwiching the droplet in an enclosed system (closed format) or lying in the same plane without a top cover (open format). These configurations, commonly termed electrowetting-on-dielectric (EWOD), rely on an applied electric potential to alter the wettability of a droplet situated on a dielectric substrate.[49] On application of a voltage potential (V), charge distribution at the liquid/dielectric interface leads to a reduction of free energy and thus interfacial surface tension, with an apparent reduction in the contact angle (see Figure 4.3(b)). The electrostatic energy stored in the dielectric, effectively a capacitor of capacitance,

$$C_d = \frac{\varepsilon_0 \varepsilon_d A_{SL}}{d} \tag{4.12}$$

where, ε_d is the relative permittivity of the dielectric, d is the dielectric's thickness and A_{SL} is the base area of the droplet, is represented by the second term on the right-hand side of Lippmann's equation (eqn (4.1)).

$$\gamma_{SL}(V) = \gamma_{SL}(0) - \frac{1}{2}C_d V^2 \tag{4.13}$$

Originally described by Berge with water droplets on hydrophobic polymer films,[49] modification of Young's equation results in eqn (4.14), from which the contact-angle reduction for the partial wetting of an electrostatically induced droplet on a dielectric surface can now be deduced. Eqn (4.14) is widely known as the Young–Lippmann equation, also referred to as the *electrowetting equation*, where the first term describes the droplet in equilibrium (*i.e.* $V = 0$) and the second term on the right-hand side is the dimensionless electrowetting number (Ew) that represents the electrically induced interfacial energy reduction at the dielectric/liquid interface relative to the interfacial surface energy of the liquid/vapour interface.[28,49]

$$\cos \theta(V) = \cos \theta_Y + \frac{C_d V^2}{2\gamma_{LV}} \tag{4.14}$$

The resulting electrowetting equation can be obtained using a thermodynamic and electrochemical approach based on early experimental observations of electrocapillarity, where an electrical double layer of the Debye–Huckel length is formed based on the surface charge density (ρ_s) at the metal/electrolyte interface due to the voltage drop (V) across the electrical double layer at the said interface.

$$\frac{d\gamma_{SL}}{dV} = -\rho_s \tag{4.15}$$

For the case of EWOD, the electrical double layer formed in series with the dielectric thin film can be ignored as the dielectric capacitance dominates that of the electrical double layer. Hence, C_{DL} from eqn (4.1) is replaced with eqn (4.12) for valid use in EWOD systems. Consequently, EWOD systems generally require much larger voltages, up to >200 V in some applications/ configurations, across the dielectric film to invoke a strong reduction in interfacial tension than would be seen across the metal–electrolyte electrical double layer at fractions of a volt, before electrolysis occurs. It is worth noting that eqn (4.14) shows that a reduction in droplet contact angle is dependent on the applied voltage (AC or DC) in order to attain its new equilibrium, though it does not describe how the electrostatic forces mechanically drive droplet contact-angle reduction or droplet displacement, as one would expect to observe for digital microfluidic platforms. Thus, as Jones postulated through electromechanical treatment of fluid volumes, the observed elec-trostatic reduction of a droplet's contact angle is not the cause of fluid motion and therefore the term electrowetting should rather be reserved for an electric field's effect on the contact angle of a fluid, with the two observable effects considered independent of each other.[50]

4.3.1   Surface Free-Energy Minimisation

Experimental observations of the apparent contact-angle reduction due to electrowetting are generally in close agreement with Berge's original deri-vation of eqn (4.14), based on an energy analysis of the system.[49] Though above a certain threshold "contact-angle saturation" is reached with no further change in contact angle observed (Figure 4.4(a)). Vallet *et al.* showed irreversible wetting of a 12-μm thick polyethylene terephthalate (PET) film by aqueous solutions was achieved by a contact-angle reduction of >30° under high-voltage conditions. Contact-angle reduction was observed to follow theory at voltages <150 V after which, droplet instabilities characterised by satellite ejection (see Figure 4.4(b)) around the 3-phase contact line

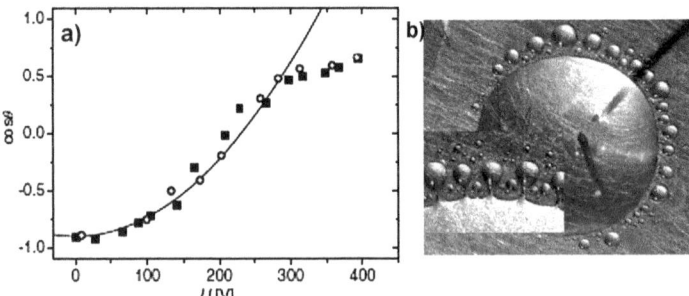

Figure 4.4   (a) Contact-angle saturation (glycerol and NaCl mixtures) occurs at >300 V (DC) for advancing (squares) and receding (circles) angles with oil as surround medium. (b) Above the saturation voltage deviation from Young–Lippmann's equation (solid line) occurs, leading to satellite ejection from the droplet. Reprinted with permission from H. J. J. Verheijen and M. W. J. Prins, *Langmuir*, 1999, **15**, 6616–6620. Copyright 1999 American Chemical Society.

becomes apparent.[28] Mugele *et al.* reported contact-angle saturation was a material-dependent quantity and the droplet morphology characterised by electric-field distribution (F_{el}) and its free energy (F) defined by eqn (4.16), do not account for stray electric-field-induced effects at the contact line resulting from instabilities of the liquid surface or local dielectric breakdown of the dielectric film.[24,51]

$$F = F_{surf} - F_{el} = A_{LV}\gamma_{LV} + A_{SV}\gamma_{LV} + A_{SL}\left(\gamma_{SL} - \frac{C_d V^2}{2}\right) - \Delta p V \qquad (4.16)$$

From eqn (4.16), a conductive droplet in a EWOD configuration is composed of voltage-independent interfacial energies (F_{surf}) and electrostatic energy (F_{el}) contributions; for which F_{el} is made up of the dielectric capacitance parallel to the base area of the droplet and constant stray capacitances along the edge of the droplet, which can be ignored for sufficiently large droplets.[51,52] The net energy gain upon charge redistribution due to the applied voltage across the dielectric drives the droplet's contact-angle reduction, thus increasing capacitance for further charge redistribution. Verheijen and Prins obtained the averaged contact angle of bulk droplets with a ~2% error for $60° < \theta < 120°$, by measuring the capacitance of the droplet and relating the droplet's base area of known volume and shape to the contact angle.[53] They reported errors in contact-angle values were due to stray capacitances originating from fringing electric fields at the contact-line localities.

4.3.2 Electromechanical: Maxwell Stress Tensor

Jones, and Kang, respectively, modelled droplet-based electrowetting in terms of electromechanical variables, showing equivalences to the energy-based and thermodynamic derivations.[48,54] The net body force of electrical origin (f^e), based on the Korteweg–Helmholtz force density, acting on a liquid volume is shown in eqn (4.17), where E is the applied electric field, ρ_f is the free charge density, ρ and ε are the fluid density and permittivity, respectively.[55]

$$f^e = \rho_f E - \frac{1}{2}E^2 \nabla\varepsilon + \nabla\left[\frac{1}{2}E^2\frac{\partial\varepsilon}{\partial\rho}\rho\right] \qquad (4.17)$$

Neglecting electrostriction (*i.e.* the 3rd term on the RHS of eqn (4.17)) for an incompressible conductive fluid, and applying a momentum conservation of the density flux, a generalisation of eqn (4.17) can be applied using the Maxwell stress tensor (T_{ij}) and the surface integral (S) of the outward-directed unit vector n normal to the surface area of the enclosed volume (see Figure 4.5).[56]

$$f^e = \oint_S T_{ij}n_j dA \qquad (4.18)$$

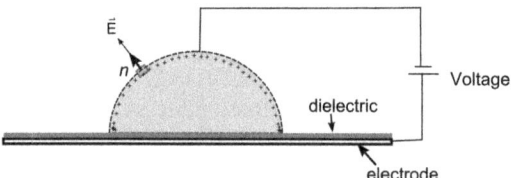

Figure 4.5 Outward-directed electrical force normal to the unbound droplet surface.

where the Einstein summation convention is employed over the indices and,

$$T_{ij} = \varepsilon_0 \varepsilon \left(E_i E_j - \frac{1}{2} \delta_{ij} E^2 \right) \tag{4.19}$$

The Kronecker delta δ_{ij} is defined as unity for $i = j$ and zero when $i \neq j$, with respect to the Cartesian coordinate system (x, y, z). From eqn (4.18) we can see that the net body force acting on a fluid volume occurs normal to the surface (with tangential forces vanishing at the surface), related to the local surface charge density $(\rho_s = \varepsilon_0 \vec{E} \cdot \Delta n)$, giving a negative (outward electrostatic pressure, P_e) contribution to the total internal pressure of the fluid. Upon integration of eqn (4.18) at the fluid/vapour interface we get,

$$\frac{\vec{F}}{dA} = P_e \vec{n} = \frac{\varepsilon_0}{2} E^2 \vec{n} = \frac{\rho_s}{2} E \tag{4.20}$$

The charge density at the apex of the contact line (also referred to as a wedged-shaped edge), where ignition of ionisation at contact-angle saturation has been reported, is extremely high relative to the dielectric/liquid interface away from the contact line,[57] thus confining the electrostatic stresses to within the contact line region. The electric field and charge distribution can be found by satisfying Laplace's equation ($\nabla^2 \varphi = 0$) with appropriate boundary conditions, and in the case of the wedge region, charge distribution is at a maximum distorting the contact line (as shown in the box of Figure 4.3(b)) at extremely high voltages. Both horizontal and vertical outward components of Maxwell stresses acting on the 3-phase contact line have commonly been thought to be the cause of wetting, with the integral of the horizontal component equal to the product of the electrowetting number and the liquid/vapour interfacial tension.[54,58] Eqn (4.21) indicates that the force pulling at the contact line is independent of droplet shape.

$$F_x = \frac{\varepsilon_0 \varepsilon_d V^2}{2d} = \gamma_{LV} Ew \tag{4.21}$$

Buehrle *et al.* investigated the effects electrostatic fields had on the deformation of the surface profile of a droplet in electromechanical equilibrium, close to the contact line, through a balancing of capillary and electrostatic forces.[59] It was found that the normalised asymptotic contact angle approached in electrowetting was in fact that of Young's angle, independent of the applied potential and material properties, suggesting electrostatic forces were not contributors to the force balances at the contact line and were more to do with interfacial tensions. Hence, for dynamic electrowetting, contact-angle reduction was believed not to be the driving mechanism.

4.3.3 Electromechanical: Capacitance-Based

Using lumped parameter electromechanics, Jones extended the theoretical understanding and alternative analysis of electrowetting, demonstrating the net force on a liquid mass in a closed EWOD configuration can be determined with no reference to fluid profile or contact angle.[48,50] This was conveniently described using the capillary height of rise set-ups as seen in Figure 4.6 (refer to ref. 60). For a system in which two dielectrically-coated electrodes, partially immersed in a volume of conductive water, are separated by a distance (D) much greater than the dielectric coating thickness (*i.e.* $d \ll D$) and the electrode length and width (w) are large enough to ignore fringing effects, the height of fluid rise in the capillary, when $V = 0$, can be found by eqn (4.22), where ρ is the fluid density.[31]

$$h_{\text{cap}} = \frac{2\gamma_{\text{LV}}\cos\theta(0)}{\rho g D} \tag{4.22}$$

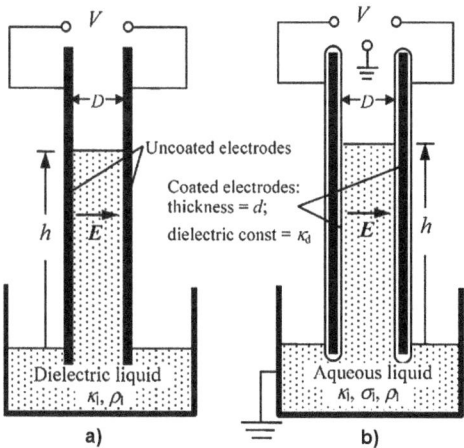

Figure 4.6 (a) Pellat's experiment and (b) modification of Pellat's experiment showing the electromechanical height of rise of a grounded aqueous fluid between two dielectrically coated parallel electrodes. Reprinted with permission from T. B. Jones, K. L. Wang and D. J. Yao, *Langmuir*, 2004, **20**, 2813–2818. Copyright 2004 American Chemical Society.

On application of a voltage potential between the two electrodes, a voltage-induced fluid rise (h_{EWOD}) proportional to a reduction in contact angle occurs, described by combining eqn (4.14) and (4.22).[61]

$$h_{EWOD} = \frac{2\gamma_{LV}(\cos\theta(V) - \cos\theta(0))}{\rho g D} = \frac{\varepsilon_d\varepsilon_0 V^2}{4\rho g D} \qquad (4.23)$$

Thus, as h_{EWOD} is proportional to the square of the applied voltage and independent of surface tension, it follows that the surface tension only determines the initial level and does not indicate the extent to which the meniscus rises or changes shape under an electrical force. Through the introduction of a coenergy function ($W_{\dot{e}}$), describing the electrical energy stored in the system,[56] the force of electrical origin to raise a liquid by a height (∂h) was formulated, with the condition that the electrode width (w) is greater than D, guaranteeing the system capacitance is based predominantly on the liquid-dielectric-covered regions between the electrodes.[50]

$$f^e = \left.\frac{\partial W_{\dot{e}}}{\partial h}\right|_V \cong \frac{\varepsilon_0\varepsilon_d w}{4d}V^2 \qquad (4.24)$$

In 2008, Jones further extended his original capacitive-based method to more practical electrode configurations for which transportation of micro-droplets are commonly encountered and that could be modelled more robustly. Using geometrical relationships of a sessile spherical cap droplet of contact radius a, spherical cap radius r, height h, and the volume conservation of the droplet (see Figure 4.7), the contact angle θ on a homogeneous surface can be expressed through (see Appendix A of ref. 62 for an analytical root expression of a droplet's height as a function of volume and a),

$$\cos\theta = \sqrt{1 - \left(\frac{a}{r}\right)^2} \qquad (4.25)$$

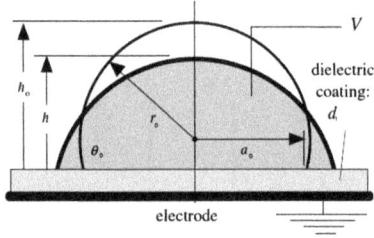

Figure 4.7   Side view of a sessile droplet resting on a dielectrically coated electrode with h, r, a, and h_o defined. The droplet, which is conducting, acts as the deformable electrode of a capacitor, subject to the constraints of constant volume and a spherical cap shape. Variables identified by the subscript "o" represent values when voltage $v = 0$. Reprinted from T. B. Jones, *Mech. Res. Commun.*, 2009, **36**, 2–9, with permission from Elsevier.

With a total surface energy (W_γ) described by the sum of each interfacial area multiplied by their respective interfacial surface tensions, such that $A_{LV}(a) = 2\pi r(a)h(a)$ and $A_{LS}(a) = \pi a^2$, any change in surface tension is related to a change in contact radius (∂a), written in the form,

$$f^\gamma(a) = -\frac{1}{2\pi a} \frac{\partial W_\gamma}{\partial a} \tag{4.26}$$

On enforcing an equilibrium constraint, *i.e.* $f^\gamma = 0$, eqn (4.26) can be found to reduce to the familiar Young's equation (eqn (4.3)).

The application of an electrostatic force to a conductive droplet on a dielectric surface will exhibit a change in shape and spreading, which can be characterised by a change in contact radius (∂a), assuming fringing fields are neglected. From eqn (4.12), the capacitance of the dielectric can be approximated by the base interface of the droplet, *i.e.*

$$C(a) \approx \frac{\varepsilon_0 \varepsilon_d \pi a^2}{d} \tag{4.27}$$

A conservation of energy $(W_e(q,a))$ condition imposed on a EWOD system from the energy gain of a constant electric potential (V) and the corresponding charge redistribution (q) acting at the contact line, is defined using the coenergy $(W_{\dot{e}}(V,a))$ of the system, such that,

$$dW_e = V dq - 2\pi a f^e da \tag{4.28}$$

Using a Legendre transformation, as charge is constrained rather than potential,

$$W_e + W_{\dot{e}} = Vq \tag{4.29}$$

And from eqn (4.28) and (4.29) we get

$$dW_{\dot{e}} = q dV + 2\pi a f^e da \tag{4.30}$$

Through integration of eqn (4.30) and rearranging for f^e along the contact line, we find the dependence of the system capacitance on the contact radius, similar in form to the second term of the Young–Lippmann equation (eqn (4.14)) and that derived through Maxwell's stress tensor (eqn (4.21)).

$$f^e = \frac{1}{2\pi a} \frac{\partial W_{\dot{e}}}{\partial a} \bigg|_V = \frac{V^2}{4\pi a} \frac{dC}{da} \tag{4.31}$$

4.4 Limitations of the Young–Lippmann Equation

4.4.1 Hysteresis

Contact-angle hysteresis (α) can occur in EWOD systems from surface irregularities such as chemical contamination, solute deposition from droplet, surface heterogeneity or simply surface roughness, affecting the free movement of the contact line leading to metastable 3-phase systems.[42,63,64] This gives rise to advancing and receding contact angles that on the macroscopic scale, if irregularities are not significant, can be adequately described by eqn (4.3) in most instances. Contact-line hysteresis (Figure 4.8) is defined as the difference in advancing contact angle and receding contact angle for a contact line moving in an opposite direction at the same velocity.[35] An excellent analysis of contact-angle hysteresis is discussed by Gao and MacCarthy in which activation energies and contact-line dynamics are viewed as key factors in droplet kinetics.[65] At a velocity of zero, differences in advancing ($\theta_A = \theta + \alpha$) and receding ($\theta_R = \theta_Y - \alpha$) contact angles can manifest as a consequence of surface irregularities, commonly termed static contact-angle hysteresis as opposed to the dynamic hysteresis we are interested in EWOD. Recall from the Young–Lippmann equation that a change in contact angle is dependent on the system capacitance and applied voltage.

$$F_{EWOD} = \gamma_{LV} \Delta \cos \theta = \frac{C_d V^2}{2} \tag{4.32}$$

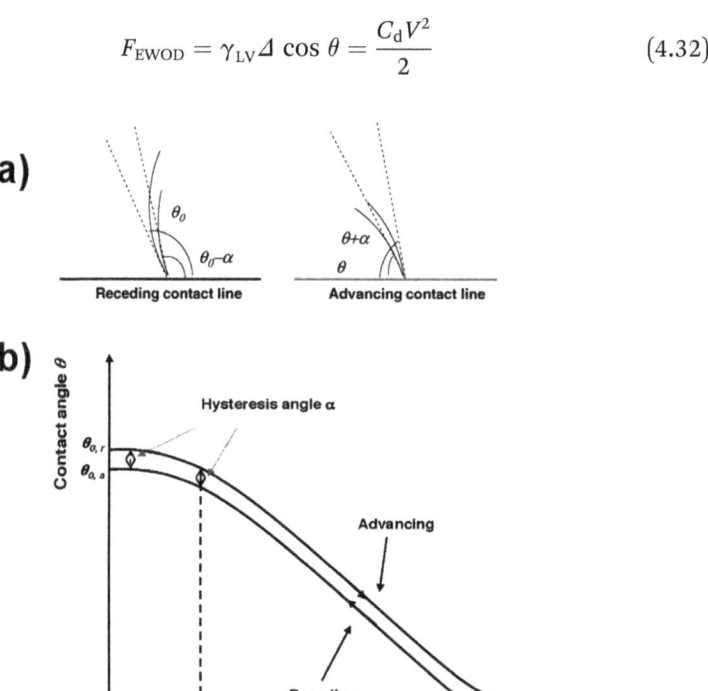

Figure 4.8   Defining contact-line hysteresis (α); (a) receding and advancing contact-line angle with and without hysteresis; (b) Electrostatic-induced contact-angle change shifted by α for advancing and receding angles. Reprinted from T. J. Berthier, P. Dubois, P. Clementz, P. Claustre, C. Peponnet and Y. Fouillet, *Sens. Actuators A*, 2007, **134**, 471–479, with permission from Elsevier.

It can be inferred from eqn (4.32) that there must be a minimum threshold potential (V_{min}) required to actuate droplet motion, dependent on the experimental conditions of the EWOD application.[66] In practice, hydrophobic materials with high dielectric strength, chemical inertness and low contact-angle hysteresis are ideal candidates for rapid and reversible motion of fluids in EWOD systems. Various insulating materials (*e.g.*, Si_3N_4, SU-8, parylene) and hydrophobic materials (*e.g.*, silicon oxycarbide, fluoropolymers, PDMS) and combinations thereof have been employed to increase the voltage tunable range of contact angles whilst minimising insulating layer thickness for effective dielectric strengths.[67-71] Mugele *et al.* employed amorphous Teflon AF® as insulating layers on indium tin oxide, prepared through annealing of two 3–5 μm layers in vacuum at 160–330 °C (depending on number of deposition steps) to increase dielectric strength to between 40–140 $V \cdot \mu m^{-1}$ and reduce surface roughness.[51] Contact-angle hysteresis was reduced for from 5–10° for a water–air system to 2–3° for a water–oil system. A combination of 10-μm thick Parylene C ($200\ V \cdot \mu m^{-1}$) or polyamide coating for insulation and a 0.1-μm Teflon AF1600 layer for hydrophobicity just above was deposited on ITO electrodes with an ITO counter electrode spaced at 250 μm.[61] With this arrangement, a 200 V potential exhibited an initial contact-angle decrease from 110° to 63° for a potassium nitrate droplet, returning to 102° at 0 V, with subsequent rapid switching of potentials showing little hysteresis, as demonstrated in capillary rise and depression experiments. Berthier *et al.* assessed contact-angle hysteresis for a range of liquids, surrounded by either air or silicone oil, and the minimum actuation potential needed for a net electrocapillary force in an open EWOD system.[66] Systems in which silicone oil was used exhibited lower hysteresis ($V_{min} = 5$ V) because of the formation of a thin homogeneous film on the substrate surface, while droplets in air required higher actuation potentials ($V_{min} = 30$–40 V) due to higher surface tensions. Based on a simple analysis of a droplet's dynamic movement between two coplanar electrodes of width w, where one electrode is actuating and the other is not, such as in Figure 4.9, the electrocapillary force in the x-direction on the actuating electrode is,

$$F_x = w\gamma \cos \theta \qquad (4.33)$$

whilst at the interelectrode gap, the constant capillary force on the 3-phase contact line is independent of the actuating voltage and droplet shape and is written as,

$$F_x = w\gamma(\cos \theta - \cos \theta_Y) \qquad (4.34)$$

Taking into consideration the advancing and receding capillary forces, due to contact-angle hysteresis, we obtain,

$$F_{x,A} = w\gamma \cos (\theta + \alpha) \qquad (4.35)$$

$$F_{x,R} = -w\gamma \cos (\theta_Y - \alpha) \qquad (4.36)$$

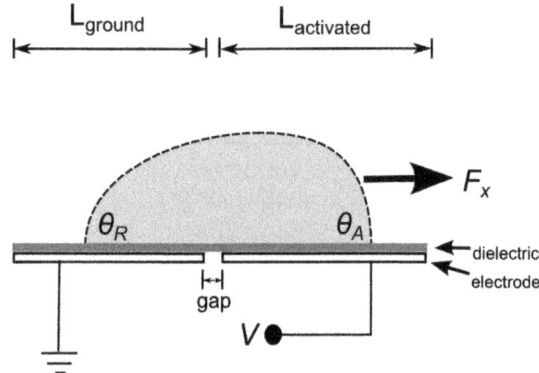

Figure 4.9 Simplified two dimensional diagram of the electromechanical force acting on a droplet above two coplanar electrodes.

which on expansion of eqn (4.33)–(4.36) gives a total capillary force dependent on Lippmann's force (1st term) and the extent of contact-angle hysteresis (2nd term) as described by eqn (4.37).

$$F_x = w\gamma[\cos\theta - \cos\theta_Y] - w\gamma\alpha[\sin\theta - \sin\theta_Y] \equiv \frac{wCV^2}{2} - w\gamma\alpha[\sin\theta - \sin\theta_Y]$$

$$(4.37)$$

Assuming no hysteresis is present within the system the total force would be entirely dependent on the actuating potential. In reality, the presence of hysteresis indicates a minimum potential is required, which from eqn (4.37) can be reduced to a simplified form based on the dependency of contact angles to voltage potential. Thus, a larger capacitance of the dielectric layer combined with a small value of hysteresis reduces the minimum potential required for EWOD droplet actuation.

$$V_{min} = 2\sqrt{\frac{\gamma\alpha\sin\theta_Y}{C}} \qquad (4.38)$$

An analytical model to predict the EWOD force on droplets in a closed system and the force reduction as a consequence of hysteresis was presented by Schertzer *et al.* as a function of the droplet's 3D shape.[72] At various applied voltages, aspect ratios, and taking into account of other parameters (*e.g.*, density, surface tensions, wall friction, *etc.*) they were able to predict the steady-state velocity and dynamic position in 1D to within 3.7% and 5.1% accuracy, respectively. For further analytical treatment of contact-angle hysteresis of purposely roughened/geometrical patterned surfaces in which droplets are distinctively classified into Cassie drops (*i.e.* drops sitting on peaks of surface with air pockets)[63] or Wenzel drops (*i.e.* drops completely wetting grooves),[73] and in which the former is more favourable to fluid manipulation particularly on super hydrophobic surfaces for electrically controlled wetting and dewetting, refer to works by Patankar and coworkers[74–77] and Krupenkin *et al.*[78,79]

4.4.2 Contact-Angle Saturation

At low voltage potentials electrically induced contact angle changes are reversible, but at a specific potential known as the maximum actuation potential (V_{max}), contact-angle reduction ceases to occur.[66] Contact-angle saturation (θ_{sat}) is not a fully understood phenomenon, and could be attributable to a number of mechanisms each acting independently and simultaneously at the contact line. Hence, Young–Lippmann's equation is not entirely descriptive across a full range of potentials, with the contact angle reaching an asymptotic value at some $V \geq V_{max}$.[59] This value varies as a function of bulk fluid properties and EWOD system design, but θ_{sat} has been observed to range between 30° and 80°. A number of varied explanations for the cause of contact-angle saturation have been put forward including charge build up and trapping in the dielectric layer which increases linearly above V_{max};[52] high field strengths causing droplet instabilities at the 3-phase contact line;[28] air ionisation, indicated by observed droplet luminescence around the contact line, though suppressed by increases in droplet salt concentrations.[57] Interestingly, Wang and Jones demonstrated contact angle limitations using height-of-rise experiments.[80] They showed that the velocity of the contact line influences the onset of saturation, through application of frequency-dependent voltages. Alternatively, as EWOD reduces the surface tension of the liquid/solid interface (γ_{LS}), it is possible to obtain V_{max} by equating $\gamma_{LS} = 0$ with the remaining surface tensions (from Young's equation) responsible for the contact angle.

$$V_{max} \approx \left(\frac{C}{2\gamma_{LS}^\circ} \right)^{0.5} \qquad (4.39)$$

At V_{max} the limit of eqn (4.13) is reached, which Peykov *et al.* showed depended on the capacitance of the system and the liquid/solid interface tension at zero point charge (referred to as the PQRS model).[81] Koopal argued, based on Fowkes equation, that it was a reasonable approximation to make based on the work of adhesion of polar surfaces such as Teflon AF1600 $(\gamma_S = 12 \text{ mJ}\cdot\text{m}^{-2})$ or polyethylene $(\gamma_S = 30 \text{ mJ}\cdot\text{m}^{-2})$ interfaced with water $(\gamma_L = 145 \text{ mJ}\cdot\text{m}^{-2})$. This creates a narrow range of interfacial tension, with strong negative values of γ_{LS} unable to occur long before the work of adhesion at the solid surface can equal the work of cohesion of the water molecules.[82] The observed linear relationship at low voltages was applied to derive a "modified" Lippmann–Young law[83] describing the contact-angle saturation effects observed experimentally (see Figure 4.10), by using Langevin's function, $L(X) = \coth(3X) - (1/3X)$.

$$\frac{\cos\theta - \cos\theta_Y}{\cos\theta_{sat} - \cos\theta_Y} = L\left(\frac{CV^2}{2\gamma_{LV}(\cos\theta_{sat} - \cos\theta_Y)} \right) \qquad (4.40)$$

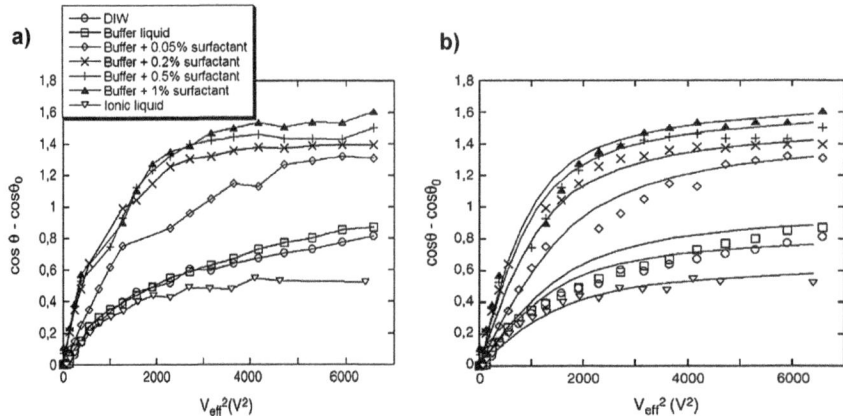

Figure 4.10   Plots of $(\cos\theta - \cos\theta_0)$ for a range of fluids immersed in silicon oil showing the (a) experimental plot and (b) the "modified" Young–Lippmann equation plot as fitted to experimental data using the Langevin function. Reprinted from J. Berthier, P. Clementz, O. Raccurt, D. Jary, P. Claustre, C. Peponnet and Y. Fouillet, *Sens. Actuators A*, 2006, **127**, 283–294, with permission from Elsevier.

4.4.3   Frequency Dependency

A limiting factor of eqn (4.14), (4.22) and (4.31) is the inability of frequency-dependent effects to be taken into consideration for predictive relationships with contact angle. Their basic assumption is that contact-angle reduction is based on a static potential for which the time-average potential (V_{RMS}) at low frequencies and infinite liquid conductivity (*i.e.* a perfect conductor) applies. Above a critical frequency ($f_c = \omega/2\pi$), eqn (4.14) no longer holds true. This critical frequency is a function of the liquid's dielectric properties, conductivity and permittivity, *i.e.* σ_m and ε_m, respectively, describing the transition between conductive currents and displacement currents. Thus, for $f \gg f_c$ various liquids will behave more as a dielectric material.[48]

$$f_c = \frac{\sigma_m}{2\pi\varepsilon_0\varepsilon_m} \tag{4.41}$$

Quinn *et al.* indicated the strong influence fluid ionic strength, pH, and thus the electrical double layer at the liquid/solid interface had on deviations from the theoretical electrowetting equation, well below V_{max}.[69] They explained their nonideal electrowetting curves obtained were due to charging inefficiencies at the interface for fixed potentials.

The application of a time-varying potential to a sessile droplet has been shown to induce internal steady electrohydrodynamic flows (Figure 4.11) through a droplet's resonant oscillation;[84,85] practically eliminating

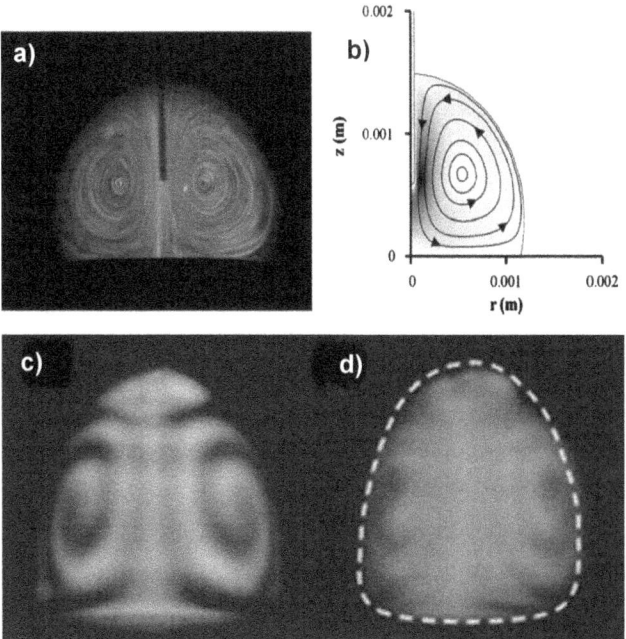

Figure 4.11 AC signals inducing (a) electrohydrodynamic vortices within a droplet; (b) Its corresponding streamline simulation. "Reproduced with permission from H. Lee, S. Yun, S. H. Ko and K. H. Kang, *Biomicrofluidics*, 2009, **3**, 044113–044112. Copyright 2009, AIP Publishing LLC.; (c) and (d) Latter stages of dye mixing within an oscillating (81 Hz) droplet showing ∼0.5 s time lapse between images (c) and (d). Reproduced with permission from F. Mugele, J. C. Baret and D. Steinhauser, *Appl. Phys. Lett.*, 2006, **88**, 204106–204103. Copyright 2006, AIP Publishing LLC.

contact-angle hysteresis for easier translation;[86] and inducing microfluidic mixing in viscous liquids.[87] These frequency-dependent phenomena arise as a result of finite liquid conductivities, and nonuniform electric fields generating gradients in fluid properties.[88] A distinction between electro-wetting and fluid displacement (referred to as *liquid dielectrophoresis* (LDEP)) has been argued to be considered independent of each other.[48,50,59] The change in "apparent" contact angle, θ_Y, is reported to be a consequence of the strong divergent electric fields close to the 3-phase contact line acting on the liquid/vapour interface within a distance (χ) above the dielectric surface, which has a different contact angle value, tending to θ_Y as V_{max} is approached. Force balances in this region have indicated electrostatic forces are not contributors to droplet dynamics.[59] The net electromechanical body force on the droplet has been described to originate from the electric field nonuniformities of the electrodes and can be deduced based solely on the knowledge of the system's capacitances (C). Chatterjee *et al.* explicitly modelled the electromechanical force on a droplet for a two-plate system (see Figure 4.12(a)) in a similar manner as Jones.[62,89] From eqn (4.31), the total x-directed electrical force on an electrically floating droplet above two coplanar electrodes of length L, such as that described in Figure 4.9, can be expressed as,

$$f_x^e = \left.\frac{\partial W_{\dot{e}}}{\partial x}\right|_V = \frac{V^2}{2}\frac{dC}{dx} = -\frac{\varepsilon_d\varepsilon_0 V^2}{2Ld}x \tag{4.42}$$

where, from eqn (4.27), the net capacitances of C_1 and C_2, equivalent to the proportion of fluid above L_{ground} and $L_{\text{activated}}$ electrodes, respectively (per unit length in the x-direction), in series is found.

$$C(x) = \frac{C_1 C_2}{C_1 + C_2} = \frac{\varepsilon_d\varepsilon_0}{2Ld}(L^2 - x^2) \tag{4.43}$$

Through RC circuit modelling, the frequency-dependent electrostatic force on a finite conductive droplet (*e.g.*, biological fluid) and thus voltage distribution of the whole system, such as those in Figure 4.12(b), can be given full consideration.[62] The phasor voltage, $v(\omega)$, across each capacitive element is obtained through impedance division rules, where $Z_i = 1/j\omega C_i x(L)$ or $Z_i = 1/(j\omega C_i + \sigma_i)L(x)$ are the complex impedance of elements (i), for dielectric layer and fluid, respectively.

$$v(\omega)_i = \frac{Z_i}{\sum_{n=1}^{k}Z_n}V(\omega) \tag{4.44}$$

a)

b)

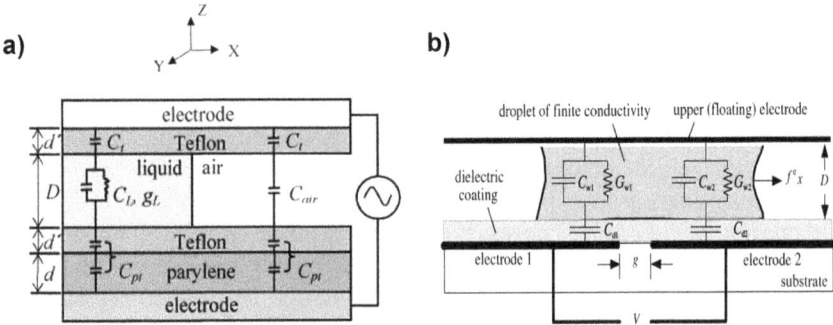

Figure 4.12   RC circuit models: (a) Two-plate equivalent circuit model. Reproduced from ref. 89; (b) Coplanar electrodes equivalent circuit model. Reprinted from T. B. Jones, *Mech. Res. Commun.*, 2009, **36**, 2–9, with permission from Elsevier.

The electric force exerted on the droplet is a sum of the energy-storing capacitor elements, and from eqn (4.42), the time-average force in the *x*-direction, per unit width of the electrode is expressed as,

$$\langle f_x^e \rangle = \frac{1}{2} \sum_n^k |\nu(\omega)_n|^2 \frac{dC_n}{dx} \tag{4.45}$$

Eqn (4.45) is an electromechanical interpretation of the frequency-dependent force on a droplet of finite conductivity, which is linearly dependent on *x*. Limiting expressions for the low-frequency (*i.e.* EWOD) and high-frequency (LDEP) values of $\langle f_x^e \rangle$ are given as:[62]

$$\langle f_x^e \rangle = -\frac{\varepsilon_d \varepsilon_0 |V(\omega)|^2}{2d} \frac{x}{L}, \quad f \ll f_c (\text{EWOD}) \tag{4.46}$$

$$\langle f_x^e \rangle = -\frac{\varepsilon_0 |V(\omega)|^2}{2} \frac{\varepsilon_d \varepsilon_m / Dd}{\varepsilon_d / d + \varepsilon_m / D} \frac{x}{L}, \quad f \gg f_c (\text{LDEP}) \tag{4.47}$$

4.5 DMF Platforms and Operations

Digital microfluidic systems emerged in the early 2000s, as scalable and easily reconfigurable 2D systems primarily fabricated from successive deposition, photolithography and etching steps with a final insulating/hydrophobic overlayer. Platforms are typically classified as either closed or opened in configuration.[90,91] In these configurations a number of fundamental operations are required for rapid, effective and reliable microfluidic processes, which Berthier *et al.* discussed in relation to CAD modelling.[92] As process complexities on a single system increase so too must the DMF architecture and droplet-routing algorithms to necessitate desired droplet operations.

4.5.1 DMF Architectures

4.5.1.1 Platform Designs

Figures 4.9 and 4.12 depict the two basic configurations of DMF systems for performing relatively simple investigations. The lower efficiency open system (Figure 4.9) consists of planar electrodes located within the same plane, *i.e.* the ground electrode is adjacent to the actuating electrode, while the closed system (Figure 4.12) consists of a single cover plate or grounded electrode a height above the electrode array of actuating electrodes. Functionally, the closed system has wider design variability[93] and is widely considered more flexible than the open system due to its ability to perform dispensing and splitting operations unlike the open system. In both configurations, electrodes are usually not in contact with the droplet but separated by one or more layers of dielectric insulator and/or hydrophobic material (see Table 4.1).[94] Design and interfacing of the DMF system is usually based

Table 4.1 Properties of commonly used dielectric coating in DMF devices (Adapted from ref. 95.)

Dielectrics	Parylene -N and -C	Teflon® 1600	Teflon (PTFE)	Cytop™	PDMS	Polyimide	SiO_2	Si_3N_4	Barium strontium titanate (BST)
Dielectric strength (kV·mm^{-1})	276 (-N) 268 (-C)	21	60	110	21.2	22	400–600	500	18–54
Dielectric constant (ε_d)	2.65 (-N) 3.15 (-C)	1.93	2.1	2.1	2.3–2.8	3.4	3.9	7.5	225–265
Applied voltage (V)	±240 DC <1 kV (50–20 kHz) AC	—	<300 DC <600 @ 1 kHz	120 DC <800 @ 2 kHz	±500 DC	<400 DC	>25 DC	>40 DC	>15 DC
Typical thickness (d)	3.5–30 µm	0.001–0.1 µm	25–50 µm	0.1–1 µm	38 µm	6–35 µm	0.1–1 µm	0.15 µm	40.8 µm
Contact angle of water (°)	126	120	114	110	120	50–80	46.7	30	40.8
Fabrication method	CVD	Spin or dip coating	Commercial	Spin coating	Spin coating	Spin coating	Thermal oxidation or PECVD	CVD	MOCVD

on the proposed application of the system, though for full on-chip operation from sample prep to analysis, compatible detection systems such as those reviewed by Malic *et al.*[21] would need appropriate integration and validation for reliable LOC systems.

An open platform all-terrain droplet actuation (ATDA) system, coined by Abdelgawad *et al.* due to the manipulation of droplets with AC signals (18 kHz, 500–700 V_{RMS}) across a range of geometries, inclines, bends and architectures was fabricated from flexible, copper-clad polyimide substrates coated with PDMS and a teflon hydrophobic layer.[95] Abdelgawad *et al.* also described a PDMS hybrid DMF-microchannel system used for sequential sample processing and chemical separation on a single substrate, as shown in Figure 4.13(a).[96] Droplets were syringe dispensed onto the open DMF platform

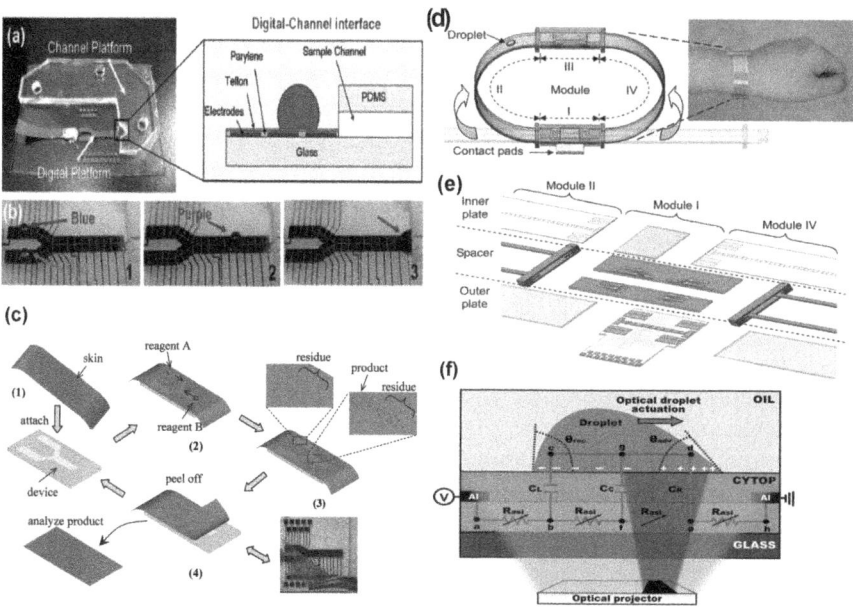

Figure 4.13   (a) Hybrid digital-channel fabricated device showing the interface of droplet translation from open DMF architecture to the microchannel inlet; (b) shows the DMF merging of two droplets before being electrokinetically driven into the microchannel. Reproduced from ref. 96; (c) Concept of "skins" strategy for removable insulating layers that can be further analysed by removing from DMF electrode architecture thus making the device reusable. Reprinted with permission from H. Yang, V. N. Luk, M. Abelgawad, I. Barbulovic-Nad and A. R. Wheeler, *Anal. Chem.*, 2009, **81**, 1061–1067. Copyright 2009 American Chemical Society; (d) Droplet on wristband (DOW) design and (e) Magnification of DOW device consisting of a three layers of modular components made from flexible PET material. Reproduced from ref. 99; (f) single-sided continuous optoelectrowetting (SCOEW) concept and equivalent circuit model. Reproduced from ref. 100.

and processed (*i.e.* mixed) before being electrokinetically driven (or *via* capillary action) into the buffer-filled microfluidic channel for micellar electrokinetic chromatographic separation of DMF labelled amino acids and enzyme digested peptides (see Figure 4.13(b)). A variation of the hybrid DMF-microchannel by Watson *et al.*, consisting of a multilayered closed DMF platform with a network of microchannels underneath the glass substrate wide-area actuating electrode array was reported to be superior in operation to the PDMS hybrid.[97] To combat contamination and interfacing problems associated with DMF systems, Yang *et al.* introduced removable polymer layers for their system in which repeated analysis could be conducted using the main electrode architecture, but only changing the polymer layer (Figure 4.13(c)).[98] Associated pitfalls with this strategy are unwanted bubble trapping between electrodes and polymer layer and material incompatibility that would routinely damage the device. A closed format chip-to-chip design by Fan *et al.*, building on the previous ATDA system, described a droplet on a wrist (DOW) band device from 4 modular flexible ITO-coated PET substrates shown in Figure 4.13(d) and (e).[99] Tests of droplet transportation against gravity were successfully performed, though issues regarding the materials and thus the voltages needed for actuation of water droplets, let alone biological fluids, to achieve reliable and safe operation are still to be investigated. An optical-based method for droplet manipulation labelled single-sided continuous optoelectrowetting (SCOEW, Figure 4.13(f)) uses a photoconductive surface instead of physical electrode features, with low-intensity light sources used to actuate droplets in an open 2D format.[100] A lateral dc bias applied across the entire device above a hydrogenated amorphous silicon dielectric layer (a-Si:H), modelled as a photoresistor, just below the hydrophobic Cytop layer enables the generation of an increased photoconductive difference between adjacent illuminated and nonilluminated regions, thus inducing droplet motion. Their optically patterned system, demonstrated to be able to perform all basic operations with a 250 pl–50 μl volume range and a 1% CV for droplet dispensation, was limited by the size of the pixilated electrodes and was shown to be compatible with a liquid-crystal display (LCD) offering future compact designs for portability using, for instance, mobile phones.

4.5.1.2 Droplet Sensing

The ability to detect or analyse samples with high sensitivity, reproducibility and reliability in microsystems can be hampered by either the design of the system or the size of the transducing element coupled to the microsystem. Integrating miniaturised sensing systems into DMF platforms has a number of critical advantages, including:

- droplet metering for accurate and reproducible size generation crucial for volume critical operations;
- determination of voltage–actuation relationship for high-fidelity droplet transportation on actuating electrodes;

- on-chip bio-/chemical analysis of reactions for real-time detection or monitoring;
- process automation and reconfigurability of droplet routing in multiplexed, surface-contaminated or high-density electrode array systems.

Continual enhancements of DMF design and development are to automate systems whilst reducing processing time and costs. These lead to innovative solutions to challenging issues, one of which is droplet contact-line pinning/stiction due to surface fouling, affecting transportation efficiencies of droplets in DMF systems. Shih *et al.* developed a simple RC based sensing with feedback system to monitor droplet motion and operations within their DMF device as a function of voltage.[101] Failure to actuate a droplet would imply a higher potential voltage is required to complete a specific operation, particularly useful if droplet hysteresis occurs as a consequence of surface fouling midoperation. The most common forms of integrated sensing mechanisms for DMF systems are electrical (*i.e.* impedance spectroscopy or capacitance sensing), with or without feedback (Figure 4.14), having been applied for both coplanar and parallel systems with applicability to droplet dispensation for uniformity, characterisation of the droplet, determining droplet position between two electrodes and more recently monitoring and analysing cell cultures.[102–109] More recent developments for DMF sensor integration include thin-film InGaAs photodetectors for chemiluminescence detection,[110,111] a customised miniaturised optical fluorimeter for PCR detection,[112] a silicon on insulator nanophotonic microring resonator for glucose measurements (Figure 4.15),[113] in-line nanoelectrospray ionisation mass spectrometry (nESI-MS) DMF coupling for quantification of sample

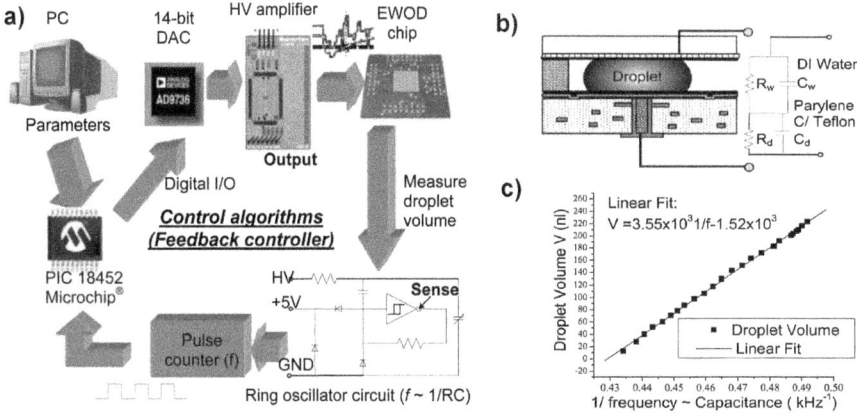

Figure 4.14 (a) Discrete-time digital feedback control system hardware; (b) Equivalent circuit model for measuring EWOD droplet capacitance; (c) Capacitance *vs.* droplet volume as measured by ring oscillator circuit showing linear relationship. Reproduced from ref. 104.

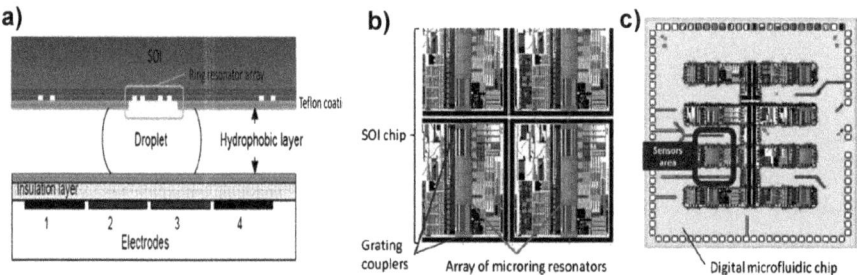

Figure 4.15 (a) Diagram of a silicon on insulator (SOI) DMF device containing arrays of microring resonator sensors on top plate; (b) Fabricated nanophotonic SOI chip layout and (c) alignment with DMF chip for label-free biosensing applications. Springer and Analytical and Bioanalytical Chemistry, 2012, 404, 2887–2984, Silicon photonic sensors incorporated in a digital microfluidic system, C. Arce, D. Witters, R. Puers, J. Lammertyn, P. Bienstman, Figures 4 and 5. With kind permission from Springer Science and Business Media.

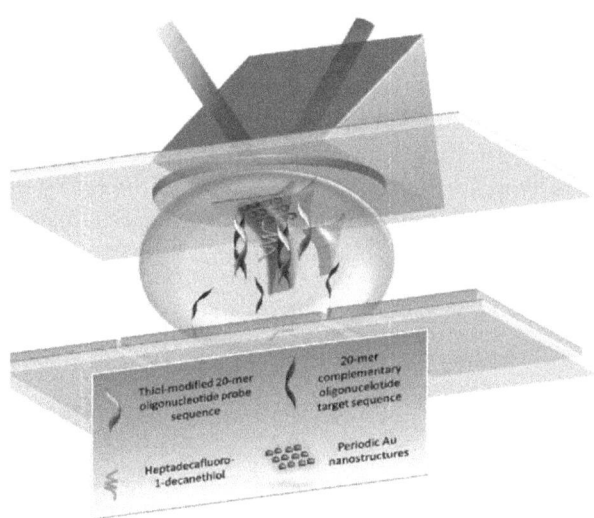

Figure 4.16 Schematic of nanostructured DMF device coupled with surface plasmon resonance imaging (SPRi). Reprinted from L. Malic, T. Veres and M. Tabrizian, *Biosens. Bioelectron.*, 2011, **26**, 2053–2059, with permission from Elsevier.

analytes,[114] and nanopost interfaces for enhanced surface plasmon resonance imaging (SPRi) biosensing of 180 nl volumes in as little as a minute, as depicted in Figure 4.16.[115]

An elaborate DMF system integrated with underlap field effect transistors (FET) was described by Choi *et al.*, in which n-channel MOSFETs were modified with a biosensing region ("underlap region") between the gate and

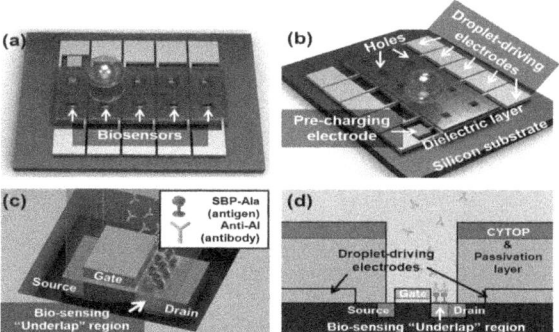

Figure 4.17   Schematics of the DMF–FET-based device: (a) a top-view, ((b) a side-view, (c) a magnified view of the embedded underlap FET biosensor and (d) a cross-sectional view of the biosensor. Reproduced from ref. 116.

drain.[116] Operating in an open format as shown in Figure 4.17, they pre-charged their droplet to reduce the driving potential to a 20-V signal, and transported and merged combinations of water, phosphate buffered saline solution and immunoglobin solutions (anti-Avian influenza and anti-rabbit IgG) close to the detection site where negligible nonspecific binding of anti-rabbit IgG was detected. When expected, high sensitivity and concentration-dependent detection of anti-AI binding to surface bound AIa (Avian Influenza antigen) on the underlap region was detected by the FET-based sensor with limits of detection (LOD) as low as 7 fM.

4.5.1.3   Routing Strategies

Reconfigurable and addressable DMF systems are an important part of the design and programming phase, facilitating automation of multiplexing/par-allelisation processes, droplet redirection from fouled surfaces (cross-contamination) and avoidance of contamination between different droplets.[91,117–119] As electrode grid arrays (*e.g.*, $M \times N$) become larger, the number of inputs and possible higher spatial resolution routes afforded for droplet operations also increases, though electrode addressability constraints with inadvertent droplet actuation also becomes an issue with scaling.[120] This can make manual protocol design optimisation and scheduling routes for multi-plexed processes a burdensome task for users of generic DMF architectures.[121] This challenge is complicated even further when small segments of the elec-trode array become nonfunctional due to dielectric breakdown (*e.g.* excessive voltages or cycle times) or production faults, further reducing device reliability and performance whilst increasing reagent consumption if dynamic sensing is not incorporated for dynamic reconfigurability of electrodes.[122,123]

To overcome these challenges a diverse set of programmable strategies have been investigated as a means for optimising device performance and facilitating reliable user-specific protocols on traditional architectural array

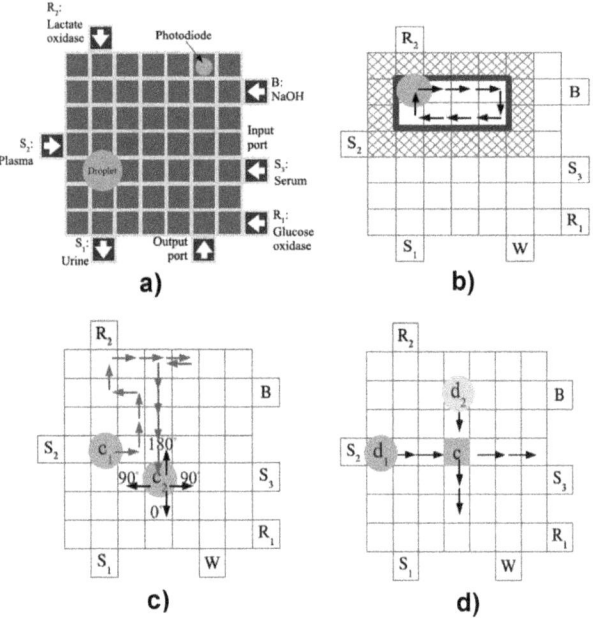

Figure 4.18 (a) Schematic of a DMF biochip architecture analogous to dynamic reconfigurable field programmable gate arrays (DR-FPGA); Possible mixing operations that can be configured for (b) module-based execution (constrained) and (c) routing-based execution (free movement); (d) If contamination at c by d_1, position *must* be avoided by d_2. Thus, routing-based strategies based on partitioning and area constrained routing, in which surfaces are regularly washed by assigned partition or area-based droplets, have been shown to be efficient routes for complex DMF operations. Springer and Design Automation for Embedded Systems, 16, 2012, 19–44, Routing-based synthesis of digital microfluidic biochips, E. Maftei, P. Pop, J. Madsen. Figure 2. With kind permission from Springer Science and Business Media.

arrangements (Figure 4.18). Böhringer described general modular algorithms based on robotic path planning (*e.g.*, prioritised $A*$ search, network flow) for droplet routing and task planning. Independent of hardware design, a balance between efficiency and optimality was necessary as complete solutions for optimised planning were difficult to attain.[124] Fouillet *et al.* employed a bottom-up architecture whereby each of the fundamental operations were standardised into a library of tools, which a user (*e.g.*, chemist or biologist) could call individually or as an integrated modular process.[125] Maftei *et al.* implemented a greedy randomised adaptive search procedure (GRASP), thus droplet routing was not just limited to between modular regions (*i.e.* dedicated regions for fundamental operations). They also incorporated a contamination-avoidance algorithm when required showing time and cost savings can be achieved through reduced-area droplet pathway-routing strategies.[126] A microelectrode dot array (MEDA) architecture, proposed by Chen *et al.* was described as being capable of achieving precise and flexible control of multiple droplets simultaneously.[127] Unlike traditional EWOD electrodes where droplet base area is a fraction larger than the underlying electrode geometric area, the MEDA architecture uses a clustered array of smaller microelectrodes each controlled by an activation circuit enabling various shapes and sizes to be activated simultaneously for droplet operations. As a consequence, the effective length of the droplet's contact line needed compensating, with more effective diagonal operations achievable. A new routing algorithm, based on priority settings, block setting, path routing extended to a 3D space (the 3D-A* search algorithm) and dynamic routing, was implemented for the MEDA system with simulation results indicating improved routing times for a droplet, from reservoir to sink, achieved in less time than previously cited results. Recently, a DMF system incorporating droplet capacitance sensing, fuzzy-based adaptability and expert routing strategies was evaluated for real-time droplet management across a small set of biologically relevant fluids (see Figure 4.19).[128] Labelled as an intelligent DMF system with fuzzy-enhanced feedback, control electronics drive individual electrodes (7–15 V_{RMS} at 1 kHz) for droplet actuation, whilst field programmable gate arrays (FPGA) scan electrode capacitances and relay information to the control unit for real-time computation. This automated system has capabilities of determining droplet volume and position, enhancement of driving voltage based on a reference charging time and real-time multidroplet tracking for process management. This high-level system represents an innovative solution to multiliquid droplet actuation in a single system where decision-based control for asynchronous operations occurs in real time, with device-lifetime extension possible as reconfigurations become a dynamic process.

4.5.2 Fundamental Operations

4.5.2.1 Dispensing

Critical to automated DMF operations is the creation or digitisation of droplets from single or multiple fluid reservoirs situated on the boundary of an electrode array grid. Considered to be one of the most difficult operations, without the use of an external pressure, droplet dispensation from an on-chip reservoir generally requires a higher V_{min} than other operations to overcome the impedance forces holding back the liquid. This higher actuating voltage gives rise to a greater probability of surface (insulating and hydrophobic layers) degeneration with long-term use, though Lin *et al.* demonstrated a low voltage actuation threshold (~7.2 V) for 5–30 pl droplet dispensation for a water-in-oil system on multilayered insulating platforms.[129,130] Upon EWOD actuation by several microelectrodes adjacent to the reservoir, the advancing meniscus exits the reservoir along with a column of liquid. To generate a droplet, microelectrodes situated close to the reservoir,

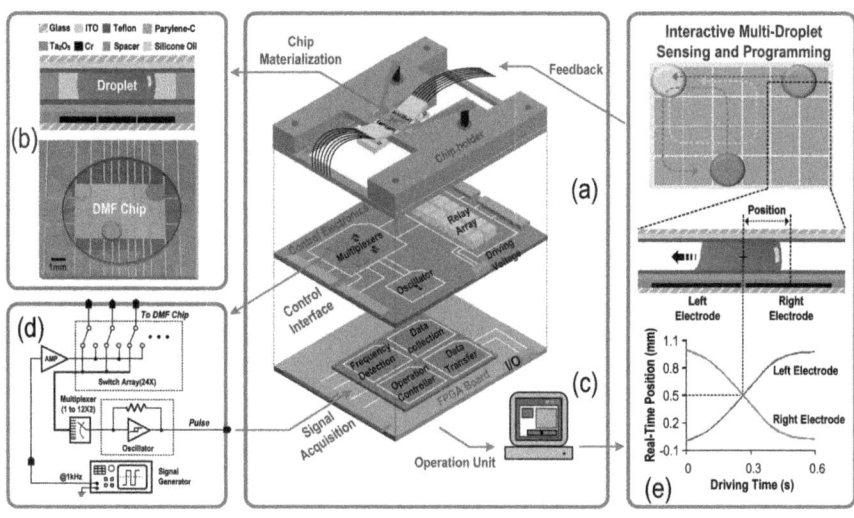

Figure 4.19 Proposed control-engaged digital microfluidic (DMF) technology. (a) The DMF module is composed of three operation layers. The top layer is an ergonomically designed chip holder that allows user-friendly operation and experimental repeatability. The midlayer is the control electronics that fits the size of the test bench. The bottom layer is the field-programmable gate array (FPGA) board. (b) The fabricated DMF chip features optimised materialisation to enhance the speed-to-voltage efficiency of droplet manipulation using the electrowetting-on-dielectric (EWOD) behaviour. (c) The operation unit powered by the inference mechanism in the computer is capable to perform (i) feedback-based position correcting, (ii) fuzzy-enhanced droplets control, and (iii) droplet transportation sequence programming. (d) The control electronics performing real-time multidroplet actuation and sensing. The presence/absence of droplet in each electrode is converted into clocked pulses *via* multiplexing each electrode to a single oscillator. (e) The operation unit features a user-friendly and upgradable human-control interface to manage the entire module. Reproduced from ref. 128.

dictated by pitch length, are energised/de-energised accordingly to create a pinch-off region along the liquid column (Figure 4.20). The actuation sequence required for droplet pinch-off, in conjunction with reservoir volume, system dimensions, possible hysteresis and variable fluid properties, makes automation a difficult task to control, particularly for rapidly generating a high density of uniform droplet volumes. Ren *et al.* attempted to overcome the issues associated with the automation process of droplet dispensing through capacitance metering of the droplet volume from a syringe pump using a ring oscillator circuit.[103] By exploiting the fluid's dielectric properties (0.1 M KCl) and the base contact area the droplet makes with the dielectric layer (Parylene C/Teflon), aligned with a buried control electrode and a ground counterelectrode an equivalent capacitance of the

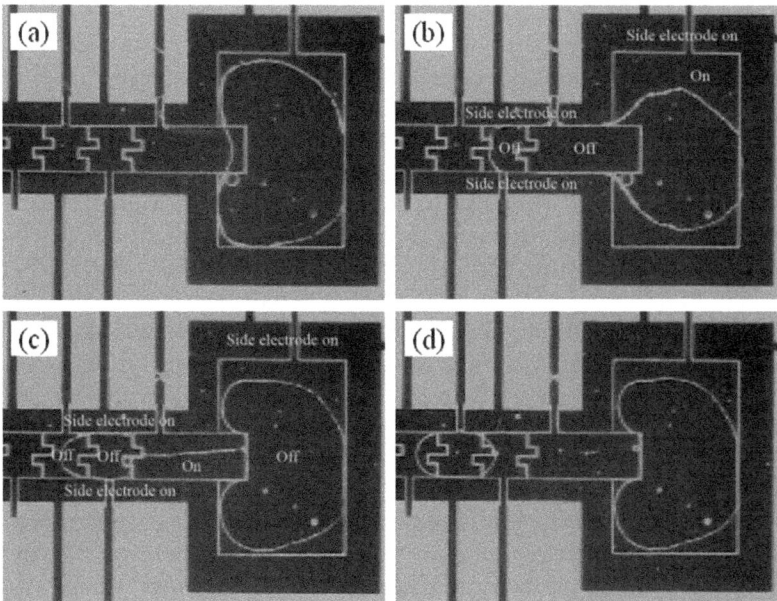

Figure 4.20 Basic operations performed on a DMF device include droplet dispensing, cutting, merging and transportation. The figure shows a sequence of video images showing the dispensing of an oil droplet of ~230 nl. The spacing between the top and bottom plates is 170 µm, and the applied voltage is 70 V_{RMS} 100 kHz AC. Note the thin film of oil in (c) that for a short time connects the dispensed droplet to the much larger reservoir volume. Reprinted with permission from ref. 193.

droplet volume was measured to ±5% reproducibility, for up to 45 droplets per min, against variable parameters such as viscosity, aspect ratio and dispensing volume. Gong and Kim employed a similar capacitance sensing technique with no ancillary pumping mechanism, thus devising an entirely electrical microfluidic system.[104] EWOD devices were fabricated on 4-layer PCB boards for seamless electronics integration and real-time proportional-integral-derivative feedback control. For a target capacitance the use of the Young–Laplace and Young–Lippmann equations gave relationships for pressure differences, droplet volume and applied voltages between the microelectrode sites wetting the reservoir. They reported a ±1% reproducibility for continuous droplet generation, at a rate of up to 30 droplets per min for low viscosity media, with an option for generating variable droplet volumes for enhanced flexibility. A recent study assessing the long term reproducibility of repetitive droplet dispensation volumes in an air-filler medium with SU-8/Teflon coating, indicated their imaging technique, testing regime and volume calculations were more robust and accurate than previously quoted literature.[131] Within a 120-µm spacer, over ~3 h 207 droplets were dispensed correlating to a 75% success rate. Though initial droplets

(<15) varied in size, as the reservoir size decreased to match the reservoir electrode size, daughter droplets dispensed were of uniform size. Further decrease in reservoir volume created a situation whereby droplet dispensing became harder and daughter volumes unpredictable again. A reduction in spacer thickness, and thus a force increase, was found to cause smaller volume variations, approximately 6% over 160 droplets.

It should be noted that the EWOD configurations above were of the closed format, hence required opposing acting forces to cut the stable liquid column. This is in contrast to droplet generation by LDEP,[102,132] in which a fluid column in open format and between two coplanar electrodes breaks up into droplets due to capillary instabilities when the applied voltage is removed.[133,134] Rayleigh's theory for capillary instability at wavelengths greater than a critical wavelength ($\lambda_c = 2\pi R$), accurately predicted droplet formation that was spatially controlled by periodic bumps according to theory for the most unstable wavelength ($\lambda^* = 9.016R = 4.5(2w + g)$), dependent on the total electrode width (w) and gap (g) employed, to gain uniformly spaced and picolitre-sized droplets. Owing to the small variations in droplet volume, a system combining droplet dispensing *via* LDEP at pre-determined intervals along the edge of a closed-format system, before entering an $M \times N$ array system for EWOD transportation is an alternative route for highly reproducible droplet volumes. EWOD and LDEP has been combined in creating double emulsion (DE) droplets of conductive (water) and dielectric (oil) liquids, respectively.[135] Although a closed system was described in DE creation, liquid tail volume (voltage dependent) close to pinch-off was prescribed as a main contributor to volume variations on droplet dispensing. This volume variation problem can be resolved through the provision of a third electrode and a T-junction gap.[132]

4.5.2.2 Splitting

Splitting or cutting is very similar to the pinch-off process described above, though instead of a liquid column, a droplet sandwich between parallel microelectrodes is forced to move in opposite directions to induce a hydro-dynamically unstable fluid. This is accomplished though elongation of the fluid droplet, where two opposing ends are wetted and the middle is non-wetted, inducing pinching at the middle. An illustrative example of this process is shown in Figure 4.21. A droplet initially situated above a de-energised/grounded control electrode, with an approximately equivalent droplet diameter, has adjacent side electrodes energised to induce necking. This gives an increase in the radii of curvature (recall eqn (4.10)),which is dependent on spacer thickness (D) being below a critical value, and a change in opposing end contact angles, such that eqn (4.48) gives the direct relationship between the principal radii of curvatures ($R_{1,2}$), Ew, and D.

$$\frac{R_2}{R_1} = 1 - \frac{R_2}{D}\frac{C_d V^2}{2\gamma_{LV}} \tag{4.48}$$

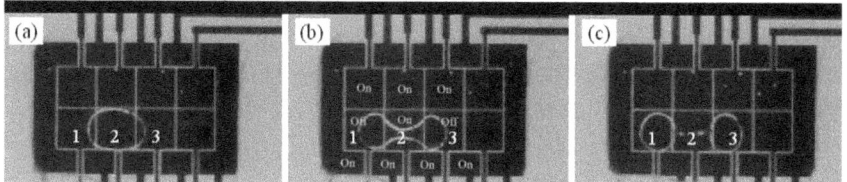

Figure 4.21   Sequence of video images of splitting and merging a mineral oil droplet. Addressable electrodes are 1 mm square, the spacing between the top and bottom plates is 170 μm and the applied voltage is 70 V_{RMS} at 100 kHz AC. Reprinted with permission from ref. 193.

Although the size of the split droplets are also sensitive to the size of the parent droplet, factors including surface properties, microelectrode size, evaporation when in air-filler medium all have a contributing part in droplet splitting.

4.5.2.3   Merging/Mixing

Merging droplets in DMF is an essential operation for mixing or diluting constituents of individual droplets in a controlled volume. This can be accomplished by moving droplets towards each other, preferably unto a single microelectrode and once merged, the new droplet is actively transported over a set number of microelectrodes to mix the constituents.[68] If the Ohnesorge number (Oh) in eqn (4.49), describing viscous effects (μ) in relation to inertial (ρ) and surface tension (γ) effects is large (Oh > 0.1), droplet coalescence may not be possible. With highly aqueous media, mixing of constituents will occur *via* diffusion mechanisms for low Reynolds number (Re) fluids, thus leading to slow mixing times.

$$Oh = \frac{\mu}{\sqrt{\rho \gamma L}} \tag{4.49}$$

To enhance mixing times of merged droplets forced convection or electrohydrodynamic flows within the droplet, through the application of frequency-dependent voltages have been demonstrated as an effective method.[85,87] The generation of fluid-property gradients and temperature gradients through electrothermal heating effects causes fluid motion within the droplet, though excessive rises in temperatures can lead to damage of heat-sensitive materials, *e.g.*, biological molecules.

4.6   Applications of DMF Technologies

The development of DMF technology over the past two decades has given rise to a better understanding of the underlying mechanisms involved in EWOD droplet actuation and a host of other microscale phenomena that can be exploited for specific applications in a range of sectors as reviewed by Jebrail *et al.*[136] EWOD system design and fabrication have become more complex to address specific needs and integration with a range of other technologies and detection systems (*e.g.*, electrochemical, fluorescence, surface plasmon resonance).[21] In the following section, examples of specific applications of DMF are discussed.

4.6.1   Biological Applications

The miniaturisation of fluidic devices for carrying out a variety bioprocesses such as cell culturing, gene expression, DNA hybridisation, *etc.*, has been seen as an enormous step for cell and molecular biologists who are now able to control the microenvironment (*e.g.*, temperature, biochemical gradients, pressures) and use significantly less reagents in their quest for new discoveries in biochemical pathways and drug discovery.

4.6.2   Cell-Based Assays

Cell-based assays on DMF platforms were first described by Wheeler's group based at the University of Toronto and continually provide innovative approaches to biological based DMF processes. Using a combination of cytotoxic assessment methods, including MALDI mass spectrometry of cell lysates, the effect of electrostatic manipulation of entrained Jurkat-T cells was found to have no adverse effects on cell viability, proliferation and physiology.[137] Actuation of variable cell densities ($\sim$10^6 cells per ml) in nl–μl-sized droplets within a closed, humidified DMF device were performed using a 15 kHz, 100–140 V_{RMS} signal with droplet conductivity approaching 5 S·m^{-1}. This represented a microenvironment of significantly high conductive media, above which mammalian cells normally require for culturing protocols. To overcome droplet stiction from contact line pinning, addition of nonionic pluronic F68 reduced cell adsorption onto the Teflon surface facilitating actuation of suspended cells. A study of varied pluronic coblock polymer (PEO$_m$–PPO$_n$–PEO^m) additives suggested that PPO with longer chain lengths and PEO of >50% content had no adverse effects on cells and reduced biofouling of DMF devices.[138] Barbulovic-Nad *et al.*, from the same group, subsequently introduced a complete mammalian culturing DMF platform (Figure 4.22) in which passive dispensing – *i.e.* a DMF platform of predominantly hydrophobic surfaces with hydrophilic patterned sites – for seeding, media exchange, subculturing or splitting of adherent cells over 6–8 days was achieved.[139] Although differences in DMF microculturing *versus* traditional macroculturing methods were found in higher seeding densities on the microscale, calling for more regular ($\sim$24 h) media exchanges, and possible surface-coating integrity issues between deposited fibronectin on Teflon compared to the plasma-treated polystyrene culture flasks, long term viability of CHO and HeLa cells were obtained enabling the demonstration of GFP gene transfection in the DMF system. Witters *et al.* demonstrated a dry lift-off (DLO) technique gave a more controlled method of biofunctionalising

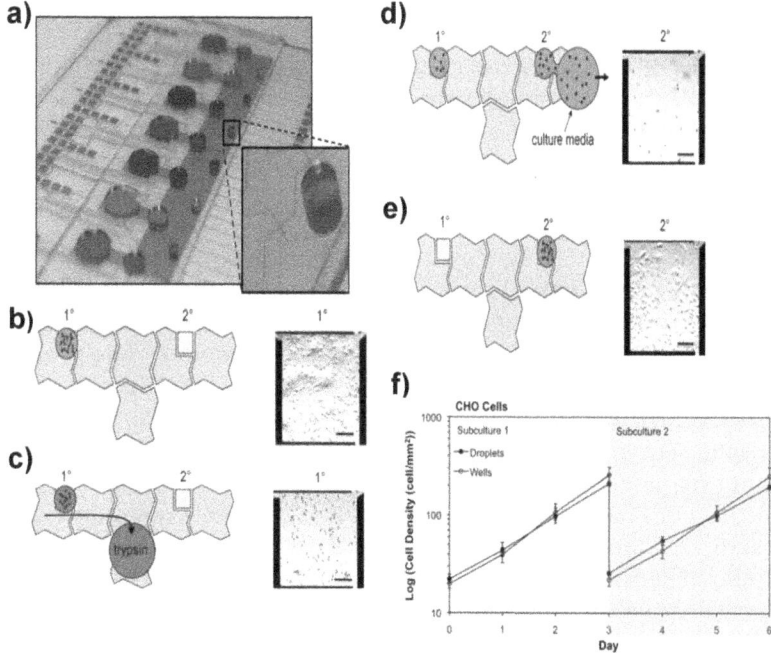

Figure 4.22 Schematic and pictures depicting (a) Used DMF device subculture of CHO–K1 cells in droplets by digital microfluidics; (b) Monolayer of the first generation of cells on the primary (1°) adhesion pad; (c) Cells dissociating from the surface of the 1° pad after delivery of trypsin; (d) trypsinised cells being harvested in a droplet of media containing serum, and then seeded on the secondary (2°) pad; (e) Monolayer of subcultured cells on the 2° pad after 72 h. Scale bars are 200 μm; (f) Figure shows comparable DMF subculturing results with traditional well plate techniques over a 6-day period. Reproduced from ref. 139.

(poly-L-lysine-FITC) Teflon-AF surfaces with "cytophilic" patches, thus leaving the remaining surface hydrophobic and promoting site-specific cell adhesion for single or small clusters cell-based assays.[140] The creation of individually addressable virtual microwells (VM) by Wheeler's group, based on hydrophilic patterning of surfaces to generate passive dispensed droplets on hydrophilic sites, was demonstrated on single-plate and two-plate DMF configurations.[141] This system showed improvements in dispensed volume reliability independent of viscosity, volume maintenance upon media exchange and cell-seeding densities, based predominantly on the device's geometrical design parameters. The use of primary cells has also been applied to DMF systems, where through the use of a fluorocarbon lift off process, cell cocultures were realised reducing sample volumes needed for analysis.[142] Recently, Shih *et al.* developed on the aforementioned studies by creating a DMF system capable of performing variable assays on mammalian cells with label-free impedance sensing occurring within the VM.[108] The development of a DMF microbioreactor for ~5 days of cell growth (bacteria, algae and yeast) with subsequent genetic transformation on bacteria by a fluorescent reporter gene, all within distinct droplets, was described by Au et al.[143] Cell-growth curves were comparable to macroscale culturing growth curves demonstrating DMF as a tool for parallel microbial culturing, with reduced volumes, for synthetic biology and pharmaceutical applications.

An advantage of DMF systems is the high-throughput capabilities afforded through assay multiplexing with appropriate routing schemes. This was demonstrated for apoptosis screening assays based on the detection of the irreversible capase-3 activation pathway at various levels of staurosporine doses (0–10 μm).[144] Using a commercial fluorescence multiwell plate reader, the effects of dose response via Nuc View 488, a bifunctional fluorogenic enzyme for capase-3 activity detection and the effect of cell washing via nuclear staining (Hoechst 33342) to assess cell loss were characterised. Low shear stresses, as a result of passive solution exchange, indicated that adherent cells did not detach from the surface during apoptotic experiments compared to a 38% loss in cells from a standard macroscale protocol. Increases in staurosporine concentration were reflected by an increase in fluorescence intensities, with lower errors for the DMF platform in comparison to standard 96-well plate protocol. This represented a 33-fold reduction in reagent consumption and superior data quality for accurate dose response curve generation for drug-discovery applications.

4.6.3 Biomolecular Assays

One of the first demonstrations of a DMF platform for on-chip processing of biological solutions involved assessing basic high-speed droplet operations and potential device contamination from DNA adsorption through hydrophobic or electrostatic interaction.[145] Yoon and Garrell described the use of a pH 4 acetate buffer and a biased asymmetric square wave for minimising biomolecular adsorption of BSA, DNA and lysozymes unto a Teflon AF surface in an air-filled system.[146] Subsequently, Luk et al. investigated the extent to which biomolecules (i.e. proteins) nonspecifically adsorbed onto the device surface and introduced the concept of pluronic additives (i.e. F127) for the first time to limit protein adsorption in DMF devices.[147] This method is seen as an important milestone for processing of biological materials in DMF systems because many biological processes are usually conducted in a static environment, for which a droplet can be interpreted as a microreactor, and the high-speed translation of droplets between electrodes to reduce contamination is not a feasible strategy for most applications. Most recently, Perry et al. have proposed using graphene oxide as a protein cargo carrier to prevent protein adsorption, though this strategy is still in its infancy.[148]

Polymerase chain reaction (PCR), a thermal cycling process important for DNA amplification and a means for analysis and detection of mutations in mRNA or DNA molecules, is a process that has been demonstrated extensively in high-throughput two-phase immiscible microdroplet systems (*e.g.*, emulsion PCR),[149-151] and also shows great compatibility and increased controllability with EWOD/DMF platforms. This was first demonstrated by Chang *et al.* in which the challenge of overcoming PCR process (heating) interference with an EWOD configuration (potential dielectric breakdown) was addressed by designing a hydrophilic/hydrophobic structure for PCR and EWOD integration with low actuation voltages.[152] Both EWOD and PCR processes were controlled by individual programmable microprocessors for droplet operations (<25 V_{RMS}) and real-time temperature sensing and heating (PCR power consumption = 0.9 W), respectively. Reagents and primers along with cDNA of Dengue II virus were dispensed from separate reservoirs at ~730 nl, mixed in 3 cycles on a 2 × 2 array and transported to the periphery of the PCR chamber where surface-tension gradients created by the hydrophobic/hydrophilic surface induces rapid droplet motion into the PCR chamber. Thermal cycles were performed 25 times, through a 120-s DNA denaturisation (95 °C), 35-s primer annealing (53 °C), 35-s DNA extension (72 °C) and 240 s for a final DNA extension (72 °C) protocol, consuming 15 µl of sample representing a 70% reduction over conventional PCR machines. Advanced Liquid Logic Incorporated (http://www.liquid-logic.com) introduced a self-contained, low-power, custom-built real-time multiplexed PCR assay DMF device, with up to 64 individually addressable channels capable of being actuated with 0–300 V potentials.[112] The two-plate DMF chip in Figure 4.23, was fabricated on a 86 × 86 mm PCB substrate (ITO-coated glass coverplate) containing electrode arrays and terminal pads, 8 loading

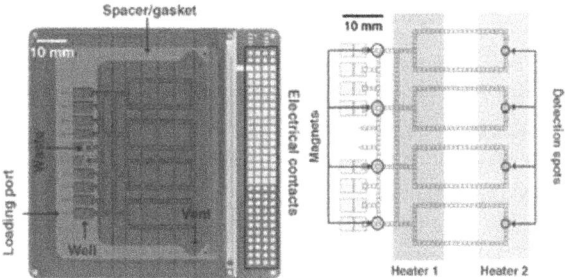

Figure 4.23 Image of an assembled microfluidic cartridge comprising a PCB chip, polymer spacer/gasket, and glass top-plate with drilled holes (left); Schematic of the PCR chip showing electrode positions relative to heaters, magnets, and detectors (right). Reprinted with permission from Z. Hua, J. L. Rouse, A. E. Eckhardt, V. Srinivasan, V. K. Pamula, W. A. Schell, J. L. Benton, T. G. Mitchell and M. G. Pollack, *Anal. Chem.*, 2010, **82**, 2310–2316. Copyright 2010 American Chemical Society.

reservoirs, two temperature zones traversed by four separate electrode loops each accommodating a single droplet that are programmed for thermocycling. Assessment of the PCR system was carried out using genome detection and analysis of methicillin-resistant *Staphylococcus aureus* (MRSA), and *Candida albicans* (at 45cycles, runtime = ~45 min),[153] and *Mycoplasma pneumonia* in blood samples. With 40 thermocycles applied to the 10-fold DNA dilutions (pentograms–femtograms), amplification efficiency based on the slope of the fluorescence curve for MRSA DNA amplification was calculated to be superior to other miniaturised PCR systems at 94.7%, while amplification speed was found by optimising the annealing/extension time over the cycle count, thus giving a total process time for optimal reaction yields of 18 min. Concurrently, work by Berthier *et al.* at Laboratoire d'Electronique et de Technologie de l'Informatique (CEA-LETI) showed that their EWOD PCR system could extract and purify target DNA from a 5-μl blood sample, thus simultaneously and in parallel detect the presence or absence of specific gene mutations *via* specific fluorescent signals.[154] A range of processes involving biomolecular interactions have been demonstrated on EWOD systems including DNA hybridisation and immunoassays,[115,155–158] surface biofunctionalisation for surface plasmon resonance imaging (SPRi) and enhanced-SPRi detection of DNA hybridisation,[115,159] sample prep and purification for proteomic analyses and DNA sequencing,[160,161] multistep protein processing for on-chip MALDI-MS analysis,[162] and employing hydrogel discs for proteolytic enzyme microreactors and tissue engineering applications.[163,164] Recently, the use of fluorinated solvents as an alternative to pluronics, for reducing protein adsorption has been reported with an added benefit of rapid *in situ* protein crystallisation for an improved signal-to-noise ratio for on-chip MALDI-MS analysis on fluoropolymer surfaces.[165]

4.7   Chemical Reactions

Compartmentalisation of components within aqueous microdroplets provides a suitable microenvironment for rapid and efficient microreactions in which crosscontamination is minimised and sample volumes are significantly reduced, such as those described in the previous section for biological entities. First reported by Washizu in the late 1990s, the idea of a "droplet microreactor" was demonstrated by manipulating water droplets in air *via* an array of addressable interdigitated microelectrodes coated with a 20-μm insulating layer.[5] The mechanism here was attributed to DEP in air, and has since been utilised by others with various microelectrode geometries (*e.g.*, tapered L-DEP structure) for biochemical assays[166,167] and enyzmatic reactions,[168] though issues pertaining to droplet evaporation and contamination have been a cause of concern with open-format configurations. Employing the use of silicone-oil-filled systems for droplet encapsulation, Taniguchi *et al.* performed alkalisation of phenolphthalein with NaOH droplet solutions and an enzyme-based reaction oxidising luciferin in the presence of ATP and luciferase on a 3/6/9-phase electrode device, similar to Washizu's set-up.[169]

A programmable fluidic processor for droplet-based chemistry was described by Gascoyne's group in which dielectric liquids were manipulated by DEP on a Fluoro-Pel® hydrophobic coating with a water immiscible hydrocarbon. Using the positive value of the complex Clausius–Mossotti factor between hydrocarbon and aqueous droplet as criterion for droplet motion in an open format, proof-of-concept experiments for a range of biochemical and chemical reactions were demonstrated.[170] As seen, chemically stable dispersant media serves a variety of functional purposes in DMF systems, not least voltage reduction for reversible droplet switching.[53,68,155,156,171,172]

A study carried out by Chatterjee *et al.* into electrostatic actuation of organic solvents, ionic liquids (ILs) and aqueous surfactants in a two-plate air-filled system revealed that a vast range of liquids can be actuated (merged, split, *etc.*) independent of surface tension value or range of voltage induced change in contact angle.[173] At operating AC voltages of 90 V, 8 kHz solvents such as DMSO, ethanol, formamide and acetone were actuated, whilst a reduction in frequency and spacer thickness, from 300 μm to >50 μm, enabled the actuation of even more solvents *e.g.*, dichloromethane, chloroform and *m*-dichlorobenzene. A general deduction made was that solvents with $\varepsilon_m > 3$, $\sigma_m > 10^{-9}$ S·m^{-1} or dipole moments greater than 0.9 D were movable with potential applications in synthesis reactions and liquid–liquid extraction operations. Classes of solvents, *i.e.* free-flowing room-temperature ionic liquids (RTILs) and task-specific ionic liquids (TSILs), were characterised by Dubois *et al.* on open-format configurations, as potential stable microreactor environments for organic synthesis of tetrahydroquinolineon electrode arrays (Grieco's reaction).[174] RTILs have a range of applications for multicomponent reactions (MCR) from organic synthesis to enzymatic catalysis,[175] with the most attractive characteristics being negligible vapour pressure and thus low rate of evaporation in open formats. IL droplets exhibited a lower contact angle and surface tension than water on Teflon AF surface, with viscosity values up to 300 times that of water. Furthermore, $18 \text{ V} < V_{min} < 30 \text{ V}$ was found as the minimum actuation potential range for ILs compared to 40 V for water, though addition of 1% Tween to water did lower V_{min} to within the IL range. The three-component reaction was conducted in a droplet microreactor by merging three droplets in sequence and incubating over an hour period at room temperature. Analysis of the reaction was conducted on-line *via* cyclic voltammetry and also validated in HPLC against macrovolume reactions. A recent study found that increasing the temperature of imidazolium-based ILs further reduces V_{min} and gives faster response times under AC voltages in alkane-filled EWOD systems.[176] Jebrail *et al.* communicated a two-plate MCR DMF system for parallel chemical synthesis of macrocylic peptides.[177] In their system, ten different solutions, including 5 amino acid substrate solutions, were handled to synchronously synthesise five different peptide-macrocycles from aziridine aldehyde with 90% reagent depletion. Further processes demonstrated on this system included, redissolution of product in trifluoroethanol for structural modification, solvent evaporation to obtain products in specified locations and reaction specificity for synthesis of a nine member Proline–Leucine-derived macrocycle. Ding *et al.* dispensed volatile solvent acetonitrile from a syringe pump in to a closed platform through a feedback mechanism, and demonstrated droplet manoeuvrability towards synthesis of radiochemical PET probe 2-[18F] fluoro-2-deoxy-D-glucose,[178] whilst Wijethunga *et al.* demonstrated increased performance capabilities of a liquid–liquid microextraction DMF system could be influenced by higher applied frequencies.[179]

Finally, a magnetic-bead-incorporated DMF platform for pyrosequencing (Figure 4.24) developed by Fair's group and Advanced Liquid Logic Incorporated implemented a three-enzyme pyrosequencing protocol on a PCB-fabricated (Figure 4.24(a)) DMF system using *Candida parapsilosis* DNA (229 bp).[180,181] DNA polymerase synthesises a complementary DNA strand using individual nucleotides (dNTPs) added to the next unpaired base, thus releasing a pyrophosphate molecule that can be converted to ATP and detected by chemiluminesence from the oxidation of luciferin by luciferase and quantifiably related to the dNTP's incorporation in a linear fashion. The proof of concept of cyclic or noncyclic order addition of the 4 dNTPs to the DNA-primer–magnetic-bead complex showed similar sequencing results, with high cycle efficiencies, though basic process cycle time optimisation, *e.g.*, washing, reaction kinetics, and system architecture is key for increasing throughput and integration of up- and downstream processes for an all-inclusive detection system.

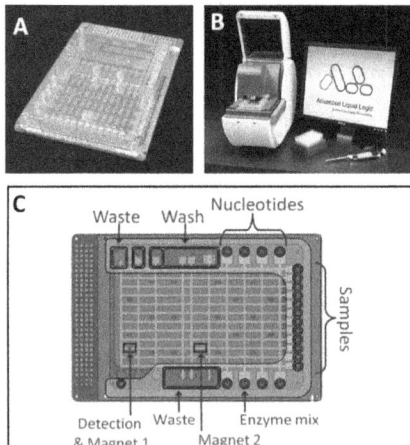

Figure 4.24   (a) Photograph of the assembled multiwell plate-sized PCB-based cartridge; (b) photograph of the sequencing instrument developed; and (c) schematic illustration of the features of the cartridge including sample and reagent wells, detectors and waste. Reprinted with permission from D. J. Boles, J. L. Benton, G. J. Siew, M. H. Levy, P. K. Thwar, M. A. Sandahl, J. L. Rouse, L. C. Perkins, A. P. Sudarsan, R. Jalili, V. K. Pamula, V. Srinivasarn, R. B. Fair, P. B. Griffin, A. E. Eckhardt and M. G. Pollack, *Anal. Chem.*, 2011, **83**, 8439–8447. Copyright 2011 American Chemical Society.

4.8 Diagnostics and Monitoring

Research into the operational limits of electrically activated DMF systems have provided researchers and engineers with a vast range of operational parameters that must be met for application to real-world samples. In particular, demonstrations of these systems in a clinical setting have shown comparable LOD with conventional methods. Along with the PCR method for whole-blood analysis for pathogens, a range of other clinically relevant diagnostic indicators are present for screening and sensing.[153] The Fair group assessed their DMF system for the determination of enzymatic activity.

Against glucose levels using a colorimetric enzyme-reaction (Trinder's reaction) over 1 min periods. The reaction involves the enzymes glucose oxidase (GOD) that forms hydrogen peroxide from the oxidation of glucose, and reacts with 4-amino antipyrine and *N*-ethyl-*N*-sulfopropyl-*m*-touidine to produce a violet quinoneimine detected by absorbance measurements *via* a noncollimated photodiode.[182] Measured absorbance over time (Figure 4.25) was related to the reaction rate of GOD and thus the concentration of glucose, with the LOD (9–15 mg·dl^{-1}) of their system found to be influenced mainly by errors relating to the photodiode's operation over time and data quantisation of the 12-bit A/D converter. Again, Srinivasan *et al.* applied their DMF system to a range of human physiological fluids to assess each fluid's operational capabilities (*i.e.* transport) on their platform and used Trinder's

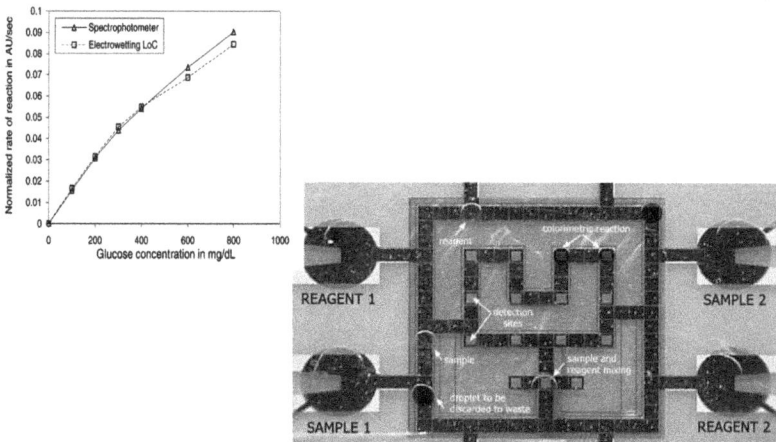

Figure 4.25 Comparison of DMF lab on a chip device with a bench-top spectrophotometer in the determination of glucose concentrations. Reprinted from V. Srinivasan, V. K. Pamula and R. B. Fair, *Anal. Chim. Acta*, 2004, **507**, 145–150, with permission from Elsevier; and the DMF platform developed for glucose assays. Reproduced from ref. 183.

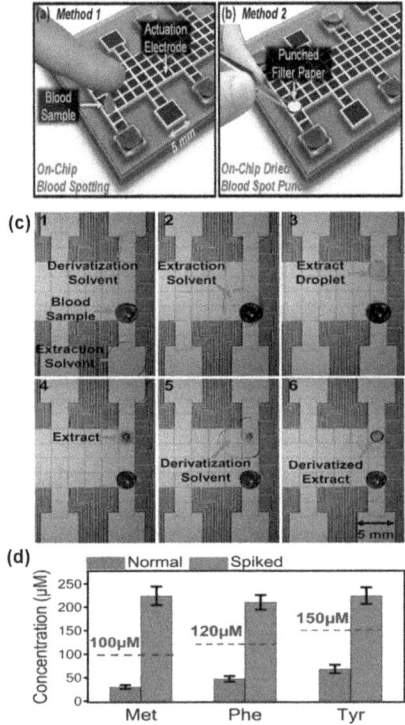

Figure 4.26 Methods of blood-sample delivery to DMF device include (a) direct spotting or (b) dried blood spot (DBS) punch from filter paper; (c) analysis of amino acids in dried blood spots by DMF with sequence of frames from a movie depicting several stages in sample processing including: (1) a dried blood sample prior to processing; (2) mixing and incubating an extractant droplet with the sample; (3) a droplet containing sample extract after translation away from the dried sample; (4) a dried extract; (5) mixing and incubating a derivatisation reagent droplet with the dried extract; and (6) the dried, derivatised product; d) Comparison of Methionine, Phenylalanine, and Tyrosine concentrations in normal (green) and spiked (red) blood samples as biomarkers for homocystinuria, phenylketonuria, and tyrosinemia, respectively. The dashed lines indicate the upper levels for normal concentrations in newborn blood samples. Each data point represents at least four replicate measurements, and error bars represent ±1 S.D. Reproduced from ref. 189.

reaction for glucose detection against each fluid type.[183] With the exception of urine, due to uric acid interference, all samples assayed compared well with their spectrophotometer reference method.

Point-of-care diagnostic systems, giving rapid sample to answer, are potential beneficiaries of DMF systems providing low cost in cartridge manufacture, facilitation of process design through software reconfigurability and modularisation for combined operations.[184] Mousa *et al.* showed the potential of DMF as a screening indicator of women at risk from developing breast cancer by extracting and measuring estradiol (active form of oestrogen) from minute tissue samples and whole blood and serum from postmenopausal cancer patients.[185] Their 20-min automated system performed lysing and liquid–liquid extraction steps on sample sizes greater than $1000\times$ less than conventional methods, which traditionally takes 5–6 h to process. Recovery of estradiol in methanol (86–119% efficiency) was achieved while unwanted constituents were extracted into isooctane, as confirmed by immunoassaying and liquid chromatography and tandem mass spectrometry analysis. Further demonstrations of DMF's capabilities in advanced automated systems have been applied to newborn screening (NBS) for treatable diseases, such as lysosomal storage diseases (LSD),[186–188] and heptorenaltyrosinemia.[114] Screening of these diseases are carried out from dried blood spot (DBS) analysis and DMF represents a remarkably apt technology for extraction/cleaning, processing and screening of samples from DBS (as in Figure 4.26) or any other solid surface for subsequent detection of diseases, pathogens or analytes in medicine and environmental monitoring.[189–192]

4.9 Summary

Beginning with Lippmann's observations over a century ago, the fundamental understanding of fluid actuation, due to electrical forces, has evolved significantly. Theoretical interpretation of electrowetting experimentation, by physicists and material chemists, has provided significant insight into the contributing mechanisms involved at the 3-phase interface. Thus, our understanding of voltage-induced fluid dynamics has enabled technologists in chemistry and biology to apply the fundamentals in the creation of low-cost efficient analytical systems. Due to advances in manufacturing technologies and miniaturisation of electronic components, electrostatic-driven droplet manipulation has undergone a significant transformation, to the point where sensor integration and droplet routing schemes are being devised for more efficient operational protocols. The functional uses of DMF systems have thus far been proven in a limited set of areas. The scope for translating this technology into other sectors will be a task accomplished by innovative multidisciplinary teams, once widely accepted as a commercially user-friendly technology in healthcare and life-sciences applications, for which a concentrated effort is currently being exerted.

References

1. G. M. Whitesides and A. D. Stroock, *Phys. Today,* 2001, **54**, 42–48.
2. Z. Jiao, N.-T. Nguyen, X. Huang and Y. Ang, *Microfluid. Nanofluid.,* 2007, **3**, 39–46.
3. A. A. Darhuber, J. P. Valentino and S. M. Troian, *Lab Chip,* 2010, **10**, 1061–1071.

4. M. A. Burns, C. H. Mastrangelo, T. S. Sammarco, F. P. Man, J. R. Webster, B. N. Johnson, B. Foerster, D. Jones, Y. Fields, A. R. Kaiser and D. T. Burke, *Proc. Natl. Acad. Sci. U. S. A.,* 1996, **93**, 5556–5561.

5. M. Washizu, *IEEE Trans. Ind. Appl.,* 1998, **34**, 732–737.

6. M. Gunji, T. B. Jones and M. Washizu, *Institue of Electrostatics Japan/ Electrostatics Society of America Joint Symposium on Electrostatics*, Japan, 2000.

7. D. H. Lee, H. Hwang and J. K. Park, *Appl. Phys. Lett.,* 2009, **95**.

8. M. K. Tan, J. R. Friend and L. Y. Yeo, *Lab Chip,* 2007, 7, 618–625.

9. Z. Guttenberg, H. Muller, H. Habermuller, A. Geisbauer, J. Pipper, J. Felbel, M. Kielpinski, J. Scriba and A. Wixforth, *Lab Chip,* 2005, **5**, 308–317.

10. E. Colgate and H. Matsumoto, *J. Vac. Sci. Technol. A,* 1990, **8**, 3625–3633.

11. K. Mohseni and E. S. Baird, *Nanoscale Microscale Thermophys. Eng.,* 2007, **11**, 99–108.

12. G. Beni and S. Hackwood, *Appl. Phys. Lett.,* 1981, **38**, 207–209.

13. H. M. Kang and J. Kim, in *MEMS 2006: 19th IEEE International Conference on Micro Electro Mechanical Systems, Technical Digest*, IEEE, New York, 2006, pp. 742–745.

14. M. A. Bucaro, P. R. Kolodner, J. A. Taylor, A. Sidorenko, J. Aizenberg and T. N. Krupenkin, *Langmuir,* 2009, **25**, 3876–3879.

15. R. K. Dash, T. Borca-Tasciuc and Asme, *Variable Focal Length Microlens by Low Voltage Electrowetting on Dielectric (EWOD) Actuation*, American Society of Mechanical Engineers, New York, 2005.

16. W. Monch, R. P. Shaik, L. Lasinger, F. Krogmann and H. Zappe, *Optical Characterization of Repositionable Liquid Micro-lenses*, IEEE, New York, 2007.

17. B. Berge and J. Peseux, *Eur. Phys. J. E,* 2000, **3**, 159–163.

18. I. Moon and J. W. Kim, *Using EWOD (Electrowetting-on-Dielectric) Actuation in a Micro Conveyor System*, IEEE, New York, 2005.

19. Y. J. Zhao and S. K. Cho, *Lab Chip,* 2007, 7, 273–280.

20. Y. Y. Lin, C. W. Lin, L. J. Yang and A. B. Wang, *Electrochim. Acta,* 2007, **52**, 2876–2883.

21. L. Malic, D. Brassard, T. Veres and M. Tabrizian, *Lab Chip,* 2010, **10**, 418–431.

22. B. T. Kelly, J.-C. Baret, V. Taly and A. D. Griffiths, *Chem. Commun.,* 2007, **0**, 1773–1788.

23. C.-X. Zhao and A. P. J. Middelberg, *Chem. Eng. Sci.,* 2011, **66**, 1394–1411.

24. F. Mugele and J. C. Baret, *J. Phys. Condens. Matter,* 2005, **17**, R705–R774.

25. J. Lee, H. Moon, J. Fowler, T. Schoellhammer and C. J. Kim, *Sens. Actuators A,* 2002, **95**, 259–268.

26. G. Beni and M. A. Tenan, *J. Appl. Phys.,* 1981, **52**, 6011–6015.

27. J. Lee and C. J. Kim, *J. Microelectromech. Syst.,* 2000, **9**, 171–180.

28. M. Vallet, B. Berge and L. Vovelle, *Polymer,* 1996, **37**, 2465–2470.

29. C. Quilliet and B. Berge, *Curr. Opin. Colloid Interface Sci.,* 2001, **6**, 34–39.

30. F. M. Fowkes, *Ind. Eng. Chem.*, 1964, **56**, 40.
31. A. W. Adamson and A. P. Gast, *Physical Chemistry of Surfaces*, John Wiley & Sons, New York, USA, 6th edn, 1997.
32. F. London, *Trans. Faraday Soc.*, 1937, **33**, 8b–26.
33. L. A. Girifalco and R. J. Good, *J. Phys. Chem.*, 1957, **61**, 904–909.
34. T. Young, *Philos. Trans. R. Soc. London*, 1805, **95**, 65–87.
35. W. A. Zisman, in *Contact Angle, Wettability, and Adhesion*, American Chemical Society, 1964, vol. 43, pp. 1–51.
36. W. A. Zisman, *Ind. Eng. Chem.*, 1963, **55**, 18–38.
37. A. Dupre, *Theorie Mechanique de la Chaleur*, Gauthier-Villars, Paris, 1869.
38. M. E. Schrader, *Langmuir*, 1995, **11**, 3585–3589.
39. W. C. Nelson and C. J. Kim, *Journal of Adhesion Science and Technology*, 2012, **26**, 1747–1771.
40. W. A. Zisman, in *Interaction of Liquids at Solid Substrates*, American Chemical Society, 1968, vol. 87, pp. 1–9.
41. J. Zeng and T. Korsmeyer, *Lab Chip*, 2004, **4**, 265–277.
42. P. G. Degennes, *Rev. Mod. Phys.*, 1985, **57**, 827–863.
43. L. Boinovich and A. Emelyanenko, *Adv. Colloid Interface Sci.*, 2011, **165**, 60–69.
44. V. Bahadur and S. V. Garimella, *Langmuir*, 2007, **23**, 4918–4924.
45. V. M. Starov, *Adv. Colloid Interface Sci.*, 2010, **161**, 139–152.
46. M. J. de Ruijter, T. D. Blake and J. De Coninck, *Langmuir*, 1999, **15**, 7836–7847.
47. S. Semenov, V. M. Starov, R. G. Rubio and M. G. Velarde, *Langmuir*, 2012, **28**, 16724–16724.
48. T. B. Jones, *Langmuir*, 2002, **18**, 4437–4443.
49. B. Berge, *C. R. Acad. Sci. Ser. II*, 1993, **317**, 157–163.
50. T. B. Jones, *J. Micromech. Microeng.*, 2005, **15**, 1184–1187.
51. F. Mugele, A. Klingner, J. Buehrle, D. Steinhauser and S. Herminghaus, *J. Phys. Condens. Matter*, 2005, **17**, S559–S576.
52. H. J. J. Verheijen and M. W. J. Prins, *Langmuir*, 1999, **15**, 6616–6620.
53. H. J. J. Verheijen and M. W. J. Prins, *Rev. Sci. Instrum.*, 1999, **70**, 3668–3673.
54. K. H. Kang, *Langmuir*, 2002, **18**, 10318–10322.
55. J. A. Stratton, *Electromagnetic Theory*, McGraw-Hill Book Company, New York, 1941.
56. H. H. Woodson and J. R. Melcher, *Electromechanical Dynamics. Part II: Fields, Forces and Motion*, Krieger Publishing Co., New York, 1985.
57. M. Vallet, M. Vallade and B. Berge, *Eur. Phys. J. B*, 1999, **11**, 583–591.
58. K. H. Kang, I. S. Kang and C. M. Lee, *Langmuir*, 2003, **19**, 5407–5412.
59. J. Buehrle, S. Herminghaus and F. Mugele, *Phys. Rev. Lett.*, 2003, **91**.
60. T. B. Jones, K. L. Wang and D. J. Yao, *Langmuir*, 2004, **20**, 2813–2818.
61. W. J. J. Welters and L. G. J. Fokkink, *Langmuir*, 1998, **14**, 1535–1538.
62. T. B. Jones, *Mech. Res. Commun.*, 2009, **36**, 2–9.
63. A. B. D. Cassie, *Discuss. Faraday Soc.*, 1948, **3**, 11–16.
64. N. A. Patankar, *Langmuir*, 2003, **19**, 1249–1253.
65. L. C. Gao and T. J. McCarthy, *Langmuir*, 2006, **22**, 6234–6237.

66. J. Berthier, P. Dubois, P. Clementz, P. Claustre, C. Peponnet and Y. Fouillet, *Sens. Actuators A,* 2007, **134**, 471–479.
67. Y. S. Nanayakkara, H. Moon, T. Payagala, A. B. Wijeratne, J. A. Crank, P. S. Sharma and D. W. Armstrong, *Anal. Chem.,* 2008, **80**, 7690–7698.
68. P. Paik, V. K. Pamula and R. B. Fair, *Lab Chip,* 2003, **3**, 253–259.
69. A. Quinn, R. Sedev and J. Ralston, *J. Phys. Chem. B,* 2003, **107**, 1163–1169.
70. X. Zeng, R. Yue, J. Wu, H. Hu, L. Dong, Z. Wang, F. He and L. Liu, *Sci. China Ser. E,* 2006, **49**, 248–256.
71. E. Seyrat and R. A. Hayes, *J. Appl. Phys.,* 2001, **90**, 1383–1386.
72. M. J. Schertzer, S. I. Gubarenko, R. Ben-Mrad and P. E. Sullivan, *Langmuir,* 2010, **26**, 19230–19238.
73. R. N. Wenzel, *J. Phys. Colloid Chem.,* 1949, **53**, 1466–1467.
74. B. He, J. Lee and N. A. Patankar, *Colloid Surf. A,* 2004, **248**, 101–104.
75. N. A. Patankar, *Langmuir,* 2010, **26**, 8941–8945.
76. N. A. Patankar, *Langmuir,* 2004, **20**, 7097–7102.
77. Y. Kwon, S. Choi, N. Anantharaju, J. Lee, M. V. Panchagnula and N. A. Patankar, *Langmuir,* 2010, **26**, 17528–17531.
78. T. N. Krupenkin, J. A. Taylor, T. M. Schneider and S. Yang, *Langmuir,* 2004, **20**, 3824–3827.
79. T. N. Krupenkin, J. A. Taylor, E. N. Wang, P. Kolodner, M. Hodes and T. R. Salamon, *Langmuir,* 2007, **23**, 9128–9133.
80. K.-L. Wang and T. Jones, *Appl. Phys. Lett.,* 2005, **86**, 054104–054103.
81. V. Peykov, A. Quinn and J. Ralston, *Colloid Polym. Sci.,* 2000, **278**, 789–793.
82. L. K. Koopal, *Adv. Colloid Interface Sci.,* 2012, **179–182**, 29–42.
83. J. Berthier, P. Clementz, O. Raccurt, D. Jary, P. Claustre, C. Peponnet and Y. Fouillet, *Sens. Actuators A,* 2006, **127**, 283–294.
84. S. H. Ko, S. J. Lee and K. H. Kang, *Appl. Phys. Lett.,* 2009, **94**, 194102–194103.
85. H. Lee, S. Yun, S. H. Ko and K. H. Kang, *Biomicrofluidics,* 2009, **3**, 044113–044112.
86. F. Li and F. Mugele, *Appl. Phys. Lett.,* 2008, **92**, 244108–244103.
87. F. Mugele, J. C. Baret and D. Steinhauser, *Appl. Phys. Lett.,* 2006, **88**, 204106–204103.
88. A. Castellanos, A. Ramos, A. Gonzalez, N. G. Green and H. Morgan, *J. Phys. D: Appl. Phys.,* 2003, **36**, 2584–2597.
89. D. Chatterjee, H. Shepherd and R. L. Garrell, *Lab Chip,* 2009, **9**, 1219–1229.
90. M. G. Pollack, R. B. Fair and A. D. Shenderov, *Appl. Phys. Lett.,* 2000, **77**, 1725–1726.
91. S. K. Cho, H. J. Moon and C. J. Kim, *J. Microelectromech. Syst.,* 2003, **12**, 70–80.
92. J. Berthier, P. Clementz, O. Raccurt, D. Jary, P. Claustre, C. Peponnet and Y. Fouillet, *Sens. Actuators A,* 2006, **127**, 283–294.
93. M. G. Pollack, A. D. Shenderov and R. B. Fair, *Lab Chip,* 2002, **2**, 96–101.

94. H. Liu, S. Dharmatilleke, D. Maurya and A. O. Tay, *Microsyst. Technol.,* 2010, **16**, 449–460.
95. M. Abdelgawad, S. L. S. Freire, H. Yang and A. R. Wheeler, *Lab Chip,* 2008, **8**, 672–677.
96. M. Abdelgawad, M. W. L. Watson and A. R. Wheeler, *Lab Chip,* 2009, **9**, 1046–1051.
97. M. W. L. Watson, M. J. Jebrail and A. R. Wheeler, *Anal. Chem.,* 2010, **82**, 6680–6686.
98. H. Yang, V. N. Luk, M. Abelgawad, I. Barbulovic-Nad and A. R. Wheeler, *Anal. Chem.,* 2008, **81**, 1061–1067.
99. S.-K. Fan, H. Yang and W. Hsu, *Lab Chip,* 2011, **11**, 343–347.
100. S.-Y. Park, M. A. Teitell and E. P. Y. Chiou, *Lab Chip,* 2010, **10**, 1655–1661.
101. S. C. C. Shih, R. Fobel, P. Kumar and A. R. Wheeler, *Lab Chip,* 2011, **11**, 535–540.
102. K.-L. Wang, T. B. Jones and A. Raisanen, *Lab Chip,* 2009, **9**, 901–909.
103. H. Ren, R. B. Fair and M. G. Pollack, *Sens. Actuators B,* 2004, **98**, 319–327.
104. J. Gong and C. J. Kim, *Lab Chip,* 2008, **8**, 898–906.
105. S. Sadeghi, H. Ding, G. J. Shah, S. Chen, P. Y. Keng, C.-J. C. Kim and R. M. van Dam, *Anal. Chem.,* 2012, **84**, 1915–1923.
106. B. Bhattacharjee and H. Najjaran, *Lab Chip,* 2012, **12**, 4416–4423.
107. M. A. Murran and H. Najjaran, *Lab Chip,* 2012, **12**, 2053–2059.
108. S. C. C. Shih, I. Barbulovic-Nad, X. Yang, R. Fobel and A. R. Wheeler, *Biosens. Bioelectron.,* 2013, **42**, 314–320.
109. T. Lederer, S. Clara, B. Jakoby and W. Hilber, *Microsyst. Technol.,* 2012, **18**, 1163–1180.
110. L. Luan, R. D. Evans, N. M. Jokerst and R. B. Fair, *IEEE Sens. J.,* 2008, **8**, 628–635.
111. L. Luan, M. W. Royal, R. Evans, R. B. Fair and N. M. Jokerst, *IEEE Sens. J.,* 2012, **12**.
112. Z. Hua, J. L. Rouse, A. E. Eckhardt, V. Srinivasan, V. K. Pamula, W. A. Schell, J. L. Benton, T. G. Mitchell and M. G. Pollack, *Anal. Chem.,* 2010, **82**, 2310–2316.
113. C. L. Arce, D. Witters, R. Puers, J. Lammertyn and P. Bienstman, *Anal. Bioanal. Chem.,* 2012, **404**, 2887–2894.
114. S. C. C. Shih, H. Yang, M. J. Jebrail, R. Fobel, N. McIntosh, O. Y. Al-Dirbashi, P. Chakraborty and A. R. Wheeler, *Anal. Chem.,* 2012, **84**, 3731–3738.
115. L. Malic, T. Veres and M. Tabrizian, *Biosens. Bioelectron.,* 2011, **26**, 2053–2059.
116. K. Choi, J. Y. Kim, J. H. Ahn, J. M. Choi, M. Im and Y. K. Choi, *Lab Chip,* 2012, **12**, 1533–1539.
117. Y. Zhao and K. Chakrabarty, *IEEE Trans. Comput-Aided Des. Integr. Circuits Syst.,* 2012, **31**, 817–830.
118. T. W. Huang, C. H. Lin and T. Y. Ho, *IEEE Trans. Comput-Aided Des. Integr. Circuits Syst.,* 2010, **29**, 1682–1695.

119. F. Su, K. Chakrabarty and R. B. Fair, *IEEE Trans. Comput-Aided Des. Integr. Circuits Syst.,* 2006, **25**, 211–223.
120. J. Nichols, C. M. Collier, E. L. Landry, M. Wiltshire, B. Born and J. F. Holzman, *J. Biomed. Opt.,* 2012, **17**.
121. J. W. Chang, S. H. Yeh, T. W. Huang and T. Y. Ho, *IEEE Trans. Comput-Aided Des. Integr. Circuits Syst.,* 2013, **32**, 216–227.
122. Y. Luo, K. Chakrabarty and T. Y. Ho, *IEEE Trans. Comput-Aided Des. Integr. Circuits Syst.,* 2013, **32**, 59–72.
123. Y. Zhao, K. Chakrabarty and B. Bhattacharya, *J. Electron. Test,* 2012, **28**, 243–255.
124. K. F. Bohringer, *IEEE Trans. Comput-Aided Des. Integr. Circuits Syst.,* 2006, **25**, 329–339.
125. Y. Fouillet, D. Jary, C. Chabrol, P. Claustre and C. Peponnet, *Microfluid. Nanofluid.,* 2008, **4**, 159–165.
126. E. Maftei, P. Pop and J. Madsen, *Des. Autom. Embed. Syst.,* 2012, **16**, 19–44.
127. Z. K. Chen, D. H. Y. Teng, G. C. J. Wang and S. K. Fan, *BioChip J.,* 2011, **5**, 343–352.
128. J. Gao, X. M. Liu, T. L. Chen, P. I. Mak, Y. G. Du, M. I. Vai, B. C. Lin and R. P. Martins, *Lab Chip,* 2013, **13**, 443–451.
129. Y. Y. Lin, E. R. F. Welch and R. B. Fair, *Sens. Actuators B,* 2012, **173**, 338–345.
130. Y.-Y. Lin, R. D. Evans, E. Welch, B.-N. Hsu, A. C. Madison and R. B. Fair, *Sens. Actuators B,* 2010, **150**, 465–470.
131. K. S. Elvira, R. Leatherbarrow, J. Edel and A. deMello, *Biomicrofluidics,* 2012, **6**.
132. K. L. Wang, T. B. Jones and A. Raisanen, *J. Micromech. Microeng.,* 2007, **17**, 76–80.
133. R. Ahmed and T. B. Jones, *J. Electrost.,* 2006, **64**, 543–549.
134. R. Renaudot, B. Daunay, M. Kumemura, V. Agache, L. Jalabert, D. Collard and H. Fujita, *Sens. Actuators B,* 2013, **177**, 620–626.
135. W. Wang, T. B. Jones and D. R. Harding, *Fusion Sci. Technol.,* 2011, **59**, 240–249.
136. M. J. Jebrail, M. S. Bartsch and K. D. Patel, *Lab Chip,* 2012, **12**, 2452–2463.
137. I. Barbulovic-Nad, H. Yang, P. S. Park and A. R. Wheeler, *Lab Chip,* 2008, **8**, 519–526.
138. S. H. Au, P. Kumar and A. R. Wheeler, *Langmuir,* 2011, **27**, 8586–8594.
139. I. Barbulovic-Nad, S. H. Au and A. R. Wheeler, *Lab Chip,* 2010, **10**, 1536–1542.
140. D. Witters, N. Vergauwe, S. Vermeir, F. Ceyssens, S. Liekens, R. Puers and J. Lammertyn, *Lab Chip,* 2011, **11**, 2790–2794.
141. I. A. Eydelnant, U. Uddayasankar, B. Y. Li, M. W. Liao and A. R. Wheeler, *Lab Chip,* 2012, **12**, 750–757.
142. S. Srigunapalan, I. A. Eydelnant, C. A. Simmons and A. R. Wheeler, *Lab Chip,* 2012, **12**, 369–375.

143. S. H. Au, S. C. C. Shih and A. R. Wheeler, *Biomed. Microdevices*, 2011, **13**, 41–50.

144. D. Bogojevic, M. D. Chamberlain, I. Barbulovic-Nad and A. R. Wheeler, *Lab Chip*, 2012, **12**, 627–634.

145. R. B. Fair, V. Srinivasan, H. Ren, P. Paik, V. K. Pamula, M. G. Pollack and *IEEE, Electrowetting-based On-Chip Sample Processing for Integrated Microfluidics*, IEEE, New York, 2003.

146. J. Y. Yoon and R. L. Garrell, *Anal. Chem.*, 2003, **75**, 5097–5102.

147. V. N. Luk, G. C. H. Mo and A. R. Wheeler, *Langmuir*, 2008, **24**, 6382–6389.

148. G. Perry, V. Thomy, M. R. Das, Y. Coffinier and R. Boukherroub, *Lab Chip*, 2012, **12**, 1601–1604.

149. V. Taly, D. Pekin, A. El Abed and P. Laurent-Puig, *Trends Mol. Med.*, 2012, **18**, 405–416.

150. N. R. Beer, B. J. Hindson, E. K. Wheeler, S. B. Hall, K. A. Rose, I. M. Kennedy and B. W. Colston, *Anal. Chem.*, 2007, **79**, 8471–8475.

151. Y. Schaerli and F. Hollfelder, *Mol. Biosyst.*, 2009, **5**, 1392–1404.

152. Y.-H. Chang, G.-B. Lee, F.-C. Huang, Y.-Y. Chen and J.-L. Lin, *Biomed. Microdevices*, 2006, **8**, 215–225.

153. W. A. Schell, J. L. Benton, P. B. Smith, M. Poore, J. L. Rouse, D. J. Boles, M. D. Johnson, B. D. Alexander, V. K. Pamula, A. E. Eckhardt, M. G. Pollack, D. K. Benjamin, J. R. Perfect and T. G. Mitchell, *Eur. J. Clin. Microbiol. Infect. Dis.*, 2012, **31**, 2237–2245.

154. J. Berthier, V. Mourier, N. Sarrut, D. Jary, Y. Fouillet, P. Pouteau, P. Caillat and C. Peponnet, *Int. J. Nanotechnol.*, 2010, **7**, 802–818.

155. D. Brassard, L. Malic, C. Miville-Godin, F. Normandin, T. Veres and *IEEE, in 2011 IEEE 24th International Conference on Micro Electro Mechanical Systems*, IEEE, New York, 2011, pp. 153–156.

156. A. H. C. Ng, K. Choi, R. P. Luoma, J. M. Robinson and A. R. Wheeler, *Anal. Chem.*, 2012, **84**, 8805–8812.

157. R. S. Sista, A. E. Eckhardt, V. Srinivasan, M. G. Pollack, S. Palanki and V. K. Pamula, *Lab Chip*, 2008, **8**, 2188–2196.

158. E. M. Miller, A. H. C. Ng, U. Uddayasankar and A. R. Wheeler, *Anal. Bioanal. Chem.*, 2011, **399**, 337–345.

159. L. Malic, T. Veres and M. Tabrizian, *Biosens. Bioelectron.*, 2009, **24**, 2218–2224.

160. A. R. Wheeler, H. Moon, C. A. Bird, R. R. O. Loo, C. J. Kim, J. A. Loo and R. L. Garrell, *Anal. Chem.*, 2005, **77**, 534–540.

161. H. Kim, M. S. Bartsch, R. F. Renzi, J. He, J. L. Van de Vreugde, M. R. Claudnic and K. D. Patel, *JALA*, 2011, **16**, 405–414.

162. D. Chatterjee, A. J. Ytterberg, S. U. Son, J. A. Loo and R. L. Garrell, *Anal. Chem.*, 2010, **82**, 2095–2101.

163. V. N. Luk and A. R. Wheeler, *Anal. Chem.*, 2009, **81**, 4524–4530.

164. L. K. Fiddes, V. N. Luk, S. H. Au, A. H. C. Ng, V. Luk, E. Kumacheva and A. R. Wheeler, *Biomicrofluidics*, 2012, **6**.

165. A. P. Aijian, D. Chatterjee and R. L. Garrell, *Lab Chip*, 2012, **12**, 2552–2559.

166. D. Chugh and K. Kaler, *Microfluid. Nanofluid.*, 2010, **8**, 445–456.
167. R. Prakash and K. Kaler, *Sens. Actuators B*, 2012, **169**, 274–283.
168. M. Kumemura, D. Collard, S. Yoshizawa, B. Wee, S. Takeuchi and H. Fujita, *ChemPhysChem*, 2012, **13**, 3308–3312.
169. T. Taniguchi, T. Torii and T. Higuchi, *Lab Chip*, 2002, **2**, 19–23.
170. J. A. Schwartz, J. V. Vykoukal and P. R. C. Gascoyne, *Lab Chip*, 2004, **4**, 11–17.
171. V. Srinivasan, V. Pamula, M. Pollack, R. Fair and *IEEE, in Mems-03: IEEE the Sixteenth Annual International Conference on Micro Electro Mechanical Systems*, IEEE, New York, 2003, pp. 327–330.
172. M. W. Royal, N. M. Jokerst and R. B. Fair, *IEEE Photon. J.*, 2012, **4**, 2126–2135.
173. D. Chatterjee, B. Hetayothin, A. R. Wheeler, D. J. King and R. L. Garrell, *Lab Chip*, 2006, **6**, 199–206.
174. P. Dubois, G. Marchand, Y. Fouillet, J. Berthier, T. Douki, F. Hassine, S. Gmouh and M. Vaultier, *Anal. Chem.*, 2006, **78**, 4909–4917.
175. N. Isambert, M. D. S. Duque, J. C. Plaquevent, Y. Genisson, J. Rodriguez and T. Constantieux, *Chem. Soc. Rev.*, 2011, **40**, 1347–1357.
176. X. D. Hu, S. G. Zhang, X. J. Lu, C. Qu, L. J. Lu, X. Y. Ma, X. P. Zhang and Y. Q. Deng, *Surf. Interface Anal.*, 2012, **44**, 478–483.
177. M. J. Jebrail, A. H. C. Ng, V. Rai, R. Hili, A. K. Yudin and A. R. Wheeler, *Angew. Chem. Int. Ed.*, 2010, **49**, 8625–8629.
178. H. J. Ding, S. Sadeghi, G. J. Shah, S. P. Chen, P. Y. Keng, C. J. Kim and R. M. van Dam, *Lab Chip*, 2012, **12**, 3331–3340.
179. P. A. L. Wijethunga, Y. S. Nanayakkara, P. Kunchala, D. W. Armstrong and H. Moon, *Anal. Chem.*, 2011, **83**, 1658–1664.
180. D. J. Boles, J. L. Benton, G. J. Siew, M. H. Levy, P. K. Thwar, M. A. Sandahl, J. L. Rouse, L. C. Perkins, A. P. Sudarsan, R. Jalili, V. K. Pamula, V. Srinivasarn, R. B. Fair, P. B. Griffin, A. E. Eckhardt and M. G. Pollack, *Anal. Chem.*, 2011, **83**, 8439–8447.
181. E. R. F. Welch, Y. Y. Lin, A. Madison and R. B. Fair, *Biotechnol. J.*, 2011, **6**, 165–176.
182. V. Srinivasan, V. K. Pamula and R. B. Fair, *Anal. Chim. Acta*, 2004, **507**, 145–150.
183. V. Srinivasan, V. K. Pamula and R. B. Fair, *Lab Chip*, 2004, **4**, 310–315.
184. R. Sista, Z. S. Hua, P. Thwar, A. Sudarsan, V. Srinivasan, A. Eckhardt, M. Pollack and V. Pamula, *Lab Chip*, 2008, **8**, 2091–2104.
185. N. A. Mousa, M. J. Jebrail, H. Yang, M. Abdelgawad, P. Metalnikov, J. Chen, A. R. Wheeler and R. F. Casper, *Sci. Transl. Med.*, 2009, **1**.
186. R. Sista, A. E. Eckhardt, T. Wang, M. Séllos-Moura and V. K. Pamula, *Clin. Chim. Acta*, 2011, **412**, 1895–1897.
187. R. S. Sista, A. E. Eckhardt, T. Wang, C. Graham, J. L. Rouse, S. M. Norton, V. Srinivasan, M. G. Pollack, A. A. Tolun, D. Bali, D. S. Millington and V. K. Pamula, *Clin. Chem.*, 2011, **57**, 1444–1451.
188. A. A. Tolun, C. Graham, Q. Shi, R. S. Sista, T. Wang, A. E. Eckhardt, V. K. Pamula, D. S. Millington and D. S. Bali, *Mol. Genet. Metab.*, 2012, **105**, 519–521.

189. M. J. Jebrail, H. Yang, J. M. Mudrik, N. M. Lafreniere, C. McRoberts, O. Y. Al-Dirbashi, L. Fisher, P. Chakraborty and A. R. Wheeler, *Lab Chip,* 2011, **11**, 3218–3224.
190. C. Delattre, C. P. Allier, Y. Fouillet, D. Jary, F. Bottausci, D. Bouvier, G. Delapierre, M. Quinaud, A. Rival, L. Davoust and C. Peponnet, *Biosens. Bioelectron.,* 2012, **36**, 230–235.
191. M. Jonsson-Niedziolka, F. Lapierre, Y. Coffinier, S. J. Parry, F. Zoueshtiagh, T. Foat, V. Thomy and R. Boukherroub, *Lab Chip,* 2011, **11**, 490–496.
192. Y. Zhao and S. K. Cho, *Lab Chip,* 2006, **6**, 137–144.
193. K.-L. Wang and T. Jones, *13th International Conference on Electrostatics, Journal of Physics: Conference Series,* 2011, **301**, 012057.

CHAPTER 5

Manipulation of Micro-/Nano-Objects via Surface Acoustic Waves

PENG LI, FENG GUO, KEVIN LIN, AND TONY JUN HUANG*

Department of Engineering Science and Mechanics, The Pennsylvania State University, University Park, PA 16802, USA
*E-mail: junhuang@psu.edu

5.1   Introduction

The ability to manipulate micro/nano-objects (such as cells, microparticles, droplets, and nanowires) in defined patterns and paths is critical for a wide variety of applications in detection science. In the past decades, researchers in the biophysics and lab-on-a-chip communities have developed several particle-manipulation methods, such as optical tweezers,[1,2] dielectrophoretic (DEP) tools,[3] magnetic tweezers,[4] and optoelectronic tweezers,[5] which each have their own advantages. More recently, surface acoustic wave (SAW)-based manipulation has drawn a lot of attention and has become an active research area.[6-11] A SAW is one type of acoustic wave that propagates along the surface of an elastic material; the major energy of a SAW is confined within one wavelength from the surface. SAW devices consist of interdigitated metallic electrodes on piezoelectric substrates that only require standard photolithography processes to fabricate. Furthermore, SAW properties can be tuned by adjusting input electric signals, which is much more convenient than most alternative techniques. Most importantly, ultrasound with appropriate

RSC Detection Science Series No. 5
Microfluidics in Detection Science: Lab-on-a-chip Technologies
Edited by Fatima H Labeed and Henry O Fatoyinbo
© The Royal Society of Chemistry 2015
Published by the Royal Society of Chemistry, www.rsc.org

intensities is safe for biological systems and has been demonstrated most commonly in the application of monitoring a human fetus during pregnancy.

In this chapter, we will discuss micro- and nano-object manipulation enabled by SAW. More specifically, both travelling surface acoustic waves (TSAW) and standing surface acoustic wave (SSAW)-based manipulation will be discussed. The mechanism of TSAW-based manipulation is mainly through the actuation of fluid when SAWs leak into the fluid, whereas SSAWs mainly utilise acoustic radiation pressure to achieve manipulation. It should be noted that the applications of SAW are not limited to micro- or nano-object manipulation. For more thorough reviews, readers are recommended to ref. 6–8 for further information.

5.2 Theoretical Considerations

5.2.1 Generation of SAW

Normally, SAWs are generated by applying an appropriate AC electric field to a piezoelectric material that is able to generate a propagating mechanical stress that reacts to the electrical signal. The electrical signal is applied through a set of metallic interdigitated transducers (IDTs) deposited on the surface of a piezoelectric substrate. IDTs consist of a series of metallic fingers interspaced with another series of metallic fingers (Figure 5.1). The period of the IDT determines the wavelengths of the SAW generated. In addition to the period, the number, shape, and aperture of IDTs can be adjusted as well to generate SAWs with unique characteristics. The applications and features of special types of IDTs will be discussed in detail in the following sections. The widely used piezoelectric material is LiNbO$_3$ due to its good piezoelectric and optical properties.

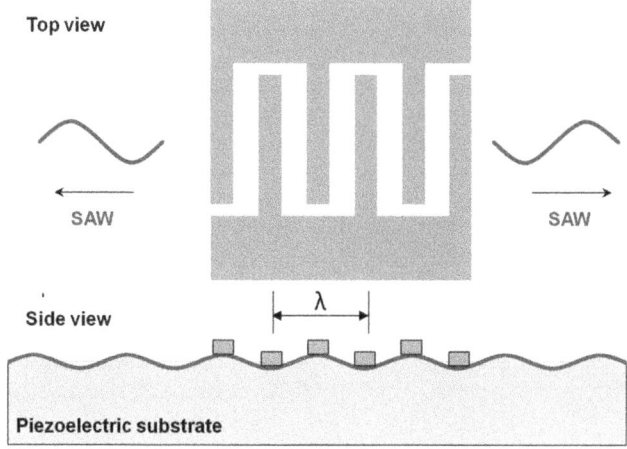

Figure 5.1 Interdigitated transducers and the generation of SAW. Reprinted with permission from ref. 6.

It should be noted that the SAW generated under the above condition is Rayleigh waves. Usually, the Rayleigh wave is composed of two components, a longitudinal and a vertically polarised shear component. When encountering the liquid, some of the energy will couple into the media. There are other types of waves that exist, *e.g.*, Love waves that only have a horizontal shear component. However, since most applications in microfluidics occur in liquids, the Rayleigh wave is the most widely used mode of SAW. In this chapter, the use of "SAW" only indicates Rayleigh SAW.

5.2.2   Formation of Standing SAW (SSAW)

The formation of SSAWs is based on the interference of surface acoustic waves. When a pair of identical IDTs generate SAWs, which travel towards each other, on a piezoelectric substrate (Figure 5.2(a)), the interference of the two waves will result in a SSAW field with parallel lines. Figure 5.2(b), shows a simulated interference pattern of a one-dimensional SSAW on a piezoelectric substrate surface. The light and dark regions indicate the amplitudes of the sound pressure field. The weakest (lightest region) and strongest regions

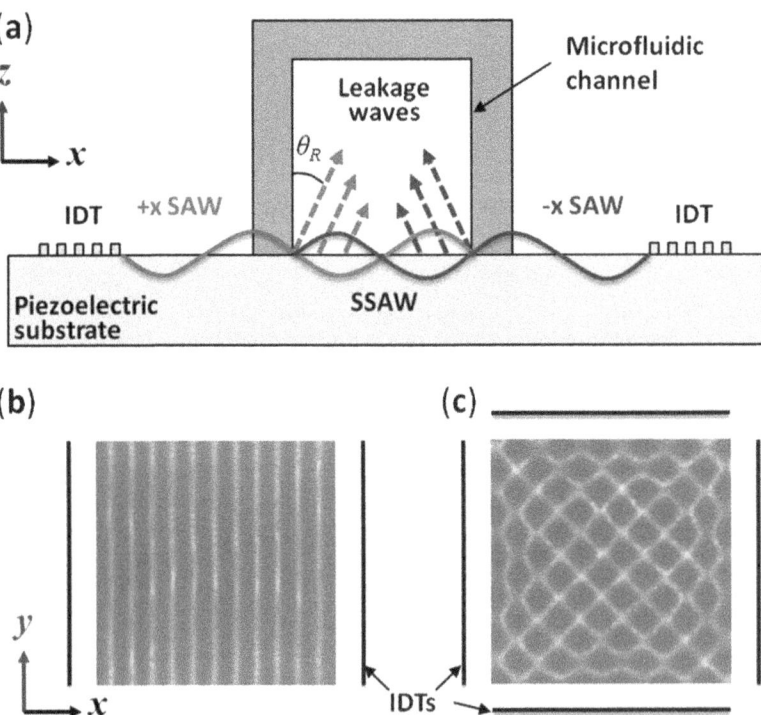

Figure 5.2   (a) Schematic of SSAW; (b) and (c) simulated interference pattern of a one-dimensional and two-dimensional SSAW field, respectively. Reprinted with permission from ref. 8.

(darkest region) of sound pressure are called pressure nodes and antinodes, respectively. The position of nodes and antinodes will not change over time as long as the input signal remains constant. The distance between two neighbouring nodes or antinodes is half the wavelength of the SAW.

Similarly, when two pairs of IDTs assemble orthogonally, a 2D interference pattern will be observed (Figure 5.2(c)). The direction of the resultant 2D interference pattern has a 45-degree angle with respect to the travelling direction of each SAW.

5.2.3   Primary Acoustic Radiation Force

When particles are present in a fluid medium with a SSAW field, the particles will experience a force pushing them to pressure nodes or antinodes depending on the properties of said particles and fluid. The force is called the primary acoustic radiation force. The primary acoustic radiation force is the foundation of all kinds of SSAW-based particle manipulation. The primary acoustic radiation force F_a is described as:

$$F_a = -\left(\frac{\pi p_0^2 V_p \beta_f}{2\lambda}\right)\varphi(\beta,\rho)\sin(2kx), \tag{5.1}$$

where p_0, V_p, λ, and k are the acoustic pressure, particle volume, wavelength and wave vector of the acoustic waves, respectively; φ is the acoustic contrast factor. When φ is positive, particles are pushed towards pressure nodes; otherwise, they are pushed towards pressure antinodes. It can be described as:

$$\varphi(\beta,\rho) = \frac{5\rho_p - 2\rho_f}{2\rho_p + \rho_f} - \frac{\beta_p}{\beta_f}, \tag{5.2}$$

where ρ_p, ρ_f, β_p, and β_f are the density of particles and fluid medium, and compressibility of particles and fluid medium, respectively. Most hard particles and biological cells have a positive acoustic contrast factor, so they will move towards the pressure nodes in a SSAW field.

5.3   Applications of SAW-Based Manipulations

In the following sections, we will discuss the examples of SAW-enabled manipulations of micro- and nano-objects. The applications range from continuous-flow-based high-throughput particle manipulation to static-flow-based precise manipulation.

5.3.1   SAW-Based Particle/Cell Concentration

Concentrating particles or cells through centrifugation is a widely used procedure in many applications in biology, chemistry, and medicine. However, concentrating particles or cells in microfluidic platforms is not a trivial task, as surface forces dominate over body forces at the microscale. It is thus difficult to apply centrifugation to microscale applications due to the lack of significant body forces. At the microscale, SAW-induced acoustic streaming has been studied to achieve particle and cell concentration for microscale applications.[8–12]

For example, Shilton *et al.* generated a TSAW to induce rotational fluid motion within a droplet.[13] The TSAW only covered a fraction of the droplet, thereby only forming circular pattern streaming. When microparticles were present in the droplet, they were concentrated to the centre due to a shear gradient between the edge and the centre of the droplet. Particles tended to move towards the region of low shear, where the linear velocities of the fluid approached zero. They demonstrated that the movement of particles from the edge to the centre of the droplet required less than 1 s. In this way, TSAWs offer a simple yet effective way of concentrating microparticles.

In addition to acoustic-streaming-induced particle concentration, SSAW has also been used as a particle/cell concentration method at the microscale. Chen *et al.* reported the enrichment of particle/cells in disposable micro-tubings by establishing a SSAW field along the flow direction (Figure 5.3).[14] Multiple pressure nodes can be formed along the length of the tubings, where the particle/cells are trapped and enriched. A 100-fold concentration increase was achieved with 99% recovery efficiency. They also reported the

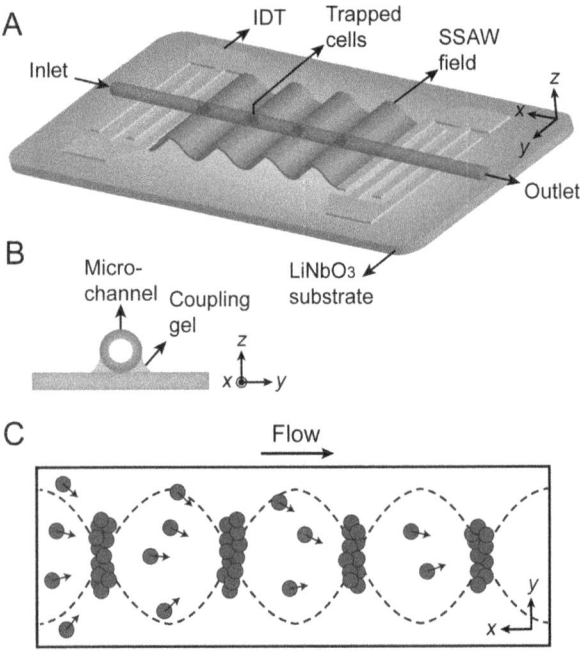

Figure 5.3   A SSAW-based cell concentration device. (a) A schematic of SSAW based cell concentration device (b) The configuration of microchannels and coupling gel (c) The mechanism of SSAW based cell concentration. Reprinted with permission from ref. 14.

enrichment process did not impact cell normal physiology in terms of viability and proliferation. Thus, the SSAW-based particle/cell enrichment method could be suitable for concentrating cells in microfluidic continuous-flow applications.

5.3.2   SAW-Based Cell/Particle Focusing

High-throughput cell analysis and treatment often requires prealignment of cells in a flow stream. An apparent example is flow cytometry that is able to achieve multiparametric, high-throughput single-cell analysis and sorting. The conventional method of cell-focusing is achieved through hydrodynamic streams. Both sample stream and sheath fluid are introduced to the flow cell. Cells can then be focused since the sample stream is compressed by the sheath flow. The sheath-flow-based method has been demonstrated to be effective and has been adopted by many applications requiring cell focusing. However, several limitations arise with this method. First, the sheath flow adds extra shear stress to the sample, increasing the risk of cell damage. Excessive shear stress is known to damage cells, particularly fragile cells.[15] Secondly, the sheath flow increases the bulkiness and complexity of the whole system. Thirdly, operational cost can be significant as sheath flow rates are much faster than sample flow rate causing a large amount of sheath fluid to be consumed on a daily basis. Therefore, focusing cells without the use of sheath flow will have unique advantages to current applications.

As discussed in the theoretical section, particles in the standing acoustic field will experience acoustic radiation forces and be pushed to the position of pressure node or antinode. Cells are normally pushed to the pressure node. Based on this mechanism, SSAW-based cell/particle focusing is possible. Shi *et al.* first reported the use of SSAW to focus polystyrene particles in a microfluidic channel.[16] The PDMS-based microfluidic channel is carefully aligned to make sure that the pressure node arises at the middle, meaning particles are focused towards the middle of the channel. Meanwhile, the microfluidic channel is designed to only cover one pressure node across the channel width. The width is usually the half-wavelength of the SAW. When particles pass through the SSAW-applied region, they will experience the acoustic radiation force and be pushed to the centre of the channel (Figure 5.4). Input amplitude, IDT lengths, and SAW frequency, are adjustable parameters for optimising the performance of SSAW focusing. Zeng *et al.* also reported that particle focusing effects were enhanced by adding Bragg reflectors inside or outside of the IDTs.[17]

More interestingly, SSAW-enabled 3D particle focusing has also been observed. 3D focusing means that particles are not only focused in the horizontal direction but also in the vertical direction, which will further reduce variations for cell manipulation and analysis. Shi *et al.* reported that when applying SSAW to a PDMS-based microfluidic channel, particles all aligned at a similar height in the channel.[18] The authors attributed this phenomenon to the effect of a primary acoustic radiation generated transverse to the particles

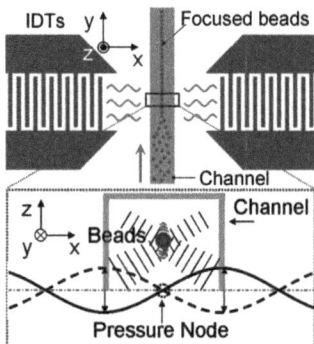

Figure 5.4   SSAW-based particle focusing. Reprinted with permission from ref. 16.

in the channel by the nonuniform acoustic field. Particles are all pushed by the force to the position of maximum acoustic kinetic energy. The authors also observed that the migration of particles in the x–y plane is faster than in the y–z plane. Although preliminary theoretical considerations have been discussed, further examinations of this phenomenon are necessary to elucidate the mechanism and improve the performance of SSAW-based focusing in the z-direction. Recently, a SSAW-focusing-based microfluidic flow cytometer has been developed by Huang and coworkers.[19]

5.3.3   SAW-Based Cell/Particle Separation

The equation of acoustic radiation force indicates that the force depends on several parameters related to a particle's own physical properties including size, density, and compressibility. As a result, particles with different physical properties experience different forces when they are in the region of SSAW. The differential forces imparted on particles translate to differential migration times to the pressure node position. Thus, by adjusting the migration time of particles in SSAW region and the position of collection channels, separation of particles with distinct physical properties can be realised.

Shi *et al.* reported a SSAW-based, particle-separation microfluidic device in continuous flow.[20] The device is capable of separating particles with different sizes. The channel width covers half a wavelength of the applied SAW, and the pressure node is at the centre of the channel (Figure 5.5). The inlet channels consist of sample inlets and sheath flow inlets. In Shi's work, there are two sample inlets to introduce samples from the side. One sheath-flow inlet is located at the centre to focus particles in the side stream where pressure antinodes are present. Thus, when particles passed through the SSAW field, they would be pushed to the centre channel. Based on the equation of primary acoustic radiation force, particles with larger size experience larger acoustic radiation force and move to the centre channel quicker than smaller particles. Larger particles were then collected from the centre outlet, whereas smaller ones were collected from side outlets. The

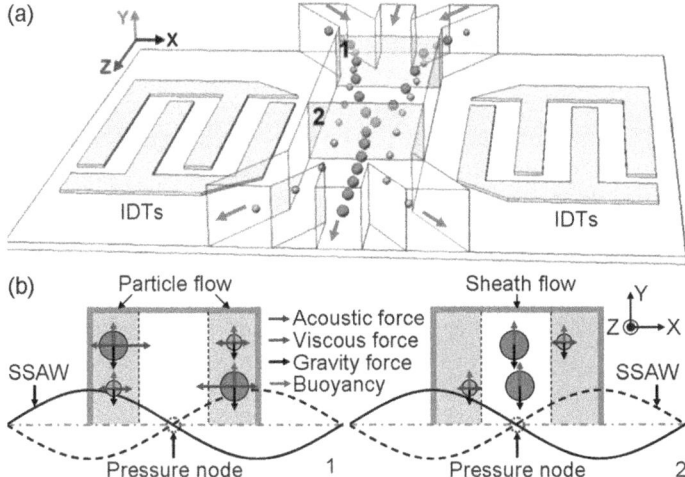

Figure 5.5 Schematic of SSAW-based particle separation. (a) A schematic of SSAW size-based particle separation device. (b) The mechanism of SSAW size-based separation. Reprinted with permission from ref. 20.

microfluidic device demonstrated that SSAW can be an efficient mechanism to separate particles of different sizes.

Based on the same principle, Nam *et al.* also reported a microfluidic device for sized-based separation.[21] In contrast to Shi's work, they changed the position of the pressure node to the channel wall, and situated the pressure antinode to the centre of the channel. Accordingly, samples are introduced from the centre inlet and focused by the side sheath flow. Larger particles were then pushed to the side outlets and collected. This setup is an alternative design for SSAW-based separation. Generally, placing the pressure node to the channel centre makes better use of space and enhances separation throughput as there are two streams containing samples. On the other hand, the design with a pressure antinode in the centre can obtain better focusing performance and avoid the interaction between channel walls and particles. Since the SSAW-based separation highly depends on the starting position and migration time, better focusing will result in better separation performance in terms of purity and efficiency. Both designs have their own pros and cons, so it is up to the specific applications and requirements to determine which design to use.

Nam *et al.* further applied the SSAW separation device to separate platelets and blood cells.[22] 99% of blood cells are red blood cells with a size range 6–8 μm in diameter. Platelets are 2–3 μm cell fragments, so there is a significant difference in size between blood cells and platelets. The reported device showed a platelet ratio of 74% in the separated sample and 99% of blood cells had been removed. Despite the low sample throughput (0.25 μl min^{-1}), this device demonstrated the feasibility of applying SSAW technique to separate biologically relevant samples. More recently, Ai *et al.* reported the use of SSAW device to separate bacteria (*E. coli*) from blood cells.[23]

The physical properties that SSAW is able to differentiate are not only limited to size, as particles with different densities can also be separated using SSAW. One such example is cell-encompassing polymer beads. Encapsulating cells in biocompatible polymers will enable many biological applications. To obtain the best performance out of cell-encompassing polymer beads, two parameters need to be well controlled: the size of polymer beads and the cell numbers in each bead. Currently, researchers have managed to generate monodispersive polymer beads efficiently. However, it is difficult to control the cell number in each bead. As a result, it is impossible to obtain highly monodispersive cell numbers in the generated polymer beads. One way to circumvent this issue is to separate these beads based on the cell number inside each bead. Although the size of polymer beads is similar, beads with different number of cells will have different densities. Thus, beads with different cell numbers can be separated using SSAW. Nam *et al.* configured the pressure node at the centre of a microfluidic channel that has five sample collection outlets: one middle outlet, two side outlets, and two in between.[24] Under this setup, beads can be categorised into three groups after they pass through the SSAW region: large cell quantity alginate beads (LQABs), small cell quantity alginate beads (SQABs), and empty beads. The more cells present in the beads, the higher acoustic radiation force they will experience. Therefore, LQABs, SQABs, and empty beads can be collected from two side outlets, two intermediate outlets, and middle outlets, respectively. 97% separation efficiency and 98% separation specificity for LQABs have been achieved.

As discussed previously, SSAW-based separation is a migration (acousto-phoresis)-based method. The separation can only be successful when different particles have different travelling distances under SSAW fields. Generally, the particles experiencing larger acoustic radiation forces will have higher travelling velocities, resulting in larger travelling distances. But the travelling distance also depends on the starting position. If the separation distance is smaller than the difference between starting positions, separation will fail. To minimise the differences in starting position, all SSAW separation devices require a sample-focusing step. Although most of the examples used sheath-flow-based hydrodynamic focusing, SSAW-based sheathless focusing can also be integrated into the separation devices. Jo and Guldiken reported a SSAW separation device with two pairs of IDTs.[25] The first pair of IDTs serves as the focusing function to align sample particles without the need of sheath fluid. The second pair of IDTs located downstream is to separate particles with different densities. 89.4% separation efficiency at a 2 µl min^{-1} flow rate has been achieved.

5.3.4 SAW-Based Particle Sorting

Particle actuation under continuous flow is a vital component of fluorescence-activated cell sorters (FACS). A range of mechanisms have been utilised to actuate cells including electrical, optical, and hydrodynamic methods.[26-33] Compared with these counterparts, acoustic-based actuation is gentler and thus has better biocompatibility with biological cells. Both TSAW and SSAW have been demonstrated to be capable of cell sorting in microfluidic channels. TSAW-based cell sorting is realised through acoustic-induced streaming, whereas SSAW is based on acoustic radiation pressure.

Franke *et al.* reported the use of TSAW to sort droplets in microfluidics under continuous flow.[34] They configured the TSAW to propagate perpendicular to the fluid flow. When the SAW is on, it generates acoustic streaming to push the fluid-containing droplet and redirects it from its original path. In this way, droplets can be sorted to a different outlet when the SAW is on. Through binary control of the SAW (on or off), droplets of interest can be selectively sorted out. They also demonstrated that the same principle can be applied to sort mammalian cells.[35]

In addition to acoustic-streaming-based sorting, SSAW is also capable of cell sorting by manipulating the acoustic radiation pressure. There are several possible ways to achieve particle actuation using SSAW. First, the most straightforward way is through controlling the presence of SSAW field. The pressure node can be positioned near the particle flow path. Once the SSAW is activated, particles will be directed to a different flow path due to the acoustic radiation force. When the SSAW is not present, particles will remain in their original flow path. Therefore, a simple two-way SSAW-based cell sorter can be achieved. Particles can also be actuated by changing the position of a pressure node. Since particles are always pushed to the position of pressure nodes, the movement of pressure nodes can be translated to the movement of particles. The advantage of this method over the first strategy is that changing position of pressure node allows multiposition actuation, adding in more flexibility to the cell-sorting process. Two approaches have been reported to change the position of pressure nodes. Several groups reported that the position of pressure nodes can be adjusted in real time by tuning the relative electronic phase of one transducer to the other.[36–38] The other way to change the position of pressure nodes is to change the frequency of activated transducers. Ding *et al.* employed chirped IDTs that consist of a series of transducers with a distance gradient to achieve this concept.[39] At each distinct frequency, a certain pair of IDTs can be resonated. Thus, when changing the applied frequency, the resonated IDTs are changed as well, leading to the change in pressure node position. Based on this design, the authors demonstrated the sorting of single cells into as many as five individual outlet channels (Figure 5.6). The channel width allows only one pressure node to exist between the channel walls. HL-60 cells were used to demonstrate the device performance. At a frequency of 9.8 MHz, the primary acoustic radiation force pushed the cell to the side outlet. On changing the frequency to 10.9 MHz, cells are pushed to the other side outlet as a result of the change of pressure node position. Cells can also be directed to the middle three outlets by switching respective frequencies. Li *et al.* further demonstrated that this principle is able to sort water-in-oil droplets into five individual outlets with a throughput of 222 droplets per second.[40]

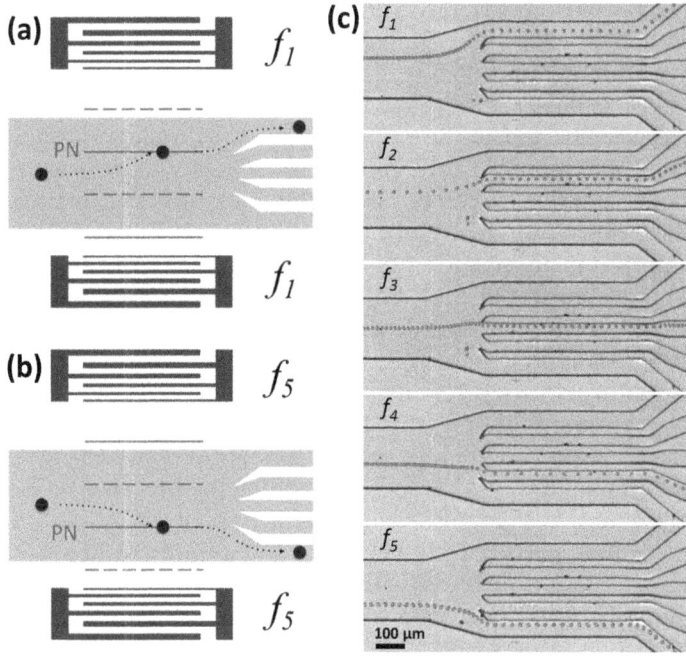

Figure 5.6   SSAW-based multichannel cell sorting. (a) The particle path with input frequency f_1 (b) the particle path with input frequency f_5 (c) Micrographs of particle paths with different input frequencies. Particles can be directed into 5 different outlets. Reprinted with permission from ref. 39.

5.3.5   SAW-Based Cell/Particle Patterning

In the above sections, we introduced the SSAW-based manipulation of cells/particles under continuous flow. These manipulations are important to high-throughput applications. Meanwhile, SSAW is also suitable to perform precise and fine manipulations under the stop flow (or batch operation). In this section, we will discuss SSAW-enabled cell patterning. Geometrically defined arrangement of cells is important in many biological applications, such as single-cell microarrays, tissue engineering, and cell–cell interactions. SSAW-based cell patterning relies on the position of pressure nodes. Since cells always tend to move to the pressure node where they experience minimal acoustic radiation force, the pattern of pressure nodes can be converted to the pattern of cells under the stop flow condition.[41,42] Depending on the number and geometry of IDTs, different types of cell patterns can be formed. When only one pair of IDTs is used, only linear shapes of pressure nodes will be formed. As a result, cells/particles can be patterned into a linear array. When two pairs of IDTs are used in an orthogonal configuration, diamond-shaped interference patterns will be formed. Under this pressure-node configuration, a two-dimensional geometry of cell patterns can be formed.

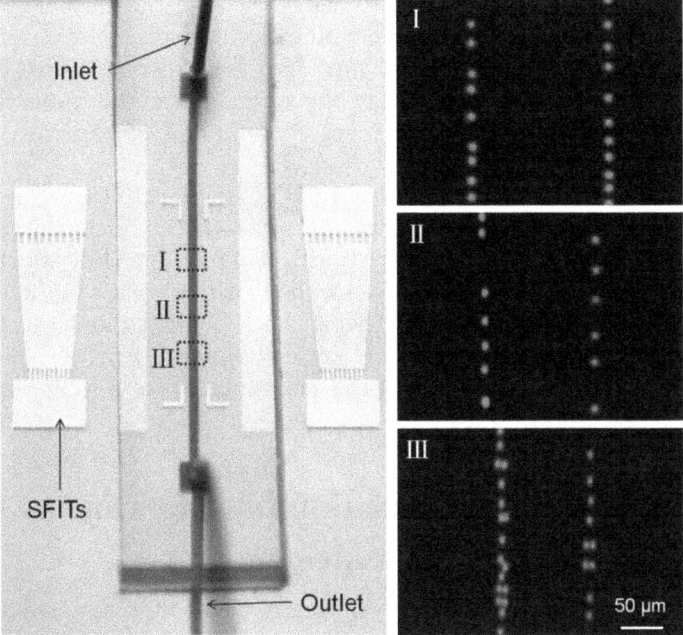

Figure 5.7   SSAW-enabled tunable particle patterning. Reprinted with permission from ref. 43.

The flexibility of SSAW-based cell patterning also allows more complex patterns. Ding *et al.* reported the use of slanted-finger interdigitated transducers (SFITs) to achieve heterogeneous distance cell patterns in one device.[43] The location of the SAW beam can be tuned by adjusting the input signal frequency. The authors showed that in one channel, HL-60 cells were patterned with a linear fashion with multiple distances between lanes (Figure 5.7). The distances were 60, 78, and 150 mm, when applying 45, 36, and 27 MHz, respectively. More complex 2D dynamic patterns are also demonstrated using two pairs of SFITs.

5.3.6   SAW-Based Single-Cell/Particle Manipulation

In addition to large-scale cell patterning, many biological applications require even finer adjustment at the single-cell level, *e.g.*, the study of cell–cell adhesion events. SSAW is also suitable to actuate single cells with μm-scale precision. The key for SSAW-based single-cell manipulation is the controllable change of pressure-node position. Several methods have been reported to be able to change the position of pressure nodes in a controllable manner. Meng *et al.* reported a SSAW-based device that is capable of moving a lipid microbubble in a 2D plane.[44] The device consists of 2 pairs of IDTs arranged orthogonally. When changing the relative phase of one IDT, the position of the pressure node will change. As a result, the movement of microbubbles is guided by the movement of the pressure-node position. The movement resolution is around 2.2 μm. Tran *et al.* reported a fast dynamic phase-shift approach by modulating the frequency of the input electrical signal. Thus, they also achieved precise movement of particles in a 2D plane.[45]

In addition to the phase-shift-based SSAW single-particle manipulation, Ding *et al.* used two orthogonal pairs of chirped IDTs.[46] As described above, chirped IDTs were able to change the position of pressure nodes by changing the input frequency. In this work, the *x* and *y* positions of the pressure nodes can be adjusted independently. The authors demonstrated the effective movement of a single bovine red blood cell along any predesigned path. Moreover, they also demonstrated the manipulation of a millimetre-sized single *C. elegan* worm.

5.4 SAW-Based Manipulation of Nano-Objects

5.4.1 Nanotube/Nanowires Patterning

Nanowires and nanotubes play important roles in many applications, including chemical/biological sensing and electronic circuits.[47] However, they often require special manipulations, such as alignment, assemblies, and patterning, to become functional. In recent years, SAW has been demonstrated as an effective tool to manipulate nanowires and nanotubes.

Strobl *et al.* showed that multiwalled carbon nanotubes (MWNTs) on a LiNbO₃ substrate can be aligned using TSAWs.[48] When a TSAW field was present in the fluid, carbon nanotubes aligned 25−45° relative to the direction of the SAW field. The resulting pattern is a combinatorial effect of SAW-induced streaming and SAW-induced piezoelectric field. The SAW-induced piezoelectric field is determined to be the major reason for the alignment of carbon nanotubes, as the patterning was not formed when the electric field is screened. Because of the SAW-induced fluid movement, the final pattern of nanotubes has an angle to the direction of the SAW. Under the same principle, Smorodin *et al.* improved the patterning of carbon nanotubes on patterned gold electrodes using thiolated single-walled carbon nanotubes (SWNTs).[49] The thiolated carbon nanotubes can be easily attached to the surface of the gold electrodes, thereby minimising the impact of acoustic streaming on the pattern.

Previous studies limited the alignment of nanotubes on piezoelectric substrates, which are not very useful in many practical applications. Seemann *et al.* overcame the limitation and achieved the patterning of carbon nanotubes on silicon substrates by applying a "flip-chip" configuration.[50] They placed a silicon substrate with patterned gold electrodes in close proximity with a LiNbO₃ substrate on which a TSAW was generated. When the SAW is on, nanotubes can be aligned and attached onto gold electrodes of the silicon substrate.

In addition to using TSAWs, SSAWs have also been demonstrated to be capable of patterning microtubes and nanowires. Kong *et al.* employed two parallel IDTs to establish a SSAW field on a LiNbO$_3$ substrate.[51] When a drop of solution containing Cr microtubes was placed in the SSAW field, the microtubes were aligned along the direction of SAW propagation. Similar to the mechanism of TSAW-based patterning, the authors determined that the alignment of microtubes was due to the SAW-induced piezoelectric field. Chen *et al.* further configured a more complex SSAW field to achieve more flexible patterning of silver nanowires.[52] Both parallel and perpendicular arrays of silver nanowires were formed. The distance between nanowire arrays can also be adjusted by tuning the frequency of the SAW. Interestingly, the authors observed 3D spark-shaped nanowire patterns in a 2D SSAW field.

5.4.2   Liquid-Crystal Reorientation

SAW is also capable of manipulating liquid crystals. Polymer-dispersed liquid crystals (PDLCs) is a transparent polymer matrix containing randomly dispersed crystal droplets. They have many applications in displays and optical elements. The light transmittance of the material can be adjusted through applying an electric field to the PDLCs. In addition to the electric-based realignment, acoustics has been exploited as an alternative controlling mechanism. Recently, Liu *et al.* reported the control of light transmittance of PDLCs using a TSAW.[53] They showed that PDLC changed from opaque to transparent after applying a TSAW as liquid crystals were aligned along the SAW propagation direction.

5.5   Summary

SAWs have demonstrated tremendous capabilities of particle manipulation in microfluidic platforms, from high-throughput cell focusing to precise manipulation of single cells. To accelerate the adoption of this technique to various kinds of manipulations, efforts on standardising current designs and operations should be encouraged. Meanwhile, it is also important to further investigate the theoretical background and technical capabilities to improve the performances of SAW-based particle manipulation. Finally, the label-free, contactless and noninvasive nature of SAW manipulation makes it an especially promising manipulation technique for biological applications. Thus, we anticipate seeing more breakthroughs and exciting applications of SAW-based particle manipulation.

References

1. A. Ashkin, J. M. Dziedzic, J. E. Bjorkholm and S. Chu, *Opt. Lett.*, 1986, **11**, 288.
2. M. B. Rasmussen, L. B. Oddershede and H. Siegumfeldt, *Appl. Environ. Microbiol.*, 2008, **74**, 2441–2446.

3. S. Park, Y. Zhang, T.-H. Wang and S. Yang, *Lab Chip,* 2011, **11**, 2893–2900.
4. A. Snezhko and I. S. Aranson, *Nat. Mater.,* 2011, **10**, 698–703.
5. A. Jamshidi, P. J. Pauzauskie, P. J. Schuck, A. T. Ohta, P.-Y. Chiou, J. Chou, P. Yang and M. C. Wu, *Nat. Photonics,* 2008, **2**, 86–89.
6. X. Ding, P. Li, S.-C. S. Lin, Z. S. Stratton, N. Nama, F. Guo, D. Slotcavage, X. Mao, J. Shi, F. Costanzo and T. J. Huang, *Lab Chip,* 2013, **13**, 3626–3649.
7. M. Gedge and M. Hill, *Lab Chip,* 2012, **12**, 2998–3007.
8. S.-C. S. Lin, X. Mao and T. J. Huang, *Lab Chip,* 2012, **12**, 2766–2770.
9. H. Li, J. R. Friend and L. Y. Yeo, *Biomed. Microdevices,* 2007, **9**, 647–656.
10. P. R. Rogers, J. R. Friend and L. Y. Yeo, *Lab Chip,* 2010, **10**, 2979–2985.
11. M. K. Tan, J. R. Friend and L. Y. Yeo, *Lab Chip,* 2007, **7**, 618–625.
12. R. Wilson, J. Reboud, Y. Bourquin, S. L. Neale, Y. Zhang and J. M. Cooper, *Lab Chip,* 2011, **11**, 323–328.
13. R. Shilton, M. K. Tan, L. Y. Yeo and J. R. Friend, *J. Appl. Phys.,* 2008, **104**, 014910.
14. Y. Chen, S. Li, Y. Gu, P. Li, X. Ding, L. Wang, J. P. McCoy, S. J. Levine and T. J. Huang, *Lab Chip,* 2014, **14**, 924–930.
15. M. Mollet, R. Godoy-Silva, C. Berdugo and J. J. Chalmers, *Biotechnol. Bioeng.,* 2008, **100**, 260–272.
16. J. Shi, X. Mao, D. Ahmed, A. Colletti and T. J. Huang, *Lab Chip,* 2008, **8**, 221–223.
17. Q. Zeng, H. W. L. Chan, X. Z. Zhao and Y. Chen, *Microelectron. Eng.,* 2010, **87**, 1204–1206.
18. J. Shi, S. Yazdi, S.-C. Steven Lin, X. Ding, I. K. Chiang, K. Sharp and T. J. Huang, *Lab Chip,* 2011, **11**, 2319–2324.
19. Y. Chen, A. A. Nawaz, Y. Zhao, P.-H. Huang, J. P. McCoy, S. J. Levine, L. Wang and T. J. Huang, *Lab Chip,* 2014, **14**, 916–923.
20. J. Shi, H. Huang, Z. Stratton, Y. Huang and T. J. Huang, *Lab Chip,* 2009, **9**, 3354–3359.
21. J. Nam, Y. Lee and S. Shin, *Microfluid. Nanofluid.,* 2011, **11**, 317–326.
22. J. Nam, H. Lim, D. Kim and S. Shin, *Lab Chip,* 2011, **11**, 3361–3364.
23. Y. Ai, C. K. Sanders and B. L. Marrone, *Anal. Chem.,* 2013, **85**, 9126–9134.
24. J. Nam, H. Lim, C. Kim, J. Yoon Kang and S. Shin, *Biomicrofluidics,* 2012, **6**, 024120.
25. M. C. Jo and R. Guldiken, *Sens. Actuators A,* 2012, **187**, 22–28.
26. B. Landenberger, H. Hofemann, S. Wadle and A. Rohrbach, *Lab Chip,* 2012, **12**, 3177–3183.
27. F. Bragheri, P. Minzioni, R. Martinez Vazquez, N. Bellini, P. Paie, C. Mondello, R. Ramponi, I. Cristiani and R. Osellame, *Lab Chip,* 2012, **12**, 3779–3784.
28. M. M. Wang, E. Tu, D. E. Raymond, J. M. Yang, H. Zhang, N. Hagen, B. Dees, E. M. Mercer, A. H. Forster, I. Kariv, P. J. Marchand and W. F. Butler, *Nat. Biotechnol.,* 2005, **23**, 83–87.
29. T.-H. Wu, Y. Chen, S.-Y. Park, J. Hong, T. Teslaa, J. F. Zhong, D. Di Carlo, M. A. Teitell and P.-Y. Chiou, *Lab Chip,* 2012, **12**, 1378–1383.

30. Y. Chen, T.-H. Wu, Y.-C. Kung, M. A. Teitell and P.-Y. Chiou, *Analyst,* 2013, **138**, 7308–7315.
31. H. Sugino, T. Arakawa, Y. Nara, Y. Shirasaki, K. Ozaki, S. Shoji and T. Funatsu, *Lab Chip,* 2010, **10**, 2559–2565.
32. S. H. Cho, C. H. Chen, F. S. Tsai, J. M. Godin and Y.-H. Lo, *Lab Chip,* 2010, **10**, 1567–1573.
33. A. Y. Fu, H.-P. Chou, C. Spence, F. H. Arnold and S. R. Quake, *Anal. Chem.,* 2002, **74**, 2451–2457.
34. T. Franke, A. R. Abate, D. A. Weitz and A. Wixforth, *Lab Chip,* 2009, **9**, 2625–2627.
35. T. Franke, S. Braunmuller, L. Schmid, A. Wixforth and D. A. Weitz, *Lab Chip,* 2010, **10**, 789–794.
36. R. D. O'Rorke, C. D. Wood, C. Wälti, S. D. Evans, A. G. Davies and J. E. Cunningham, *J. Appl. Phys.,* 2012, **111**, 094911.
37. N. D. Orloff, J. R. Dennis, M. Cecchini, E. Schonbrun, E. Rocas, Y. Wang, D. Novotny, R. W. Simmonds, J. Moreland, I. Takeuchi and J. C. Booth, *Biomicrofluidics,* 2011, **5**, 044107.
38. L. Meng, F. Cai, Z. Zhang, L. Niu, Q. Jin, F. Yan, J. Wu, Z. Wang and H. Zheng, *Biomicrofluidics,* 2011, **5**, 044104.
39. X. Ding, S.-C. S. Lin, M. I. Lapsley, S. Li, X. Guo, C. Y. Chan, I. K. Chiang, L. Wang, J. P. McCoy and T. J. Huang, *Lab Chip,* 2012, **12**, 4228–4231.
40. S. Li, X. Ding, F. Guo, Y. Chen, M. I. Lapsley, S.-C. S. Lin, L. Wang, J. P. McCoy, C. E. Cameron and T. J. Huang, *Anal. Chem.,* 2013, **85**, 5468–5474.
41. J. Shi, D. Ahmed, X. Mao, S.-C. S. Lin, A. Lawit and T. J. Huang, *Lab Chip,* 2009, **9**, 2890–2895.
42. C. D. Wood, S. D. Evans, J. E. Cunningham, R. O'Rorke, C. Wälti and A. G. Davies, *Appl. Phys. Lett.,* 2008, **92**, 044104.
43. X. Ding, J. Shi, S. C. Lin, S. Yazdi, B. Kiraly and T. J. Huang, *Lab Chip,* 2012, **12**, 2491–2497.
44. L. Meng, F. Cai, J. Chen, L. Niu, Y. Li, J. Wu and H. Zheng, *Appl. Phys. Lett.,* 2012, **100**, 173701.
45. S. B. Q. Tran, P. Marmottant and P. Thibault, *Appl. Phys. Lett.,* 2012, **101**, 114103.
46. X. Ding, S.-C. S. Lin, B. Kiraly, H. Yue, S. Li, I.-K. Chiang, J. Shi, S. J. Benkovic and T. J. Huang, *Proc. Natl. Acad. Sci. U. S. A.,* 2012, **109**, 11105–11109.
47. R. H. Baughman, A. A. Zakhidov and W. A. de Heer, *Science,* 2002, **297**, 787–792.
48. C. J. Strobl, C. Schäflein, U. Beierlein, J. Ebbecke and A. Wixforth, *Appl. Phys. Lett.,* 2004, **85**, 1427–1429.
49. T. Smorodin, U. Beierlein, J. Ebbecke and A. Wixforth, *Small,* 2005, **1**, 1188–1190.
50. K. M. Seemann, J. Ebbecke and A. Wixforth, *Nanotechnology,* 2006, **17**, 4529–4532.

51. X. H. Kong, C. Deneke, H. Schmidt, D. J. Thurmer, H. X. Ji, M. Bauer and O. G. Schmidt, *Appl. Phys. Lett.,* 2010, **96**, 134105.
52. Y. Chen, X. Ding, S.-C. Steven Lin, S. Yang, P.-H. Huang, N. Nama, Y. Zhao, A. A. Nawaz, F. Guo, W. Wang, Y. Gu, T. E. Mallouk and T. J. Huang, *ACS Nano,* 2013, **7**, 3306–3314.
53. Y. J. Liu, X. Ding, S.-C. S. Lin, J. Shi, I. K. Chiang and T. J. Huang, *Adv. Mater.,* 2011, **23**, 1656–1659.

Introduction to Optofluidics for LOC Systems

HENRY O. FATOYINBO

University of Surrey, Faculty of Engineering and Physical Sciences,
Department of Mechanical and Engineering Sciences, Centre for Biomedical
Engineering, Guildford, Surrey, GU2 7XH, UK
E-mail: henry.fatoyinbo@uclmail.net

6.1 Introduction

The synergistic coupling of light with microfluidics, though recurrent in sensing systems, has recently undergone a revolutionary transformation with regards to the development of miniaturised integrated system architectures for all-optical particle and fluid manipulation capabilities.[1,2] Electromagnetic radiation interacting with matter and its use in instrumentation and detection is widely referred to as optics. It has a significant role in medical, chemical and biological sciences where commercially available clinical diagnostic equipment and detection systems such as glucose monitoring systems, surface plasmon resonance (SPR), surface-enhanced Raman spectroscopy (SERS), flow cytometry and fluorescence-activated cell sorting (FACS) are established and indispensable tools for clinicians and researchers.[3-9]

Miniaturised optical elements integrated into LOC systems (*e.g.*, micro-optical electromechanical systems or MOEMS) has long been realised as a desirable goal for creating low-cost, highly sensitive and space saving data processing, imaging and detection systems.[10] Integrating microfluidic networks with discrete optical elements such as microlenses, mirrors,

RSC Detection Science Series No. 5
Microfluidics in Detection Science: Lab-on-a-chip Technologies
Edited by Fatima H Labeed and Henry O Fatoyinbo
© The Royal Society of Chemistry 2015
Published by the Royal Society of Chemistry, www.rsc.org

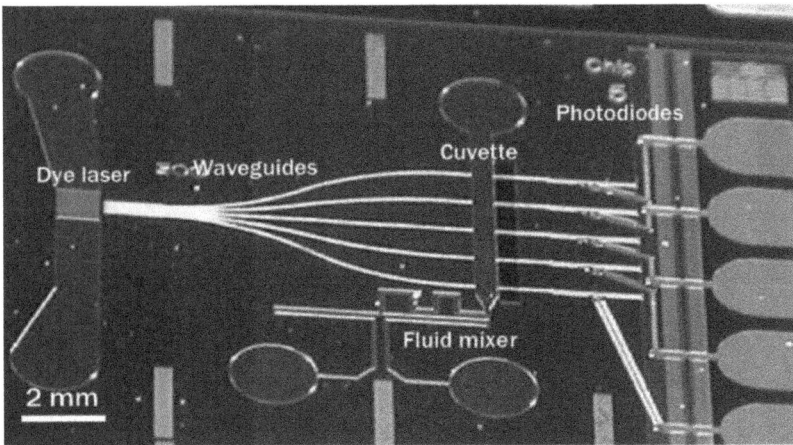

Figure 6.1   Photograph of the lab-on-chip device with integrated microfluidic dye
laser, optical waveguides, microfluidic network and photodiodes. The
metallic contact pads for the photodiodes are seen on the far right.
The chip footprint is 15 mm by 20 mm. The photograph was taken
before a lid was bonded to the structures. Reproduced from ref. 11.

waveguides and diffractive elements in monolithic substrates have been
accomplished using a range of materials including PMMA, glass, quartz and
semiconductors (see Figure 6.1).[11] Cheap and rapid prototyping of elastomeric microdevices through soft lithography and cofabrication techniques,[12-16] and the incorporation of nanomaterials and nanostructures,
optical elements through the functionality of nano/microfluidics in complex
channels has improved the limits of detection (LOD) in gas, micro/nano-
organisms, biomolecular and nanoparticulate sensing and analysis.[17,18]
In addition, the precise and dynamic reconfigurability of various optical
properties from the tuning of fluid parameters (*e.g.*, refractive index) to the
localisation of optical forces for particle manipulation, are just a few of the
reasons why optofluidic technologies are rapidly evolving as an exciting and
integral prospect to the processing tool-kit of LOC systems.

6.1.1   Optofluidics Defined

When light interacts with materials that deform under shear stresses (*e.g.*,
fluids), a number of possible outcomes exist that adequately encompasses
this nascent field of optofluidics. On the one hand, light can be used to
manipulate fluids and/or the dispersed particulates within, alternatively,
fluids can be used to accurately manipulate light over μm and nm scales. This
broad definition of optofludics has been further segmented into 3 device-
based categories by Psaltis *et al.*[19] They include structured solid–liquid
hybrids where the optical properties of both materials are relevant; pure
fluid-based systems where the material properties of the fluids involved are

important; and finally, colloidal-based systems in which particles are manipulated or the dispersed particle's optical properties are of particular relevance. It should be noted that although light has been used in various detection systems, such as in fluorescently tagged cells and biomolecules for optical imagining and luminescence detection in microfluidic systems,[20] they fall short of the definition of optofluidics. In this scenario, for example, fluorophor emissions are merely detected by the optical system and the fluorescence does not alter the fluid in a manner consistent with the afore-mentioned categories. This does not preclude the idea that fluorescence detection is separate from optofluidics. As recently demonstrated by Chen *et al.*, a protein laser produced from a optofluidic ring resonator operating in whispering-gallery mode (WGM) was used to amplify the FRET (Förster resonance energy transfer) signal of genetically encoded fluorescent protein interactions within a laser cavity.[21] Applying this type of technique in microfluidic systems could potentially reduce the amount of proteins needed to achieve superior results over conventional FRET detection systems.

In this chapter, the reader is introduced to some of the more recent innovations of optofluidic devices for particle manipulation in LOC systems, from nanophotonic tweezers to the hybrids of optoelectrofluidic devices. The latest applications of the technologies developed to date are extremely varied and will not be covered in its entirety, though readers are referred to *"The Handbook of Optofluidics"* where a comprehensive treatment of this techno-logical field is given.[22] Before discussing these latest developments, an overview of the basic physics employed in classical optical tweezers for particle trapping and manipulation *via* electromagnetic radiation is first summarised in the following section.

6.2   Radiation Pressures

Since Ashkin and coworkers first demonstrated the manipulation and trap-ping (also termed "optical tweezers") of micrometre and submicrometre particles using continuous-wave (cw) laser beams, there have been numerous developments in the use of optical forces to manipulate and probe single cells, micro-organisms, viruses, biomolecules and nanoparticles (NPs).[23-31] The radiation pressures on Mie ($d_p \gg \lambda$) and Rayleigh ($d_p \ll \lambda$) regime dielectric particles, where d_p and λ are the particle diameter and wavelength of incident light, respectively, is of the order of pN and arises from the momentum of light itself. These radiation pressures, commonly composed of scattering (F_{scat}) and gradient (F_{grad}) forces, arise from absorption, refraction (*i.e.* transmitted) and reflection of light (*i.e.* photons) momentum between media of differing refractive indices ($n = c/v$); where $c(= 1/\sqrt{\varepsilon_0 \mu_0})$ is the speed of light in a vacuum and v is the phase velocity in media. Assuming negligible reflection and ignoring diffraction, when parallel rays of light interact at a dielectric sphere's surface of higher refractive index than the surrounding medium, *i.e.* $n_1 < n_2$, light is refracted through the sphere (Figure 6.2(a)). Thus, from the conservation of momentum, forces on the

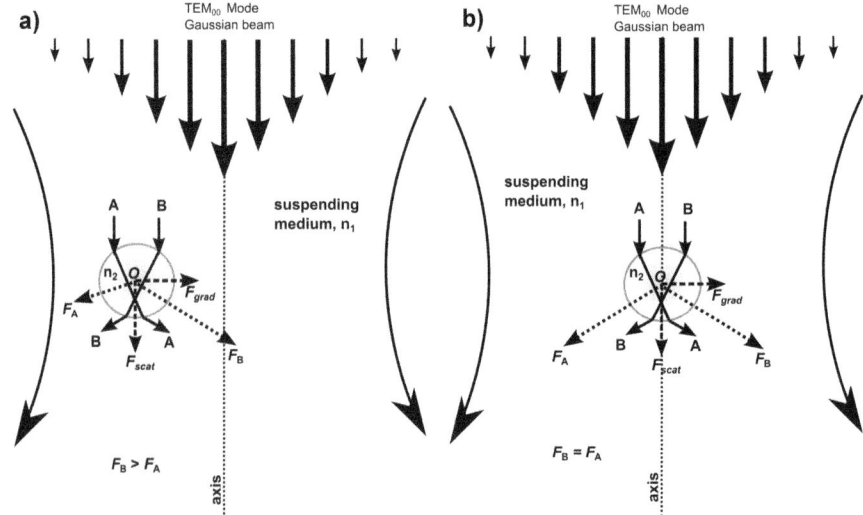

Figure 6.2 Beam of light rays travelling through the suspending medium of refractive index $= n_1$ and interacts with a dielectric sphere of refractive index $= n_2$. When $n_1 < n_2$, light is refracted through the dielectric sphere generating forces F_A and F_B. For particles situated off-centre of the beam (a) the forces generated on the side of the sphere closer to the beam's axis pulls the sphere towards the beam $(F_B > F_A)$. Once aligned with the beam's axis (b), the forces are equal $(F_B = F_A)$ and light momentum accelerates the sphere in the direction of the incident light only.

sphere $(F_A$ and $F_B)$ are generated in the direction of momentum change. As ray intensities of a Gaussian laser beam (I_0) decrease away for the beam's axis, the sum of ray intensities interacting with the sphere closer to the beam axis will generate a force greater than that on the farther side of the sphere, away from the beam axis. These forces are made up of 2 components: F_{scat}, which is directed in the same direction as the propagating light beam and F_{grad}, which is the transverse component directed towards the beam axis. The gradient force tends to pull the sphere into the beam's axis where upon alignment, $F_A = F_B$ and the particle is accelerated in the direction of the incident light only (Figure 6.2(b)).[32] For incident light acting on a sphere with a lower refractive index than the surrounding medium $(n_1 > n_2)$, the sphere will be pushed away from the beam axis, though in the same direction as the propagating light. In a dual-beam configuration, the use of a second beam (I_1) of equal intensity, directed in the opposite direction of I_0 exerts an opposing force on the particle, trapping it in an optical potential well.[24] Uses of acoustic and electrical forces to balance out the optical force of a single beam have also been applied to create this simple optical trapping zone.[29,33] Ashkin and Dziedzic also demonstrated the stability of optically levitated transparent dielectric hollow[34] and solid[35] spheres in air using vertically directed TEM_{01}^{*}-mode (annular light) and TEM_{00}-mode laser beams respectively. In the case of the solid spheres, axial scattering forces ($F_{scat} \gg F_{grad}$) were balanced by gravitational forces to create a stable levitated sphere. These seminal investigations opened up a range of potential applications using optical forces as accurate nondestructive optical manipulation tools for nanoscale to microscale particles in biomolecular, chemical and biophysical studies.

6.2.1 Optical Tweezers

Optical tweezers use a single tightly focussed beam of light with a stable 3D backward axial gradient force ($F_{grad} \gg F_{scat}$), commonly produced by a high numerical aperture of a microscope objective, to attract and trap Mie- and Rayleigh-sized particles in the beam's high-intensity focus.[25] For a more detailed review of optical force quantification, trapping theory, design considerations of traps and force measurements practicalities refer to works by Svoboda and Block,[36] Ashkin *et al.*,[25,37] Wright *et al.*,[38] Mazzolli, *et al.*[41] and Neuman and Block.[39] Theoretical treatment of the optical forces arising on trapped particles are traditionally limited to the particle's size, where for the Mie regime (≥ 10 μm) ray optics will suffice while in the Rayleigh regime (≤ 1 μm) particles are considered as point dipoles. The general relation between the optical force (F_{opt}) and the intensity power (P) is shown in eqn (6.1). The dimensionless parameter (Q), describes the proportion of incident momentum per second exerted at the interface. It is the most readily tuneable parameter, affecting the trap's design configuration (*e.g.*, convergence angle, spot size, wavelength, beam profile), thus enhancing optical trapping forces, while other parameters remain constant, such as in the limiting case of biological cell trapping where higher P or changes to media properties could potential adversely affect cell viabilities.

$$F_{opt} = \frac{n_1 P Q}{c} \tag{6.1}$$

Particles of diameter comparable to that of a laser of wavelength (λ) ≈ 0.1 to ~5 μm, (*e.g.*, microsphere or nanoparticle handles used to indirectly manipulate other particles) are not adequately covered by either the Mie or Rayleigh regimes and though generally treated using classical electrodynamics, theoretical development in this regime is still nascent.[40] The force of a focused laser beam incident on a homogeneous particle ($2r \leq \lambda$) of arbitrary geometry (*e.g.*, spheroid, ellipsoid) can be obtained using Maxwell's stress tensor, eqn (6.2), where the electromagnetic fields at the particle surface and within the particle volume are determined for the incident and scattered fields and are superpositioned external to the particle.[41-44]

$$F = \oint_{s} T_{ij} n_j \mathrm{d}S \tag{6.2}$$

From eqn (6.2), S is any arbitrary closed surface containing a volume of system charges and the Maxwell stress tensor T_{ij} is defined by eqn (6.3), where δ_{ij} is the Kronecker delta, H is the magnetic field and E is the electric field.

$$T_{ij} = \frac{1}{4\pi}\left[\varepsilon_0 E_i E_j + \mu_0 H_i H_j - \frac{1}{2}\delta_{ij}(\varepsilon_0 E^2 + \mu_0 H^2)\right] \tag{6.3}$$

6.2.2 Rayleigh Regime

The force on a homogeneous dielectric sphere of diameter much smaller than the wavelength (λ) of an incident plane wave can be approximated using the Lorentz force,[42,45]

$$F = (\mathbf{p} \cdot \nabla)E + \mathbf{p'} \times B \tag{6.4}$$

where E and B ($= \mu_0 H$) are the electric field and magnetic flux density respectively, $\mathbf{p} = \alpha E$ is the induced dipole moment and $\mathbf{p'}$ is the time deriv-ative. The polarisability (α), also known as the Clausius–Mossotti factor (CMF), depends on the particle volume and includes the relation of dielectric constants of the two differing media,

$$\alpha_0 = \left[\frac{\varepsilon_1 - \varepsilon_2}{\varepsilon_1 + 2\varepsilon_2}\right] \tag{6.5}$$

Eqn (6.5) can also be described using the Lorentz–Lorenz (eqn (6.6)) in terms of the real part of the complex relative refractive index (*i.e.* $m = n_2/n_1$),

$$\alpha_0 = \left(\frac{m^2 - 1}{m^2 + 2}\right) \tag{6.6}$$

In general, eqn (6.5) is frequency dependent in which case the material's complex refractive index ($\tilde{n}_b$) is defined as $\tilde{n}_b = (n_b + i\kappa_b)^2$, where n_b is the refractive index and κ_b is the extinction coefficient, related to the absorption coefficient (a') of the Beer–Lambert's law, *i.e.* $a' = 4\pi\kappa_b/\lambda$. This is analogous to the complex permittivity of a material ($\varepsilon_b^* = \varepsilon_1 + i\varepsilon_2$), which from empirical refractive index studies gives, $\varepsilon_b^* \approx \tilde{n}_b^2$, where,

$$\varepsilon_1 = n_b^2 - \kappa_b^2 \text{ and } \varepsilon_2 = 2n_b\kappa_b \tag{6.7}$$

In the simplified case of a nonabsorbing dielectric the imaginary compo-nent is ignored, giving $\varepsilon_b \approx n_b^2$. With nonmagnetic field contributions to the optical trapping dynamics and for simplicity ignoring any absorption by the dielectric, the electric dipole scattering force on a Rayleigh particle eqn (6.8) is a function of optical intensity (I_0), the inverse fourth power of the wave-length and the relative refractive indices of the particle and medium, with larger particles giving a more intense scattering force.[46]

$$F_{\text{scat}} = \frac{128\pi^5 r^6 I_0 n_1}{3c\lambda^4}\alpha_0^2 \tag{6.8}$$

The gradient force, a function of the nonuniform optical intensity field ∇I_0 and the sphere's polarisability, is expressed as,

$$F_{\text{grad}} = \frac{2\pi n_1 r^3 \alpha_0 \nabla I_0}{c} \qquad (6.9)$$

6.2.3 Mie Regime

The forces on a homogeneous dielectric sphere of diameter larger than the wavelength (λ) of an incident plane wave can be approximated using simple ray-optics, as quantitatively demonstrated by Ashkin[37] and followed by Wright *et al.*[38] As shown in Figure 6.3(a), parallel beams of light enter a high numerical aperture (NA = 1.2 to 1.4), where individual rays converge to a diffraction-limited focus point (f) on the axis of the sphere. The incident momentum of each ray through f is computed from the radial position (r), with respect to the Z-axis, of the ray entering the aperture and the respective

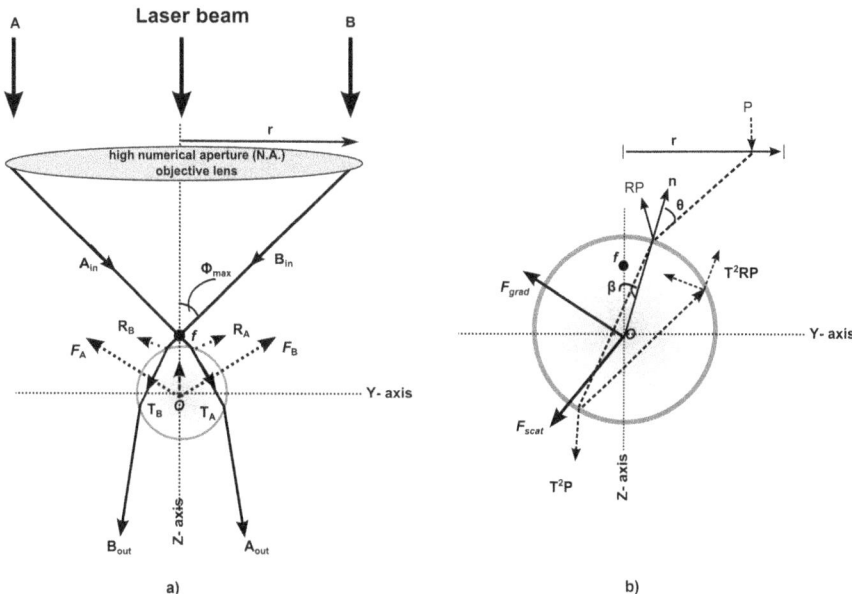

Figure 6.3 (a) Classical optics with a diffraction-limited focus point on a dielectric sphere's axis for optical trapping. Optical forces are generated from the reradiated rays at the particle surface with trapping occurring when gradient forces are more dominant than scattering forces. (b) Shows how a single ray of power P incident at an angle θ is successively refracted and reflected, internal of the sphere, with decreasing power giving rise to gradient and scattering forces made up of all incident rays integrated over the numerical aperture radius.

angle made against the *Y*-axis. The rays are reradiated (*i.e.* reflected and refracted) at the surface of the sphere giving rise to optical forces. The refracted portion is reflected internal of the sphere an infinite amount of times, along with the associated infinite number of refracted rays emerging out of the sphere *e.g.*, A_{out} and B_{out}.

For a single ray propagating in a straight line of a homogeneous media, the total force on a dielectric sphere of power *P* with an angle of incident, θ, is due to the sum of the successive power decreasing reflected and refracted rays, as depicted in Figure 6.3(b). At the sphere's origin (*O*) the orthogonal force components in terms of eqn (6.1), of the convergent rays (integrated over *r*, where $r = \Phi_{max}$) are exactly described as functions of the incident ray angle (θ), the angle of refraction (β), and the Fresnel reflection (*R*) and transmission (*T*) coefficients, respectively, due to the respective perpendicular and parallel polarisation at the plane of incidence of the dielectric.

$$F_{grad} = \frac{n_1 P}{c} Q_{grad} = \frac{n_1 P}{c} \left(R \sin 2\theta - \frac{T^2 [\sin(2\theta - 2\beta) + R \sin 2\theta]}{1 + R^2 + 2R \cos 2\beta} \right) \quad (6.10)$$

and,

$$F_{scat} = \frac{n_1 P}{c} Q_{scat} = \frac{n_1 P}{c} \left(1 + R \cos 2\theta - \frac{T^2 [\cos(2\theta - 2\beta) + R \cos 2\theta]}{1 + R^2 + 2R \cos 2\beta} \right) \quad (6.11)$$

Efficient axial trapping occurs when Q_{grad} is more dominant than Q_{scat} over the sphere's cross section. Although independent of sphere size, high convergence angles, small spot sizes, spherical aberrations, focal depth and higher particle refractive index with respect to the surrounding medium are contributing factors of *Q*. In the more common usage of TEM_{00} beams with profiles $I(r) = I_0 e^{(-2r^2/w_0^2)}$ at the entrance of a microscope objective's high NA, where w_0 is the beam's waist radius, it was found that the higher the waist radius with respect to the aperture radius, the more uniformly filled the aperture producing a greater trapping efficiency. This efficiency rapidly degrades with underfilling of the aperture, though it has been reported that a combination of lower NA with small levels of overfilling provide optimal trapping efficiencies, including maximum trap stiffness in the *z*-direction at large trapping depths.[47] Stevenson *et al.* give an excellent review on the uses of optical forces to manipulate biological materials, from single molecules (*e.g.*, DNA, RNA polymerase, kinesin, myosin, proteins) in their various experimental configurations for studies into protein folding, RNA translation, and DNA transcription and cleavage, to single-cell sorting, spectroscopy and trapping.[48] Interestingly, for the first time, optical trapping experiments were recently demonstrated on photopolymerised cholesteric liquid-crystal (CLC) microspheres dispersed in water using circularly polarised light from a linear polarised Ar ion beam laser with $\lambda = 488$ nm.[49]

Variable dynamic behaviours were exhibited based on the internal architecture of the particles and the handedness of the incident light. Figure 6.4(a)–(c) show the propagation of rays of light on to the anisotropic CLC microsphere with various orientations in a ray optics system. The ability to form stable trapping is thought to be strongly influenced by the propagation direction of the symmetric incident rays with respect to the orientation of the particle. Depending on the helical internal configurations of the microparticles, obtained through chemical control of the external environment of the LC droplet doped with reactive mesogens, they can exhibit either optical isotropic or anisotropic chirality (Figure 6.4(d)).

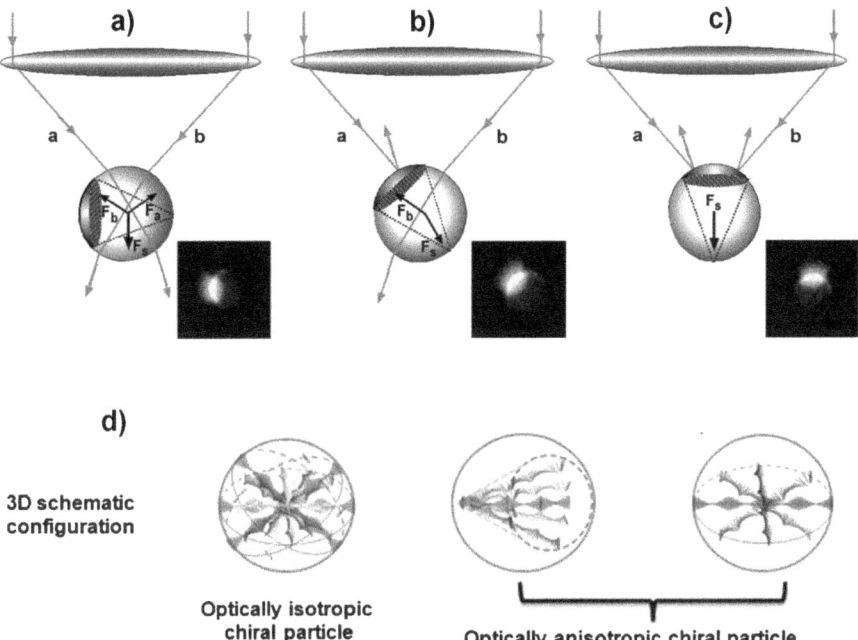

Figure 6.4 Qualitative representation of a pair of light rays impinging on an optically anisotropic particle with different orientations, and the consequent optical forces exerted by these rays. (a) The rays refract symmetrically inside the particle and thus there is a balance of forces in the transverse plane resulting in a stable optical trap. (b) The case of broken symmetry on forces and torque, depending on the orientation and position of the particle with respect to the optical trap. (c) The extreme situation when the reflecting part of the surface is oriented such that the incident light is reflected and the particle is repelled due to the scattering force. Insets show the microscope images of the solid CLC sphere with optically anisotropic structure in reflection mode; (d) Shows the 3D schematic configuration of the CLC helicoidal structures; optically isotropic (radial), and optically anisotropic. Reproduced from ref. 49.

6.3 Planar Integration of Optofluidic Elements

Light incident on metallic surfaces or nanostructures (including NPs) can be coupled to conduction electrons, coherently excited near the metal surface (*e.g.*, surface plasmon polaritons, localised surface plasmons) thus typically enhancing the optical near-field at subwavelength dimensions. Electromagnetic radiation at a dielectric/metallic interface can be reflected, absorbed or attenuated (frequency dependent) by the metal, but unlike dielectrics cannot propagate through the metal, making them useful in a range of LOC applications such as cladding for waveguides and resonators or sensor enhancement surfaces.[50] Furthermore, irradiation of fluids (*e.g.*, liquid crystals, Newtonian fluids) or between fluids of differing refractive indices with differential flow rates can give rise to thermal gradients inducing fluid motion,[51–54] liquid microlenses,[55,56] or liquid–liquid waveguides and dye lasers.[57,58] This section will describe recent developments in planar optofluidic elements, including waveguides, nanostructures for particle manipulation and hybrid systems, with a general description of their operation within the context of LOC systems.

6.3.1 Optical Guiding

Propagation of light to and from regions within a microfluidic system has been the objective of on-chip waveguide designs, where full translation of macroscale optical system functionalities are miniaturised with negligible attenuation in transmission and precision. Conventional waveguide assemblies interfaced with microfluidic channels tend to consist of a dielectric core (*e.g.*, glass fibre) in which light propagates, surrounded by a lower index of refraction cladding material to reduce optical loss with longitudinal (*z*-direction) wave propagation. A significant number of optical waveguides operate in the regime of total internal reflection (TIR) at the core interface. This condition is met when the angle of incidence (θ_i) is larger than the critical angle meaning, from Snell's law, the ray angles must be closer to the axis than θ_c, where from eqn (6.12), n_{clad} and n_{core} are the refractive indices of the cladding and core, respectively.

$$\theta_c = \sin^{-1}\left(\frac{n_{clad}}{n_{core}}\right) \qquad (6.12)$$

A distinction is made between reflection at a metallic surface and at a dielectric surface due to the different boundary conditions. In the former case the tangential electric field is zero, whereas in the latter case field propagation into the lower refractive index medium (*evanescent wave*) along with a shift in phase of the reflected wave is possible.

Micro- and nanofabrication processes have delivered a multitude of complex designs for planar waveguides, such as material coated (*i.e.* Teflon, metal) on silicon microchannels[59] and antiresonant reflecting optical waveguides (ARROWs)[60–62] to photonic crystal waveguides,[63] and liquid–liquid (L^2)

waveguides.[64,65] Categorised into index-guiding or interference-based wave-guides, fabrication and implementation of these on-chip structures, which include multilayered designs and interfacing of solid and liquid waveguides are discussed in significant depth elsewhere,[66,67] hence we limit our discussion to the more common structures with promise in LOC system integration.

6.3.2 Index-Guided Structures

6.3.2.1 Liquid Core/Liquid Cladding (L^2)

These waveguide configurations provide a number of distinct advantages over liquid core/solid cladding (LCW) or solid core/liquid cladding wave-guides fabricated in solid-based microfluidic channels.[2,68] First, L^2 waveguides are highly suitable for dynamic reconfiguration of the optical system's properties through a number of means, from laminar flow-rate differentials between core and cladding for single or multimode waveguides, to the tuning of physical and optical properties of the different fluids continually introduced into the system.[69,70] Secondly, utilising soft lithography technologies to create complex microfluidic networks, multiple streams of laminar fluid flows can be applied to switch the path of the L^2 waveguide, or couple multiple liquid core streams through manipulating the dimensions and optical properties of the liquid cladding.[71,72] Wolf *et al.* developed a number of nondeformable PDMS microfluidic networks, where light from an optical fibre was coupled into and out of the waveguide system (5 M CaCl$_2$ ($n_{core} = 1.445$) solution as the core sandwiched between deionised water ($n_{clad} = 1.335$)). The PDMS acted as a cladding layer ($n_{PDMS} = 1.40$) in the region where the optical fibre was coupled to the microfluidic channel, before the fluids are introduced. They demonstrated a variety of capabilities including mode switching through diffusive broadening at the core/clad interface; optical switching through significant variations in cladding flow rates from initial flow rate; and an evanescent coupler where parallel liquid cores, separated by a suitably thin inner liquid cladding, had light transferred from one waveguide to the other through overlapping of the evanescent fields. It was further shown that by controlling the flow rates of L^2 waveguides in parallel, but in the opposite direction of the coupled illumination, refractive-index gradients based on diffusion-controlled concentration gradients at the core-clad interface created an optical splitter, with the capacity to filter and generate multicolour light sources from a single white light source.[73] Chung and Erickson demonstrated an adaptable high-performance L^2 waveguide integrated with a fibre in and fibre out reconfigurable optofluidic system for a 1×2 optical switch and a tuneable attenuator.[70] They reported a coupling efficiency of 3.1 dB, with low crosstalk over a broad range of wavelengths. To limit diffusion across liquid streams, though compromising scattering losses, Conroy *et al.* introduced dielectric nanoparticles (30–300 nm diameter at ≤10% volume fraction) into the liquid core.[74] A symmetrical core cross section (100 μm × 100 μm), based on core and cladding flow-rates of 1 ml h^{-1} and 2 ml h^{-1}, respectively, gave maximal guiding efficiencies that were highly dependent on the NA of the input light and the volume fraction of the particles, decreasing as particle size increased.

The principles of L^2 waveguide operation have been applied in the creation of dye lasers and broadband and fluorescent light sources using single, cascaded and arrays of L^2 with an ethylene glycol (EG) or methanol core stream doped with mM concentrations of fluorophore such as rhodamine 6G ($\lambda_{max} = 596$ nm), and cladded with water or pure methanol to enhance optical tuning parameters.[64,69,75] Perpendicular irradiation (also termed light pumping), such as that shown in Figure 6.5(c), of a section of the waveguide microchannel enables light capture and guiding by the core stream, with an efficiency dependent on the contrast in refractive index between the core and the cladding ($\Delta n = n_{core} - n_{clad}$). Again, the issue of diffusion was found to affect the efficiency of the system, particularly below a threshold core flow-rate. By introducing the immiscible fluid silicone oil as the cladding

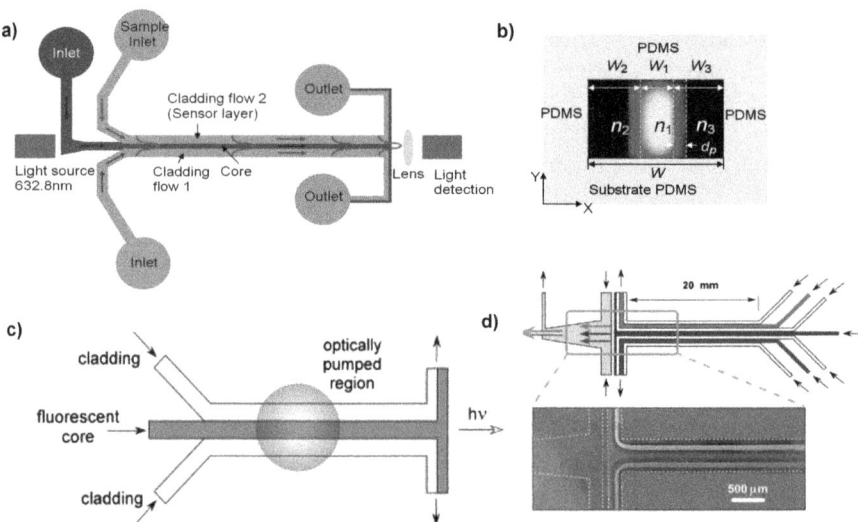

Figure 6.5   (a) Schematic of a L^2 waveguide sensor and (b) a cross-sectional view of the transparent window where $n_1 > n_3 \geq n_2$. Reprinted with permission from X. C. Li, J. Wu, A. Q. Liu *et al.*, *Appl. Phys. Lett.*, 2008, **93**, 193901. Copyright 2008, AIP Publishing LLC. (c) Illustration of a fluorescent L^2 waveguide with an optical pumping region situated downstream of the core and cladding channels; (d) Top-view scheme for the array of L^2 fluorescent light sources, consisting of parallel L^2 waveguides in a single PDMS microchannel. An end-coupled, tapered, liquid-core waveguide filled with DMSO collected the total fluorescence output. Inset: Optical micrograph of the T-junction. Dotted lines outline the walls of the PDMS channels. Reprinted with permission from B. T. Mayers, D. V. Vezenov, V. I. Vullev and G. M. Whitesides, *Anal. Chem.*, 2005, **77**, 1310–1316. Copyright 2005 American Chemical Society.

stream, smooth and distinguishable boundaries between core and clad were realised and flow rate tuning to limit diffusion was no longer warranted. Notable advantages reported with this fluorescent L^2 waveguide included the ability to couple light to the system without the need for alignment, whilst also generating intensity outputs comparable to standard fibre-optic spectrophotometer-based light sources. Although the dyes could be continually replenished, external pressure-driven flow techniques of the fluids was cited as a drawback of the system.

6.3.2.2 Liquid Core/Air Cladding (LA)

Lim *et al.* further demonstrated the ability to eliminate diffusional mixing through coupling light in a liquid core–air cladding system (LA), thus achieving stronger optical confinement due to a larger refractive index contrast in comparison to L^2 waveguide systems ($\Delta n_{L^2} \approx 0.1$, $\Delta n_{LA} \approx 0.432$).[58] The stratified laminar flow within this T-junction system consisted of fluorescently doped (*i.e.* rhodamine, fluorescein sodium salt and coumarine) EG liquid core pumped through using a syringe pump, while the air-cladding stream was driven by a vacuum suction with vacuum pressures ranging between 400–600 mmHg. To ensure a stable 2-phase stratified flow, the PDMS channel surfaces were treated with (heptadecafluoro-1,1,2,2-tetrahydrodecyl) trichlorosilane or FDTS, which reduced EG's affinity for PDMS. They reported capture fractions ($\eta = \Delta n/2n_{core}$)[75] two- to three-fold larger than traditional L^2 waveguides configurations, with propagation losses as low as 0.14 dB cm^{-1}. To overcome issues related to fluid resupplies and move a step closer to more practical uptake of optofluidic technologies, Jung and Erickson designed a reconfigurable hybrid optofluidic system (Figure 6.6) wherein

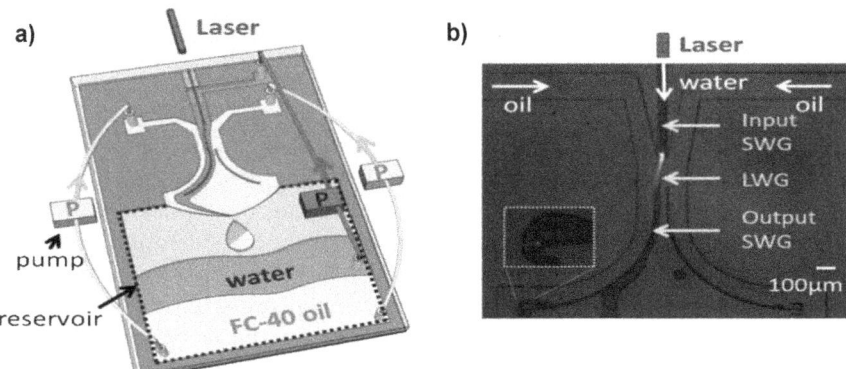

Figure 6.6 (a) Schematic showing the operation of the fluid recirculation system for the reconfigurable hybrid optofluidic system; (b) Microscopic view of the actual chip performing end-fire coupling between the liquid and solid waveguides with an inset showing the coupled light. (Input and output SWG = input and output solid-core waveguides, LWG = liquid-core optical waveguide). Reprinted from ref. 76.

liquid-state and solid-state on-chip photonic elements were end-fire coupled, and a fluidic recirculation system on a silicon substrate allowed for continuous (>20 h) optical switching at speeds of 1–5 s, with minimal degradation in performance and only 200 µl of fluidic consumption.[76] This milestone was achieved through the use of immiscible fluids and on-chip separation columns reducing fluid volumes typically needed over this period of operation by a factor of 200, and by demonstrating evanescent and efficient contact coupling between liquid and solid waveguides. The vast majority of liquid core waveguide systems within optically transparent PDMS microchannels have been demonstrated to be dynamically reconfigurable based on differential flow velocities of the cladding streams. This tends to produce liquid cores capable of size alterations in either the vertical or horizontal direction (*i.e.* 2D), but not both simultaneously. Lee *et al.* introduced a graded-index (GRIN) 3D L^2 waveguide, reducing optical losses between the liquid core and the microchannel walls through the application transverse secondary flow (*i.e.* Dean vortex) and parallel sheath flow for lateral and vertical focusing, respectively.[65] This system's core was reported to guide 5430 modes, accurately controlled by both sheath flow rates and the Dean number, thus providing reduced modal dispersions with improved constant intensity distribution along the axis of propagation for excitation and detection applications.

6.3.2.3   Liquid Core Waveguides (LCW)

A variety of approaches have been devised to improve the performance of integrated LCWs over traditional hollow glass fibres (n_{clad} = 1.52) with a bromobenzene liquid core (n_{core} = 1.56) system.[77] To reduce the refractive index of the capillary wall, characterisation, simulation and analyses of the chemically stable and optically transparent amorphous copolymer Teflon AF 2400 (n_{Teflon} = 1.29), coated on the inner wall of hollow capillary tubes (>5 µm thick), showed improved performance over uncoated capillary waveguides.[78–80] Subsequent efforts led to the fabrication of a bonded LCW device on silicon, with Teflon AF coated on the etched channel walls by Datta *et al.*[81] The two methods described for promoting adhesion of Teflon to silicon and glass were either plasma-enhanced chemical vapour deposition (PECVD) of thin-film fluorocarbon or the spin coating of perfluorooctyltrimethoxysilane solution followed by multiple spin coatings of Teflon layers. Although optical loss was observed in these etched microchannels compared to those of the smoother extruded tubes, the LOD of this silicon-based LCW was estimated to be of the order of 3 nM and represented a good level of detection compared to the cost of fabrication. Cho *et al.* demonstrated a novel process for coating PDMS-based (n_{PDMS} = 1.41) microfluidic channels with Teflon AF, thus eliminating the elasticity mismatch between the two materials while preserving optical transparency, and thus increasing the applications of LCW in highly complex microfluidic networks for on-chip integrated biosensing, µFACs and flow cytometry.[82–85] PDMS substrates were spin coated with an adhesion promoter (1*H*,1*H*,2*H*,2*H*-perfluorodecyltriethoxysilane) for Teflon AF and then bonded together *via* UV/ozone treatment, prior to a 6% Teflon AF solution being pumped through the channel. Excess Teflon AF solution was removed with a vacuum pump leaving a smooth cladding layer of thickness 5–15 μm, dependent on the applied pressure and Teflon AF solution concentration.

6.3.3 Interference-Guided Structures: ARROW Technology

In the context of optofluidic planar integration, Bragg fibres,[86] hollow-core photonic crystals fibres (HC-PCF),[87,88] Omni Guide fibres,[89] and 1D or 2D photonic crystals (PC)[63,90] are all systems designed such that the electromagnetic waves are localised through the use of wave interferences. In a somewhat similar fashion, antiresonant reflective optical waveguides (ARROWs) work on the principle that light is confined and propagates in low refractive index media (*i.e.* liquid or air core) *via* high-reflectivity Fabry–Perot-like resonators made up of multiple higher index cladding dielectric layers surrounding the core (*e.g.*, SiO_2–Si; SiO_2–SiN_4).[91] Crucially, unlike Bragg fibres in which the surrounding cladding layers are composed of periodic alternating lower and higher indexed material around a low-index core giving rise to complicated multilayer structures,[92] ARROWs do not rely upon periodicity of dielectric cladding material layers to attain low propagation losses but rather on refractive-index contrast and thickness, as modelled by Litchinitser *et al.*[87] Instead, any losses associated with the transverse component of the wave vector (k_T) are significantly reflected at the spectrally broad antiresonant wavelength at each successive cladding interface. This enables exceptionally high-efficiency propagation at the fundamental mode over large distances. Significant developments in planar ARROW design and fabrication has been accomplished by Hawkins and Schmidt and colleagues, some of which include hybrid micro- and nanopore ARROWS for electrical and fluorescence nanodetection applications; integration of low-loss single-mode liquid-core ARROWs (LC-ARROW) on a semiconductor chip for subnanometre-volume light guidance; developing smooth-walled, evenly distributed layer deposits and structurally more resilient hollow-core ARROWs possessing arch-shaped designs; integration of LC-ARROWs with liquid reservoirs and solid-core optical waveguides to collect and transport light in fluorescence correlation spectroscopy; improving interface transmission design using a self-aligning pedestal technique (SAP) for integration with silicon chip for sensing platforms; and on-chip surface enhanced Raman scattering for molecular detection and spectral filtering for biomolecular FRET detection resulting in an ~18-fold increase in *S*/*N* ratio.[60,61,93–99]

Although relatively young, ARROW technologies have evolved at a significant pace with a variety of novel systems fabricated for integration on to compact microfluidic systems. Based on rectangular LC-ARROWs (Figure 6.7), Bernini *et al.* designed and characterised a silicon-based integrated optofluidic asymmetric Mach–Zehnder interferometer (MZI), with

a)   b)

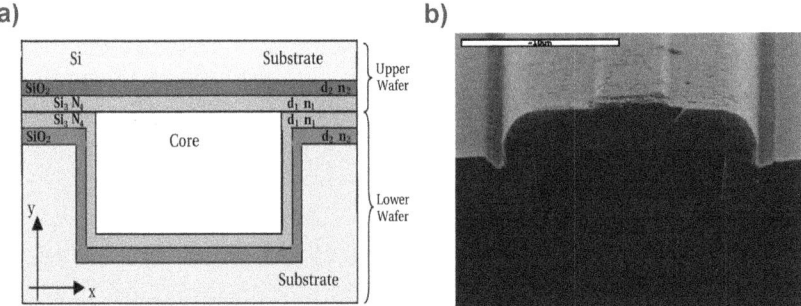

Figure 6.7 (a) Schematic cross section of a 2D hollow core arrow waveguide. Reprinted with permission from R. Bernini, G. Testa, L. Zeni and P. M. Sarro, *Appl. Phys. Lett.*, 2008, **93**, 011106. Copyright 2008, AIP Publishing LLC. (b) SEM image of hollow-core ARROW waveguide (core dimensions: 3.5 by 12 μm). Reprinted with permission from D. Yin, D. W. Deamer, H. Schmidt, J. P. Barber and A. R. Hawkins, *Appl. Phys. Lett.*, 2004, **85**, 3477–3479. Copyright 2004, AIP Publishing LLC.

a peak antiresonance at 633 nm, demonstrating strong interactions between the sample and the monomodaloptical field compared to solid-core configurations.[100,101] In addition to MZ interferometers, Testa *et al.* also demonstrated the application of this technology in fabricating 90°-bend rectangular waveguides on silicon, cladded with layers of TiO_2 and SiO_2 for enhanced reduction in bending losses, thus creating an optical ring resonator and a LC-ARROW multimodal interference (MMI) coupler for a balanced splitting ratio between bus waveguide and ring.[102] By using the waveguide to introduce a 0.11 nl sample of dimethylformamide ($n_{DMF} = 1.43$) as the liquid core, they showed an overall improvement in system integration and resonator performance in terms of both extinction ratio and quality factor.

More recent developments in practical ARROW integration on LOC systems have been accomplished by Ozcelik *et al.* for particle detection and spectral filtering.[103] Optically connected but fluidically isolated, LC-ARROWs (×2) in serial alignment on a planar optofluidic chip (see Figure 6.8) were fabricated to provide fluid-based tuneable spectral filtering capabilities for high-sensitivity fluorescence detection of nanobeads, in each section, respectively. Through a combination of pH and refractive-index tuning of the filter section's core liquid (EG and water mixtures), multiple-wavelength spectral ARROW filtering was achieved with a range of 90 nm. Nonpermanent interfacing of hybrid layers has also been reported for potential applications in fully integrated detection systems. Parks *et al.* vertically combined a PDMS-based microfluidic layer with a silicon-based liquid-core ARROW optical detection layer creating a single hybrid LOC optofluidic system.[104] In this simple yet reconfigurable system, capabilities in sample processing (*i.e.* mixing, distribution and filtering) and optical bioanalysis of single molecules in flow were demonstrated on a single chip, whilst also successfully

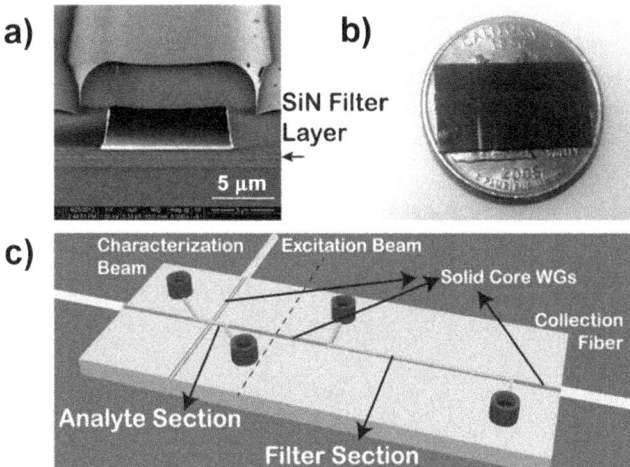

Figure 6.8 (a) SEM image of the liquid-core ARROW waveguide cross-section. Top and bottom ARROW layers are surrounding the hollow core. The thicker SiN filter layer is marked; (b) Photograph of complete chip; c) Schematic of an integrated optofluidic dual-channel chip with two different liquid-core sections (blue) for detection and tuneable filtering, optically coupled by a solid-core waveguide section. For characterisation of the filter section, the chip is cleaved along the dashed line and transmission spectra are collected from the other end of the liquid-core filter. For particle detection in the analyte section, the excitation beam is introduced with a single-mode fibre (white) from the solid-core waveguide perpendicular to the collection axis. The fluorescence signal is collected from either end of the chip. Reprinted from ref. 103.

maintaining high interlayer fluidic pressures. A recent demonstration of a 100% particle sorting system, based on the physical properties (*i.e.* size) of the inherent particles was reported by Leake *et al.*[105] Employing a network of LC-ARROWs configured in a "H" formation, particle mixtures were sorted into various streams using a combination of pressure-driven flow and optical forces, with relative particle size cut-offs being dictated by the applied flow speed and/or the power of the laser. This optofluidic scheme potentially lends itself to upstream processing techniques, such as sample pretreatment, where for instance, a separation operation of analytes in mixture can be effectively sorted based on size prior to entering a detection unit for further optical or electrochemical sensing.

6.4 Near-Field Photonics

The invention of optical tweezers by Ashkin and coworkers was a significant step in the development of advanced tools and techniques for the acceleration, trapping, detection and characterisation of macromolecules, colloidal systems and bioparticles using optical forces.[25,26] Although optical tweezers are highly advantageous at the microscale, limitations arise when NPs are the subject of manipulation. They are diffraction limited based on the fact that light cannot be confined to linear dimensions much smaller than $\lambda/2$.[106] Hence, the lower resolvable limit of an optical system can be calculated from the intensity distribution of the light beam (k_1), the numerical aperture of the objective (NA) and the wavelength. This critical dimension (CD) has lower limits of about 100 nm, based on commercially available objectives at near-UV irradiation.

$$CD = k_1 \frac{\lambda}{NA} \qquad (6.13)$$

In addition, from eqn (6.9), we see that the trapping force (F_{grad}) is proportional to the radius cubed of the particle of interest and the gradient of the optical field intensity; this makes handling many nanometre particles (*e.g.*, biomolecules, viruses, and metallic/dielectric nanoparticles) difficult with far-field optics.

Based on TIR of incident light propagating in a waveguide, the use of evanescent waves for nanomanipulation and detection is one approach employed to overcome the limitations associated with classical optics. Electromagnetic fields on the order of light wavelengths propagate close to the vicinity of interfaces, nanometre-sized structures and apertures, decaying exponentially with distance from the interface. Use of these evanescent waves is well evolved in optical-based sensor applications such as SERS, FRET and SPR systems (see Chapter 8). For a detailed discussion of near-field photonic forces experienced by subwavelength particles and for maximising the optical trapping force (to overcome Brownian motion) in planar waveguide designs refer to works by Nieto-Vesperinas *et al.* and Ng *et al.*[107,108] The use of evanescent waves from a laser beam to manipulate microscale particles and biological cells (*i.e.* RBC and yeast) were demonstrated by Kawata and Sugiura, and Gaugiran *et al.*, but we will focus our discussion here to optical-based particle nanomanipulation and advances in novel nanophotonic devices, for which Erickson *et al.* have provided a comprehensive review across experimental implementation to potential applications in microfluidics.[109–111]

6.4.1 Optofluidic Nanoslots

Experiments conducted by Kawata and Tani described the lateral trapping and longitudinal motion of Mie-scattering polystyrene particles and metallic microspheres in a channelled waveguide parallel to a glass surface.[112] Illumination of the waveguide with a cw Nd:YLF laser (@ ~80 mW), with an incident angle parallel to the glass surface, attracted dispersed particles (*via* gradient forces) just above the waveguide region towards the centre of the waveguide and then along the channel at speeds of 14 μm s^{-1} in water. Approximately 4 years later, colloidal Au-NPs (10 nm diameter) were reported to have been transversely trapped and then propelled on the glass surface of 3.5–5.5 μm channel width waveguides at speeds that were linearly dependent on the launch power.[113] With a maximum velocity of 4 μm s^{-1} observed, decreases were influenced by increases in channel width, while particle velocities were greater for TM polarisation than TE polarisation. By equating the Navier–Stokes drag force to the sum of the forward scattering ($F_{scat,z}$) and absorption ($F_{ab,z}$) forces, such that $F_z = F_{scat,z} + F_{ab,z}$, the forward velocity of the NP (v_p) on the waveguide surface can be approximated using eqn (6.14).[113] Compared to the static case, further developments saw Schmidt *et al.* perform optohydrodynamic trapping and transportation (@ ~28 μm s^{-1}) of dielectric spheres 3 μm in diameter from externally pumped crossmicroflows using evanescent waves of a planar SU-8 photonic structure (0.560 × 2.8 μm) combined with PDMS microfluidics on fused-silica substrate.[114]

$$v_p = \frac{F_z}{6\pi\eta r} \tag{6.14}$$

In contrast to guiding light within high index cores, researchers from Cornell University demonstrated light confinement and propagation in nanometre wide (slot) low-index materials, between high index slabs (*i.e.* SiO$_2$) on SOI wafers.[115,116] Baehr-Jones *et al.* showed high optical Q-factors (~27 000) could be obtained for ring resonators based on these nanoslot geometries, and the potential of these waveguide nanostructures as important candidates for use with novel low-index materials, circuit integration (*i.e.* electronic and photonic) on monolithic silicon substrates with applications in chemical and biological sensing applications.[117] These nanophotonic structures, as typically depicted in Figure 6.9 possess high E-field amplitudes and light intensities (∇I_0) higher than conventional waveguides, and the field enhancement provided by these devices showed an ~10^2-factor of improvement in trapping stiffness over plasmonic and optical tweezers. Whilst optical propulsion velocities for low refractive index contrast NPs (<50 nm) were extremely low, various system parameters were identified that would improve the operational efficiency of both trapping and transport speeds. Through 3D finite-difference time-domain simulations of a cavity-based PC nanoslotted waveguide, described by Lin *et al.*, a 1300-times enhancement of the trapping field resulted for NPs of radii down to 10 nm.[118] Consisting of tapered periodic 1D hole arrays of 100 nm radius on a SOI photonic wire waveguide, and a 20-nm nanoslot cavity at the centre, when combined enhanced optical trapping force by a factor of 6 compared to conventional waveguides. Although the ability of trapped DNA and macromolecules to overcome Brownian motion in the nanoslot cavity was a significant advantage for potential simultaneous biodetection applications, the enhanced intensity confined in the nanoslot gives rise to heightened damage of biological material over short time frames. Subwavelength slot waveguides (60–120 nm wide) were analysed and practically demonstrated by Yang *et al.*, in which tagged λ-DNA (48 kilobases) and dielectric NP were optically captured, trapped and transported using <300 mW, TE-polarised

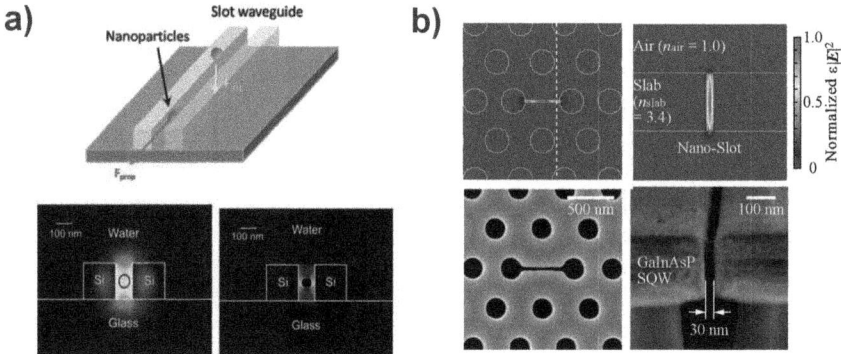

Figure 6.9   (a) Scheme of forces acting on nanoparticles in a slot waveguide (top) with optical field-intensity simulations on a 65-nm gold (bottom left) and 65-nm polystyrene particle. Reprinted with permission from A. H. J. Yang, T. Lerdsuchatawanich and D. Erickson, *Nano Lett.*, 2009, **9**, 1182–1188. Copyright 2009 American Chemical Society. (b) Simulation of the modal energy distribution within a nanoslot (top), with the fabricated device on a GaInAsP/InP single quantum-well wafer (SQW). Reprinted with permission from S. Kita, S. Hachuda, K. Nozaki and T. Babab, *Appl. Phys. Lett.*, 2010, **97**, 161108. Copyright 2004, AIP Publishing LLC.

light ($\lambda = 1550$ nm) within the high-intensity slot region.[119,120] Upon removal of the fibre-coupled optical power, the extended tagged λ-DNA was shown to be released from the extended length trap, as opposed to a focal point trap where supercoiling of the trapped DNA had previously been observed.

In addition to nanomanipulation, the enhancement of the field in the nanoslot has been utilised for sensing activities. Strong localised laser mode confinement in a 30-nm nanoslot device (Figure 6.9), fabricated in a 180-nm thick GaInAsP/InPPC wafer, showed its presence enhanced the sensitivity to the environmental index ($\Delta\lambda/\Delta n_{\text{env}}$), from air to liquid, with a sensitivity of ~350 nm per refractive index unit (RIU) and a spectral linewidth as narrow as 18 pm, giving an index resolution of approximately 4.4×10^{-5} RIU.[121] Kita *et al.* subsequently described the use of their nanoslot nanolaser in the processes of accelerating the adsorption of bovine serum albumin (BSA) unto the device for ultralow LOD label-free biosensing. The proposed mechanisms for observed trapping at the nanoslot included optical trapping forces from the enhanced localised field, fluid convection, or thermal gradient forces (also termed thermophoresis) from heating, though reports have suggested particle accumulation or depletion is correlated to the sign of the thermophoretic coefficient in addition to the particular system's operating parameters.[122,123]

A novel compact, chip-scale development for mid-IR spectrometry *via* an optonanofluidic slot waveguide (100 nm) system was applied for the label-free and surfacefunctionalisation-free sensing of chemicals and biomolecules (Figure 6.10).[124] The engineered Si-liquid-Si ($\Delta n \approx 2$) fluidic slot waveguide, connected at each end with a Si–SiO$_2$–Si slot waveguide and

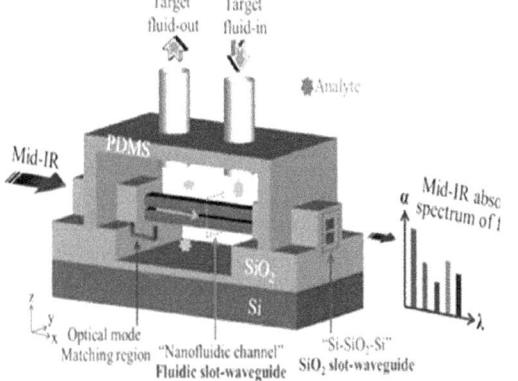

Figure 6.10 Schematic of a mid-IR optonanofluidic sensor composed of (i) a nanofluidic channel with a gap in which fluidic target analytes can fill (labelled as a fluidic slot-waveguide), (ii) a Si–SiO$_2$–Si structure, which acts as an optical mode matching region as well as a mechanical support (labelled as a SiO$_2$ slot-waveguide), and (iii) a PDMS fluidic chamber with an inlet and outlet for delivery of target analytes into and out of the fluidic slot waveguide. Liquid analyte is injected in/out of the chamber through plastic tubes connected to the PDMS chamber. This liquid fills the nanofluidic channel and it is where the optical absorption for spectrum scanning takes place. Mid-IR light is initially coupled into the front SiO$_2$ slot-waveguide and then passes through the optical mode-matching region, the fluidic-slot waveguide (nanofluidic channel), and finally into the SiO$_2$ slot-waveguide on the exit end. Light transmitted from the waveguide edge encodes the absorption spectrum of the analytes. Reprinted with permission from P. T. Lin, S. W. Kwok, H.-Y. G. Lin, V. Singh, L. C. Kimerling, G. M. Whitesides and A. Agarwal, *Nano Lett.*, 2013, **14**, 231–238. Copyright 2014 American Chemical Society.

embedded in the centre of a PDMS microfluidic chamber, was able to differentiate between several organic liquids; *n*-bromohexane (R–Br), iso-propanol (R–OH) and toluene (Ar–CH$_3$), and determine the molar fractions of binary mixtures from the characteristic absorption wavelength of the presented functional group, detected by a mid-IR camera at the Si–SiO$_2$–Si slot waveguide exit. Functionally, the fluidic slot waveguide provides lossless propagation of the mid-IR light waves, whilst simultaneously serving as the detection zone for which there is a maximal overlap with the optical field and the analyte, enhancing sensitivity by up to 50× relative to strip-waveguide evanescent wave sensing.

6.4.2 Whispering-Gallery Mode Resonators

Over the past two decades, high-Q optical microresonators have evolved as a powerful technology in chemical and biochemical sensing to cellular and biomolecular assay developments, due in large part to advancements in a multitude of micro- and nanofabrication capabilities for on-chip integration.[125–127] Serpengüzel *et al.* demonstrated the ability to externally excite polystyrene microspheres at resonance, with negligible off-resonance scattering, *via* a single-mode optical fibre coupled to a tuneable linearly polarised cw dye laser (linewidth = 0.025 nm).[128] They proposed prospective applications of these resonant microspheres in diverse fields from microphotonics to adsorption and chemical reaction sensing between biofunctionalised species on the microsphere's surface; with surface additions at the molecular level producing detectable shifts of a given resonance frequency.[129]

Resonators operating in whispering-gallery mode (WGM) are characterised by high Q factors; exhibiting low optical losses and enhanced confinement of recirculating photons. The average time (τ) a photon takes to circulate in a mode is proportional to Q and inversely proportional to the resonant frequency (*i.e.* $\tau = Q/\omega_r$).[130] Evanescent coupled light from an optical fibre to a silica microsphere circulates around the microsphere in WGM (Figure 6.11). The resonant mode is detected from dips in the transmittance through the optical fibre (Figure 6.11). Biomaterials (*e.g.*, DNA, bacteria, protein) adsorbed on the surface of the microsphere interact with the WGM

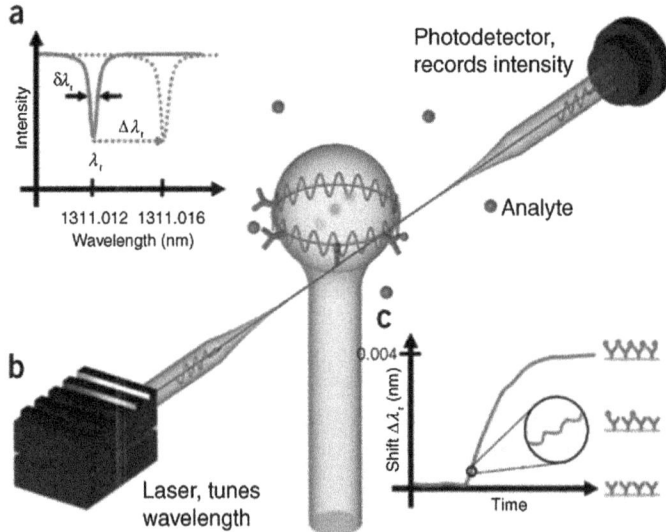

Figure 6.11   WGM biosensing: (a) Dips in transmission spectrum at a specific wavelength from a tuneable laser indicate the resonance wavelength (λ_r). Whereas shifts in λ_r are associated with molecular binding. (b) WGM in a dielectric sphere driven by evanescent coupling to a tapered optical fibre, with the light wave in red circumnavigating the surface on the glass sphere (blue) where binding of analytes to the surface immobilised antibodies is detected by shifts in λ_r, as depicted in (c). Reprinted by permission from Macmillan Publishers Ltd: Nature Methods (F. Vollmer and S. Arnold, *Nat. Methods*, 2008, **5**, 591–596), copyright (2008).

evanescent field, polarising the molecule and shifting the frequency of the mode, making these devices exceptionally sensitive to binding of single molecules or analytes detected through shifts in resonant wavelength, $\Delta\lambda_r$.[129,131,132] The excess polarisation, α_{ex}, of a bound molecule, due to its in-phase reactivity with the WGM, is proportional to the shift in resonance frequency and inversely proportional to the size of the microsphere (eqn (6.15)), where σ is the average surface density of bound material, ε_0 is the permittivity of free space, R is the sphere radius, n_s and n_m refer to the refractive indices of the sphere and medium, respectively.

$$\frac{\Delta\lambda_r}{\lambda_r} = \frac{\alpha_{ex}\sigma}{\varepsilon_0\left(n_s^2 - n_m^2\right)R} \tag{6.15}$$

For a more quantifiable approach to the sensitivity of the microsphere with binding materials, the Q factor ($= \lambda_r/\gamma_r$), where γ_r is the resonance linewidth, plays an important role in the determination of the smallest measurable frequency shift or "measurement acuity factor" ($F = \Delta\lambda_r/\gamma_r$) to changes in surface density. Eqn (6.16) describes this limit of detection (LOD) due to surface coverage, and it can be seen that to attain a high sensitivity, the R/Q ratio must be minimised.[133]

$$\sigma_{LOD} = \frac{R\left(n_s^2 - n_m^2\right)F}{(\alpha_{ex}/\varepsilon_0)Q} \tag{6.16}$$

One of the first reports for fabricating planar microresonators, with cavity Q factors in excess of 100 million was published in 2003 by Armani *et al.*[134] (Figure 6.12). Through a combination of lithography, dry etching and a selective reflow process, they managed to produce ~100 μm diameter toroid-shaped microresonators on a silicon wafer, with high Q factors only surpassed by microspherical resonators with attainable Q factors of ~10^{10} in WGM.[135] Although light coupling to the toroidal cavity was achieved through a freely movable tapered optical fibre waveguide, similar to that of the microsphere,[128] significant achievements in planar waveguide–ring resonator integration have since been accomplished, with variable degrees of Q values.[116,117,136–138] In general, within resonant cavities, the Q of the cavity gives a measure of the sharpness of the cavity to external excitation and can be defined as in eqn (6.17), where ω_r is the resonance frequency assuming no losses.[42]

$$Q = \omega_r\frac{\text{stored energy}}{\text{power loss}} \tag{6.17}$$

Similar to microsphere resonators, other resonator designs such as liquid-core optical ring resonators (LCORR),[139,140] microrings,[136,141] microdisks,[142,143] and microtoroids,[144] used predominantly as optofluidic WGM sensors, have been fabricated and demonstrated at chip-level integration for biological and chemical applications, including real-time DNA hybridisation monitoring *via* fluorescent-label excitation of ssDNA from the evanescent field of the

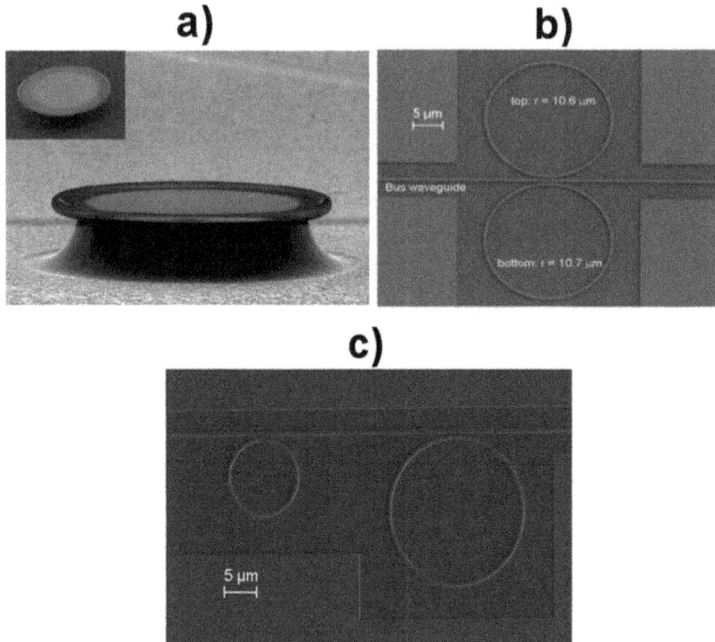

Figure 6.12   Planar ring microresonators: (a) Scanning electron micrograph of a silica microdisk toroidal microresonator with intrinsic cavity Q of 1.00×10^8. Reprinted by permission from Macmillan Publishers Ltd: Nature (D. K. Armani, T. J. Kippenberg, S. M. Spillane and K. J. Vahala, *Nature*, 2003, **421**, 925–928) copyright (2003). SEM images of two-way particle storage chain; (b) Opposite double-ring microresonator structures, and (c) adjacent microrings. Reprinted from ref. 158.

microtoroidal resonator.[145] Luchansky and Bailey have given an excellent review on these micro/nanostructures and their applications in bio/chemical optical sensing research and development between 2009 and 2011.[146] The following section focuses on the use of resonator structures in particle-handling operations using optical fields.

6.4.3   Optical Nanomanipulation

Active optical particle manipulation towards sensing regions enhances rates of detection over passive processes such as adsorption. This was demonstrated by Arnold *et al.*, who described the active transportation of dielectric NPs to the sensing volume of a WGM oblate microspherical silica resonator, *via* optical evanescent attraction forces, enhancing detection rates by about a factor of one hundred.[147] Furthermore, it was observed that these prebound NPs circumnavigated the microsphere along the radial equatorial plane, through multiple orbits ("carousel phenomenon"), and in the same direction as the light in the WGM. To obtain a bound state, altering system properties to reduce the electrostatic repulsion through increasing solution conductivity would enable closer interactions with the microsphere's surface. This short-range repulsive interaction (U_s) is independent of power and influenced by the surface properties. A distance from the microsphere's surface the polarisation potential, U_p, describes the long-range attractive interactions of the evanescent field with the NP, similar to the gradient forces of optical tweezers that is proportional to the negative of the NP's excess polarisability.[147] Empirical studies suggested that at the surface the polarisation potential ($U_p(0)$) is independent of solution conductivity, but proportional to the power (P) entering the mode that can be directly calculated from the maximum shift in resonance wavelength (eqn (6.18)). The minimum power (P_{min}) observed for particle trapping from rearranging eqn (6.18), assuming, $|U_p(0)| \approx k_B T$, was approximately 7.3 µW. This corresponded to a reliable trapping operation, for several minutes, when $U_p(0) > 2k_B T$. This trapping limit enabled the NPs to be assessed from analysis of fluctuations in the microcavity's resonance frequency, thus providing information relating to the size and mass of the trapped particle.[147]

$$U_p(0) = -(\Delta \lambda_r)_{max} \frac{PQ}{2\pi c} \tag{6.18}$$

Nanomanipulations on planar resonator structures have recently been demonstrated. Mandal and Erickson designed an optical trap based on standing-wave optical fields generated from a 1D silicon photonic crystal (PC), evanescently coupled to a single-mode waveguide.[148] Apart from using PCs as tools for enhancing optical manipulation of NPs, they are also being increasingly applied for potential uses in the amplification of label-free biomolecular recognition strategies.[149] Briefly, 1-, 2- and 3-dimensional PCs rely upon the patterned periodicity of high refractive index contrasts, which can create a range of forbidden photon modes called a photonic bandgap (PBG) or stop bands where light of a specific frequency range falling within the PBG cannot propagate through the PC; this is analogous to the electronic bandgap of semiconductors.[150] The introduction of an imperfection or defect to the PC attempts to localise and trap light, which for the case of a line defect enables practically lossless guiding of light, even around extremely tight bends. On the other hand, a regional defect such as a vacancy in an array of holes on the 1D PC can lead to a strongly localised state in the bandgap. This effectively creates a microcavity and confines the supported mode to within the bandgap, with frequency tuning achieved through varying the size of the defect (Figure 6.13).[151]

Mandal *et al.* demonstrated the tightly confined optical traps generated by the microcavity enabled trapping of NPs as small as 48 nm in diameter.[152] Along with the simultaneous transportation of particles along the waveguide surface through light propagation, the ability to tune the nanotweezer's functionalities by altering the light mode or wavelength, or on/off status allowed for particles to be released into the microfluidic flow. Due to the stronger optical field of the resonator over the adjacent waveguide, particles

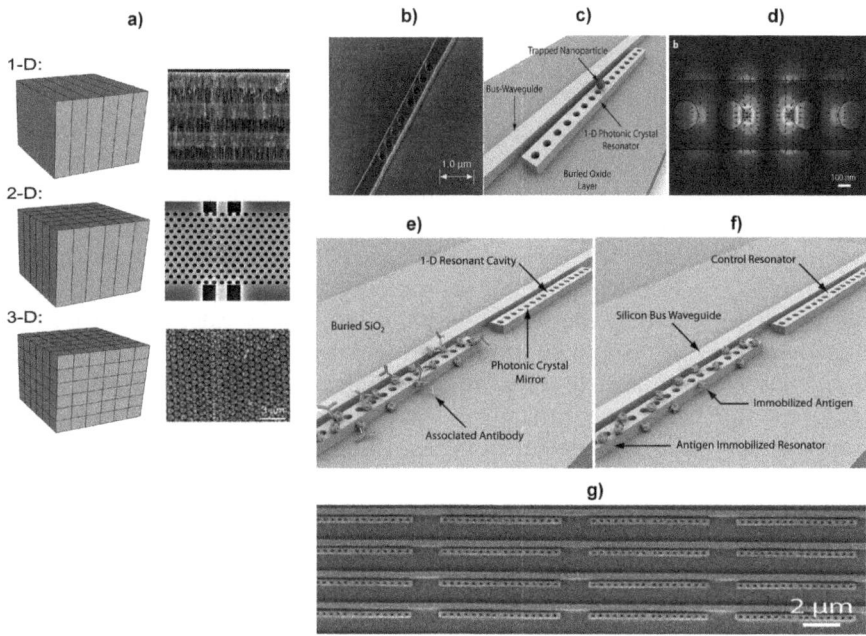

Figure 6.13 (a) 1-, 2-, and 3D photonic crystals (PC). Schematics on the left indicate the alternating patterns of high and low refractive index materials, while scanning electron micrographs of representative examples of each type of PC are on the right (top: a Bragg mirror in porous silicon; middle: a 2D PC created on a silicon-on-insulator chip *via* E-beam lithography; bottom: top surface of an inverse 3D copper(I) oxide PC. Reprinted with permission from S. Pal, P. M. Fauchet and B. L. Miller, *Anal. Chem.*, 2012, **84**, 8900–8908. Copyright 2012 American Chemical Society. (b) Scanning electron micrograph of a PBG waveguide microcavity fabricated by X-ray lithography. Reprinted by permission from Macmillan Publishers Ltd: Nature (J. S. Foresi, P. R. Villeneuve, J. Ferrera, E. R. Thoen, G. Steinmeyer, S. Fan, J. D. Joannopoulos, L. C. Kimerling, H. I. Smith and E. P. Ippen, *Nature*, 1997, **390**, 143–145) copyright (1997). (c) 3D schematic of the one-dimensional photonic crystal resonator optical trapping architecture. (d) 3D FEM simulation illustrating the strong field confinement and amplification within the one-dimensional resonator cavity. The black arrows indicate the direction and magnitude of the local optical forces. Reprinted with permission from S. Mandal, X. Serey and D. Erickson, *Nano Lett.*, 2010, **10**, 99–104. Copyright 2010 American Chemical Society. Nanoscale Optofluidic Sensor Arrays. (e) 3D rendering of the NOSA device showing two 1D photonic crystal resonators evanescently coupled to a silicon bus waveguide. The first resonator is immobilised with an antigen whereas the second resonator acts as a control. (f) 3D rendering illustrating the association of the corresponding antibody to the antigen immobilised resonator. (g) SEM image demonstrating the 2-dimensional multiplexing capability of the NOSA architecture. Reprinted from ref. 153.

"hop" over from the waveguide surface to the resonator when on-resonance is operated for trapping, and back to the waveguide through microfluidic flow, when off-resonance is operating. Characterisation of the optical trap's stiffness suggested at least an order of magnitude improvement over all other reported optical trapping systems for a 100-nm NP. The application of this optofluidic system, for multiplexed label-free sensing of immunological entities (IL-4, IL-6 and IL-8), was described using a nanoscale optofluidic sensor array (NOSA).[153] Through SOI wafer fabrication processes, the array architecture consisted of multiple resonators of unique resonance wavelengths (based on the size of the cavity spacing) adjacent to a single-mode waveguide. Targeted binding to the biofunctionalised resonator increased the refractive index of the optical cavity, thereby altering the resonant wavelength in the cavity. This is detected in real time by a shift in the bus waveguide output frequency spectra, with LODs as low as 63 attograms reported with this system. Optofluidic devices based on the 1D PC designs are being commercialised under the name of NanoTweezers,[154] by Optofluidic, Inc. (http://www.optofluidicscorp.com), for protein aggregate analysis and submicrometre particle imaging.

Yang and Erickson described an optical switch employed to manipulate particles travelling along the bus waveguide's evanescent fields and a microring resonator through on/off resonance tuning of the microring (Figure 6.14).[155] Similarly, Cai and Poon demonstrated the ability to optically route 1-μm-sized polystyrene particles *via* travelling waves, on and off silicon nitride waveguides evanescently coupled with a microring resonator, showing the ability of the integrated device as a highly effective circuit particle filter.[156] Furthermore, a planar microring cavity trapping system capable of transporting a trapped particle round the circumference of 5 and 10 μm radii ring resonators at constant velocity was described by Lin *et al.*[157] Fabricated on a SOI wafer combined with a PDMS microfluidic channel bonded to the resonator structure, a bus waveguide excited into TM-mode coupled light into the microring over a separation distance of 150 nm. The field-intensity enhancement in the microring, derived from the transmission coefficient of the resonant frequency and the finesse, were 1.76 and 1.50 for the 10 μm and 5 μm rings, respectively. In addition, a 5–8 times enhancement in optical force kept the 500 nm diameter NP stably trapped in a potential well of $9.3k_BT$ ($P = 9$ mW), with a maximum velocity of 180 μm s^{-1} over several minutes. The system was further developed to demonstrate a switchable particle storage chain for trapping and sensing applications using multiple microrings of variable resonance wavelengths, enabling particle counting without the need of imaging systems.[158] Through integration of microring resonator and a tapered PC cavity, a reusable trapping and sensing platform for green fluorescent protein (GFP) detection was created, eliminating the need for surface biofunctionalisations.[159] Instead, functionalised carrier particles coated with antibodies create clusters in solution that are subsequently trapped and then sensed by the microcavity through histogram analysis of concentration dependent resonant shifts. Although

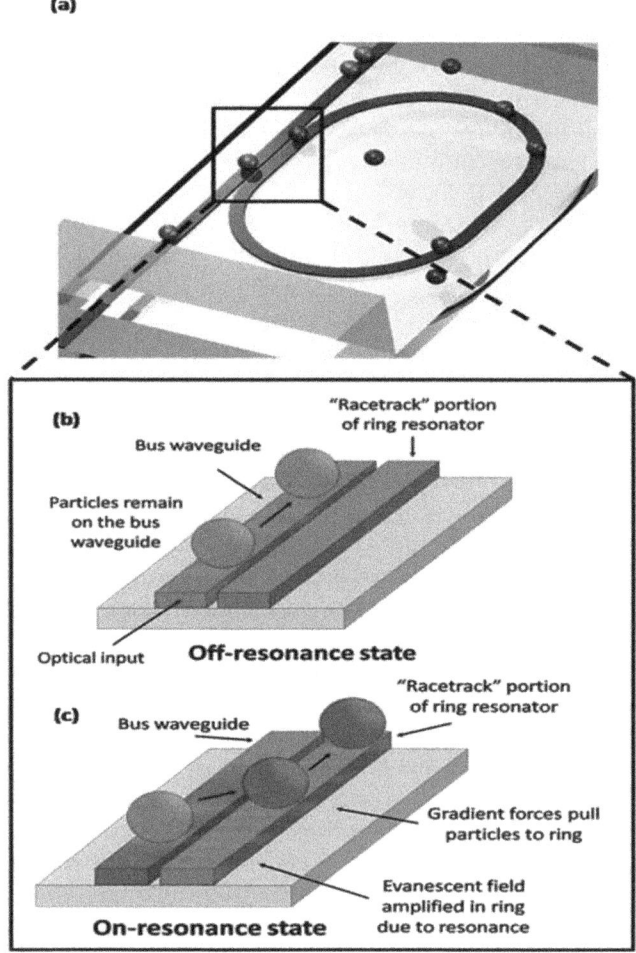

Figure 6.14   Schematic of optofluidic ring resonator switch. (a) Rendered picture of device with PDMS microfluidics. (b) and (c) Illustration of switching mechanism due to optical gradient forces when the ring is strongly coupled at the resonant wavelength. Reprinted from ref. 155.

this system does demonstrate multiple advantages over other individual resonator systems, particularly for areas such as environmental monitoring, its drawback is the lack of information given with regards to binding kinetics, as clusters are formed prior to transportation to the sensing region.

6.5   Optically Induced Electrokinetics

A relatively new class of optical particle manipulation technology is rapidly evolving, in part due to the coupling of other effective manipulation micro-technologies and the integration of advanced photosensitive materials in the microsystem design. Broadly termed *optoelectrofluidics*, it generally encompasses one of two operations, namely, (1) direct alteration of the fluid property by light, or (2) alteration of a surface's conductivity by light. Although optically induced thermal gradients have been shown to induce electrothermal fluid microvortices aiding particle trapping,[160] and in photothermal NP-based colloidal systems effecting fluid motion selectively driving suspended particles around microfluidic networks at controllable speeds and direction,[161] an overwhelming focus on programmable particle manipulation has been realised through surface conductivity alterations by light in platforms known as optoelectronic tweezers (OET), sometimes also referred to as optically induced dielectrophoresis (ODEP), introduced by Chiou *et al.*[162] Although optically induced electrophoretic, electrowetting, AC electro-osmotic, and electrothermal strategies can be designed, based on the operating parameters and setup inherent to each system, dielectrophoretic principles have been the most commonly used to date.[163,164]

6.5.1 Optoelectronic Tweezers (OET)

High-resolution virtual electrodes optically patterned on feature-less photoconductive materials, in dynamically reconfigurable systems, has enabled the generation of high field gradients from low-light power sources. These electric-field gradients act upon the suspended particles in a manner similar to dielectrophoresis (see Chapter 3), such that the field gradients interacting with a particle generate either an attractive (positive OET) or a repulsive (negative OET) force, of which the observed magnitude is dependent on the relative polarisability of the particle and surrounding medium, and the magnitude of the imposed electric field gradient. This can be explained by the increased voltage drop across the fluidic volume, relative to that across the illuminated photoconductive material, consequently generating the nonuniform electric field. A range of photoactive alternatives have been reported, including lithium niobate crystals,[165] doped and undoped hydrogenated amorphous silicon,[162,166] bismuth silicon oxide crystals,[167] and a thin-film bulk heterojunction polymer.[168,169]

The basic components of an OET system generally consists of transparent electrodes (usually AC biased) such as indium tin oxide (ITO) and a photoconductive layer deposited onto the ITO, a display device and a light source (Figure 6.15). Depending on the photoconductor used, fabrication of the OET system, planar or 3D, can vary in complexity and cost also. For instance the use of single-crystalline bipolar junction transistor (BJT) as the photoconductor layer has shown improved operational capabilities, particularly with physiologically relevant conductive fluids, but the fabrication process is more complex than the spin-coating process of the light-sensitive organic dye titanium oxide phthalocyanine (TiOPc) that has been shown to be effective at patterning HepG2 cells in a regular array (see Figure 6.16).[170,171] Light projection of the virtual electrodes, onto selective areas of the photoconductor, is accomplished through the use of a photomask and objective lenses for size

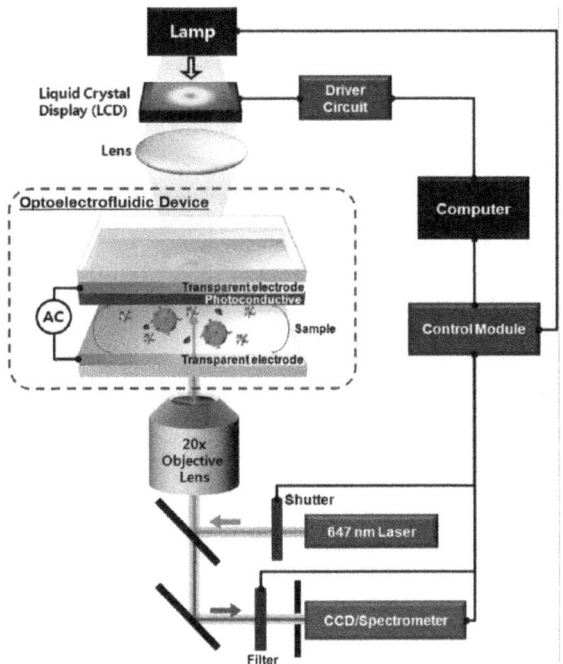

Figure 6.15 An experimental setup for optoelectrofluidic sandwich immunoassay.
A conventional inverted fluorescence microscope was applied to
construct the optoelectrofluidic immunoassay system. An LCD
module was mounted under the illumination lamp to generate
a programmed image pattern for controlling an electric field in
a conventional optoelectrofluidic device. Reprinted with permission
from H. Hwang, H. Chon, J. Choo and J.-K. Park, *Anal. Chem.*, 2010,
82, 7603–7610. Copyright 2010 American Chemical Society.

reduction. Spatial light modulation of the image, for programmable particle
manipulation, can be achieved through display devices such as a digital
micromirror display (DMD),[162] a beam projector,[169] or a liquid-crystal display
(LCD).[166,172] The advantages of an LCD-OET based system are the direct wide-
area patterning of images onto the photoconductor, eliminating the need for
lenses and thus misalignment issues. Also, the compactness of LCDs and
redundancy of optical elements lends this type of configuration as a portable
system since the alignment issues are no longer presented. A limiting issue
with the LCD-based system is light diffraction causing blurred virtual electrodes. To overcome this issue and enable more effective particle manipulation
Hwang *et al.* developed a lens-integrated LCD-OET system (such as that
depicted in Figure 6.15), demonstrating its improved particle handling
capabilities on microbeads and red blood cells (RBCs).[173]

    With regards to energising the system, Chiou *et al.* reported a $\times 10^5$
reduction in power requirements compared to optical tweezers, to enable
particle motion.[162] They used light-emitting diodes to turn on the virtual

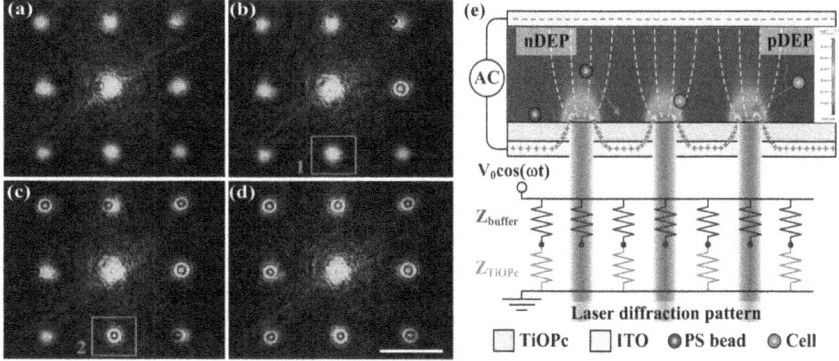

Figure 6.16 (a) to (d) laser diffraction pattern generates virtual electrode spots on the TiOPc substrate. Nine HepG2 cells are attracted toward each spot centre due to positive DEP forces, recorded at 0 s (a), 7 s (b), 17 s (c) and 26 s (d). (e) A schematic diagram of DEP electric-field simulation and the relative alternating current bias model for the sandwiched liquid medium and the organic photoconductive layer. Without illumination, most of the voltage drops across the TiOPc layer. As the light is projected onto the photoconductive layer, its impedance is reduced and the charges pass through the TiOPc layer within the illuminating region to form a virtual electrode on the bottom substrate. Then, most voltage drops across the medium. The nonuniform electric field is generated *via* the large ITO electrode on the top cover and the small virtual electrode at the bottom. The yellow dotted lines illustrate the electric-field line. The colour simulation reveals the AC electric-field (E^2) distribution that induces the DEP phenomenon. The positive DEP force attracts the HepG2 cell (red) toward the illuminating spot and the negative DEP force repels the polystyrene particle (blue) to the darker regions of the laser diffraction pattern. Reprinted from ref. 171.

electrodes, though others have used lasers and halogen lamps of conventional microscopes with success. The vast array of demonstrated OET applications, ranging from nanoparticulates to biological cells handled has been succinctly summarised by Hwang and Park.[174] Due to its inherently simple structure, hybrid-OET systems are becoming more prominent, overcoming earlier challenges (*e.g.*, closed microchannel integration) once encountered through integration of various structures and microelements to create systems which can be of real significance in detection processes, with low manufacturing costs across a range of industries.

6.6 Summary

The role of optofluidic technologies in laboratory-on-a-chip systems, with a focus on particle-manipulation methods and the integration of planar elements for light guidance and resonance-based sensing has been introduced. The developments in the field of optofluidics over the past two decades is wide ranging, with scaled-down devices demonstrating cost efficiencies in various processes ranging from energy production screening in photocatalytic microreactors,[175,176] to in-field detection of pathogens and hazardous substances in the healthcare, environment and food industries.[177] Factors that have contributed to this field's prominent rise include, on-chip integration of optofluidic elements such as tuneable lenses for beam manipulation,[55,178] switches for optical routing in microfluidic circuits through differing mechanisms (*e.g.*, acoustic, hydrodynamic),[179,180] and optical filters in detection systems.[97,103,181] They have played a major role in realising the potential of inexpensive all-optical microfluidic systems. Secondly, the advances in the use of a range of materials, including clear elastomeric (*e.g.*, PDMS) materials combined with variable actuation mechanisms to create more complex and reconfigurable optofluidic devices,[182] whilst also exhibiting the capacity to be integrated with traditional semiconductor substrates (*e.g.*, silicon) has spurred on innovation in applications, compactness and technology coupling. In addition, nanoplasmonic-based optofluidic devices are showing extremely high promise in a number of areas including using nanohole arrays to concentrate and detect analytes,[183–185] nanosensing of cancer markers and proteins using a nanoplasmonic-photonic hybrid microcavity,[186] immunological assays of live viruses from biological media,[187] and a variety of surface plasmon nanoparticle manipulation systems recently reviewed by Juan *et al.*[188] The costly fabrication processes of these complex metallic-based nanosensing platforms may pose greater challenges than the method for incorporation into microfluidic devices. But with potential LODs significantly greater than current limits, and with the reduced volume of sample required, cost *versus* benefits of these systems will play a key role in deciding the future uptake of these devices as commercial ventures start emerging.

References

1. J. Gluckstad, *Nat. Mater.,* 2004, **3**, 9–10.
2. P. Domachuk, M. Cronin-Golomb, B. J. Eggleton, S. Mutzenich, G. Rosengarten and A. Mitchell, *Opt. Exp.,* 2005, **13**, 7265–7275.
3. A. Abbas, M. J. Linman and Q. A. Cheng, *Biosens. Bioelectron.,* 2011, **26**, 1815–1824.
4. K. D. Patel and T. D. Perroud, *Miniaturized Sorters: Optical Micro Fluorescence Activated Cell Sorter*, Pan Stanford Publishing Pte Ltd, Singapore, 2010.
5. E. Verpoorte, *Lab Chip,* 2003, **3**, 60N–68N.
6. K. Kneipp, Y. Wang, H. Kneipp, L. T. Perelman, I. Itzkan, R. Dasari and M. S. Feld, *Phys. Rev. Lett.,* 1997, **78**, 1667–1670.
7. A. Campion and P. Kambhampati, *Chem. Soc. Rev.,* 1998, **27**, 241–250.
8. A. Y. Fu, C. Spence, A. Scherer, F. H. Arnold and S. R. Quake, *Nat. Biotechnol.,* 1999, **17**, 1109–1111.

9. Y. Huh, A. Chung and D. Erickson, *Microfluid. Nanofluid.*, 2009, **6**, 285–297.

10. E. Verpoorte, *Lab Chip*, 2003, **3**, 42N–52N.

11. S. Balslev, A. M. Jorgensen, B. Bilenberg, K. B. Mogensen, D. Snakenborg, O. Geschke, J. P. Kutter and A. Kristensen, *Lab Chip*, 2006, **6**, 213–217.

12. J. C. McDonald, M. L. Chabinyc, S. J. Metallo, J. R. Anderson, A. D. Stroock and G. M. Whitesides, *Anal. Chem.*, 2002, **74**, 1537–1545.

13. J. C. McDonald, D. C. Duffy, J. R. Anderson, D. T. Chiu, H. K. Wu, O. J. A. Schueller and G. M. Whitesides, *Electrophoresis*, 2000, **21**, 27–40.

14. G. M. Whitesides, E. Ostuni, S. Takayama, X. Y. Jiang and D. E. Ingber, *Annu. Rev. Biomed. Eng.*, 2001, **3**, 335–373.

15. A. C. Siegel, S. K. Y. Tang, C. A. Nijhuis, M. Hashimoto, S. T. Phillips, M. D. Dickey and G. M. Whitesides, *Acc. Chem. Res.*, 2009, **43**, 518–528.

16. H. Becker and L. E. Locascio, *Talanta*, 2002, **56**, 267–287.

17. M. Li, H. Gou, I. Al-Ogaidi and N. Wu, *ACS Sustain. Chem. Eng.*, 2013, **1**(7), 713–723.

18. X. D. Fan and I. M. White, *Nat. Photonics*, 2011, **5**, 591–597.

19. D. Psaltis, S. R. Quake and C. H. Yang, *Nature*, 2006, **442**, 381–386.

20. J. Wu, G. Zheng and L. M. Lee, *Lab Chip*, 2012, **12**, 3566–3575.

21. Q. Chen, X. Zhang, Y. Sun, M. Ritt, S. Sivaramakrishnan and X. Fan, *Lab Chip*, 2013, **13**, 2679–2681.

22. *Handbook of Optofluidics*, CRC Press, 2010.

23. A. Ashkin, J. M. Dziedzic and T. Yamane, *Nature*, 1987, **330**, 769–771.

24. A. Ashkin, *Phys. Rev. Lett.*, 1970, **24**, 156.

25. A. Ashkin, J. M. Dziedzic, J. E. Bjorkholm and S. Chu, *Opt. Lett.*, 1986, **11**, 288–290.

26. A. Ashkin and J. M. Dziedzic, *Science*, 1987, **235**, 1517–1520.

27. N. Neve, S. S. Kohles, S. R. Winn and D. C. Tretheway, *Cell. Mol. Bioeng.*, 2010, **3**, 213–228.

28. E. Eriksson, J. Scrimgeour, A. Graneli, K. Ramser, R. Wellander, J. Enger, D. Hanstrop and M. Goksor, *J. Opt. A: Pure Appl. Opt.*, 2007, **9**, S113–S121.

29. G. Thalhammer, R. Steiger, M. Meinschad, M. Hill, S. Bernet and M. Ritsch-Marte, *Biomed. Opt. Express*, 2011, **2**, 2859–2870.

30. S. Chu, *Science*, 1991, **253**, 861–866.

31. A. Ashkin, *Proc. Natl. Acad. Sci. U. S. A.*, 1997, **94**, 4853–4860.

32. A. Ashkin, *IEEE J. Sel. Top. Quantum Electron.*, 2000, **6**, 841–856.

33. Q. Lu, A. Terray, G. E. Collins and S. J. Hart, *Lab Chip*, 2012, **12**, 1128–1134.

34. A. Ashkin and J. M. Dziedzic, *Appl. Phys. Lett.*, 1974, **24**, 586–588.

35. A. Ashkin and J. M. Dziedzic, *Appl. Phys. Lett.*, 1971, **19**, 283.

36. K. Svoboda and S. M. Block, *Annu. Rev. Biophys. Biomol. Struct.*, 1994, **23**, 247–285.

37. A. Ashkin, *Biophys. J.*, 1992, **61**, 569–582.

38. W. H. Wright, G. J. Sonek and M. W. Berns, *Appl. Opt.*, 1994, **33**, 1735–1748.

39. K. C. Neuman and S. M. Block, *Rev. Sci. Instrum.*, 2004, **75**, 2787–2809.

40. S. M. Block, L. S. B. Goldstein and B. J. Schnapp, *Nature*, 1990, **348**, 348–352.
41. A. Mazolli, P. A. M. Neto and H. M. Nussenzveig, *Proc. R. Soc. London, Ser. A*, 2003, **459**, 3021–3041.
42. J. D. Jackson, *Classical Electrodynamics*, John Wiley & Sons, Inc., USA, 3rd edn, 1999.
43. J. P. Barton, D. R. Alexander and S. A. Schaub, *J. Appl. Phys.*, 1988, **64**, 1632–1639.
44. J. P. Barton, *Appl. Opt.*, 1995, **34**, 5542–5551.
45. J. P. Gordon, *Phys. Rev. A*, 1973, **8**, 14–21.
46. M. Kerker, *The Scattering of Light*, Academic Press, London, 1969.
47. E. Fallman and O. Axner, *Appl. Opt.*, 2003, **42**, 3915–3926.
48. D. J. Stevenson, F. Gunn-Moore and K. Dholakia, *J. Biomed. Opt.*, 2010, **15**, 041503.
49. R. J. Hernandez, A. Mazzulla, A. Pane, K. Volke-Sepulveda and G. Cipparrone, *Lab Chip*, 2013, **13**, 459–467.
50. S. Maier, *Plasmonics: Fundamentals and Applications*, Springer Science + Business Media LLC, New York, 2007.
51. C. W. Liu, C. P. Hsu, J. A. Yeh, Y. C. Sun, Y. F. Huang, B. H. Chu, F. Ren and Y. L. Wang, *Microsyst. Technol.*, 2013, **19**, 245–251.
52. K. T. Kotz, K. A. Noble and G. W. Faris, *Appl. Phys. Lett.*, 2004, **85**, 2658–2660.
53. D. Baigl, *Lab Chip*, 2012, **12**, 3637–3653.
54. D. C. Zografopoulos, R. Asquini, E. E. Kriezis, A. d'Alessandro and R. Beccherelli, *Lab Chip*, 2012, **12**, 3598–3610.
55. P. Fei, Z. He, C. Zheng, T. Chen, Y. Men and Y. Huang, *Lab Chip*, 2011, **11**, 2835–2841.
56. X. Mao, J. R. Waldeisen, B. K. Juluri and T. J. Huang, *Lab Chip*, 2007, **7**, 1303–1308.
57. X. Li, Y. C. Seow, J. Wu, K. Xu, J. Lin and A. Q. Liu, *Photonics and Optoelectronics Meetings (POEM) 2008: Fiber Optic Communication and Sensors*, 2008, 72780U.
58. J.-M. Lim, S.-H. Kim, J.-H. Choi and S.-M. Yang, *Lab Chip*, 2008, **8**, 1580–1585.
59. M. P. Duggan, T. McCreedy and J. W. Aylott, *Analyst*, 2003, **128**, 1336–1340.
60. D. Yin, D. W. Deamer, H. Schmidt, J. P. Barber and A. R. Hawkins, *Appl. Phys. Lett.*, 2004, **85**, 3477–3479.
61. J. P. Barber, E. J. Lunt, Z. A. George, Y. Dongliang, H. Schmidt and A. R. Hawkins, *IEEE Photonics Technol. Lett.*, 2006, **18**, 28–30.
62. U. Hakanson, P. Measor, D. Yin, E. Lunt, A. R. Hawkins, V. Sandoghdar and H. Schmidt, in *Silicon Photonics II*, ed. J. A. Kubby and G. T. Reed, Spie-Int Soc Optical Engineering, Bellingham, 2007, vol. 6477.
63. M. Loncar, T. Doll, J. Vuckovic and A. Scherer, *J. Lightwave Technol.*, 2000, **18**, 1402–1411.

64. D. V. Vezenov, B. T. Mayers, R. S. Conroy, G. M. Whitesides, P. T. Snee, Y. Chan, D. G. Nocera and M. G. Bawendi, *J. Am. Chem. Soc.,* 2005, **127**, 8952–8953.

65. K. S. Lee, S. B. Kim, K. H. Lee, H. J. Sung and S. S. Kim, *Appl. Phys. Lett.,* 2010, **97**, 021109.

66. A. Hawkins and H. Schmidt, *Microfluid. Nanofluid.,* 2008, **4**, 17–32.

67. H. Schmidt and A. Hawkins, *Microfluid. Nanofluid.,* 2008, **4**, 3–16.

68. A. Hanning, J. Westberg and J. Roeraade, *Electrophoresis,* 2000, **21**, 3290–3304.

69. B. T. Mayers, D. V. Vezenov, V. I. Vullev and G. M. Whitesides, *Anal. Chem.,* 2005, **77**, 1310–1316.

70. A. J. Chung and D. Erickson, *Opt. Exp.,* 2011, **19**, 8602–8609.

71. R. Conroy, D. Wolfe, P. Garstecki, B. Mayers, D. Vesenov, M. Fischbach, K. Paul, M. Prentiss and G. Whitesides, *Conference on Lasers and Electro-Optics/Quantum Electronics and Laser Science and Photonic Applications Systems Technologies,* Baltimore, Maryland, 2005.

72. D. B. Wolfe, R. S. Conroy, P. Garstecki, B. T. Mayers, M. A. Fischbach, K. E. Paul, M. Prentiss and G. M. Whitesides, *Proc. Natl. Acad. Sci. U. S. A.,* 2004, **101**, 12434–12438.

73. D. B. Wolfe, D. V. Vezenov, B. T. Mayers, G. M. Whitesides, R. S. Conroy and M. G. Prentiss, *Appl. Phys. Lett.,* 2005, **87**, 181105.

74. R. S. Conroy, B. T. Mayers, D. V. Vezenov, D. B. Wolfe, M. G. Prentiss and G. M. Whitesides, *Appl. Opt.,* 2005, **44**, 7853–7857.

75. D. V. Vezenov, B. T. Mayers, D. B. Wolfe and G. M. Whitesides, *Appl. Phys. Lett.,* 2005, **86**, 041104.

76. E. E. Jung and D. Erickson, *Lab Chip,* 2012, **12**, 2575–2579.

77. J. Stone, *IEEE J. Quantum Electron.,* 1972, **8**, 386–388.

78. P. Dress, M. Belz, K. F. Klein, K. T. V. Grattan and H. Franke, *Sens. Actuators B,* 1998, **51**, 278–284.

79. B. Schelle, P. Dress, H. Franke, K.-F. Klein and J. P. Slupek, *Proc. SPIE 3912, Micro- and Nanotechnology for Biomedical and Environmental Applications,* 2000, 150–156.

80. P. Dress and H. Franke, *Appl. Phys. B: Lasers Opt.,* 1996, **63**, 12–19.

81. A. Datta, I. Y. Eom, A. Dhar, P. Kuban, R. Manor, I. Ahmad, S. Gangopadhyay, T. Dallas, M. Holtz, F. Temkin and P. K. Dasgupta, *IEEE Sens. J.,* 2003, **3**, 788–795.

82. S. H. Cho, J. Godin and Y. H. Lo, *IEEE Photonics Technol. Lett.,* 2009, **21**, 1057–1059.

83. C. H. Chen, S. H. Cho, H. I. Chiang, F. Tsai, K. Zhang and Y. H. Lo, *Anal. Chem.,* 2011, **83**, 7269–7275.

84. P. Fei, Z. T. Chen, Y. F. Men, A. Li, Y. R. Shen and Y. Y. Huang, *Lab Chip,* 2012, **12**, 3700–3706.

85. X. Lu, D. R. Samuelson, Y. Xu, H. Zhang, S. Wang, B. A. Rasco, J. Xu and M. E. Konkel, *Anal. Chem.,* 2013, **85**, 2320–2327.

86. A. Y. Cho, A. Yariv and P. Yeh, *Appl. Phys. Lett.*, 1977, **30**, 471–472.
87. N. M. Litchinitser, A. K. Abeeluck, C. Headley and B. J. Eggleton, *Opt. Lett.*, 2002, **27**, 1592–1594.
88. S. Mandal and D. Erickson, *Appl. Phys. Lett.*, 2007, **90**, 184103.
89. Y. Fink, D. J. Ripin, S. H. Fan, C. P. Chen, J. D. Joannopoulos and E. L. Thomas, *J. Lightwave Technol.*, 1999, **17**, 2039–2041.
90. G. Barillaro, S. Merlo, S. Surdo, L. M. Strambini and F. Carpignano, *Microfluid. Nanofluid.*, 2012, **12**, 545–552.
91. M. A. Duguay, Y. Kokubun, T. L. Koch and L. Pfeiffer, *Appl. Phys. Lett.*, 1986, **49**, 13–15.
92. P. Yeh, A. Yariv and E. Marom, *J. Opt. Soc. Am.*, 1978, **68**, 1196–1201.
93. E. J. Lunt, B. S. Phillips, J. M. Keeley, A. R. Hawkins, P. Measor, B. Wu and H. Schmidt, *Proc. SPIE 7591, Advanced Fabrication Technologies for Micro/Nano Optics and Photonics III,* 2010, 759109.
94. D. L. Yin, E. J. Lunt, M. I. Rudenko, D. W. Deamer, A. R. Hawkins and H. Schmidt, *Lab Chip,* 2007, **7**, 1171–1175.
95. A. R. Hawkins, E. J. Lunt, M. R. Holmes, B. S. Phillips, D. Yin, M. Rudenko, B. Wu and H. Schmidt, *Proc. SPIE 6462, Micromachining Technology for Micro-Optics and Nano-Optics V and Microfabrication Process Technology XII,* 2007, 64620U.
96. H. Schmidt, D. L. Yin, D. W. Deamer, J. P. Barber and A. R. Hawkins, in *Nanoengineering: Fabrication, Properties, Optics, and Devices*, ed. E. A. Dobisz and L. A. Eldada, Spie-Int Soc Optical Engineering, Bellingham, 2004, vol. 5515, pp. 67–80.
97. P. Measor, B. S. Phillips, A. Chen, A. R. Hawkins and H. Schmidt, *Lab Chip,* 2011, **11**, 899–904.
98. P. Measor, L. Seballos, D. L. Yin, J. Z. Zhang, E. J. Lunt, A. R. Hawkins and H. Schmidt, *Appl. Phys. Lett.,* 2007, **90**, 21107.
99. M. R. Holmes, M. Rudenko, P. Measor, T. Shang, A. R. Hawkins and H. Schmidt, *MOEMS,* 2010, **9**, 023004–023006.
100. R. Bernini, G. Testa, L. Zeni and P. M. Sarro, *Appl. Phys. Lett.,* 2008, **93**, 011106.
101. R. Bernini, G. Testa, L. Zeni and P. M. Sarro, in *Sensors and Microsystems*, ed. P. Malcovati, A. Baschirotto, A. d'Amico and C. Natale, Springer, Netherlands, 2010, vol. 54, pp. 373–376.
102. G. Testa, Y. Huang, P. M. Sarro, L. Zeni and R. Bernini, *Appl. Phys. Lett.,* 2010, **97**, 131110.
103. D. Ozcelik, B. S. Phillips, J. W. Parks, P. Measor, D. Gulbransen, A. R. Hawkins and H. Schmidt, *Lab Chip,* 2012, **12**, 3728–3733.
104. J. W. Parks, H. Cai, L. Zempoaltecatl, T. D. Yuzvinsky, K. Leake, A. R. Hawkins and H. Schmidt, *Lab Chip,* 2013, **13**, 4118–4123.
105. K. D. Leake, B. S. Phillips, T. D. Yuzvinsky, A. R. Hawkins and H. Schmidt, *Opt. Exp.,* 2013, **21**, 32605–32610.
106. D. W. Pohl, *Philos. Trans.: Math., Phys. Eng. Sci.,* 2004, **362**, 701–717.
107. M. Nieto-Vesperinas, P. C. Chaumet and A. Rahmani, *Philos. Trans.: Math., Phys. Eng. Sci.,* 2004, **362**, 719–737.

108. L. N. Ng, B. J. Luff, M. N. Zervas and J. S. Wilkinson, *J. Lightwave Technol.,* 2000, **18**, 388–400.
109. S. Kawata and T. Sugiura, *Opt. Lett.,* 1992, **17**, 772–774.
110. D. Erickson, X. Serey, Y. F. Chen and S. Mandal, *Lab Chip,* 2011, **11**, 995–1009.
111. S. Gaugiran, S. Gétin, J. Fedeli, G. Colas, A. Fuchs, F. Chatelain and J. Dérouard, *Opt. Exp.,* 2005, **13**, 6956–6963.
112. S. Kawata and T. Tani, *Opt. Lett.,* 1996, **21**, 1768–1770.
113. L. N. Ng, M. N. Zervas, J. S. Wilkinson and B. J. Luff, *Appl. Phys. Lett.,* 2000, **76**, 1993–1995.
114. B. S. Schmidt, A. H. J. Yang, D. Erickson and M. Lipson, *Opt. Exp.,* 2007, **15**, 14322–14334.
115. V. R. Almeida, Q. F. Xu, C. A. Barrios and M. Lipson, *Opt. Lett.,* 2004, **29**, 1209–1211.
116. Q. F. Xu, V. R. Almeida, R. R. Panepucci and M. Lipson, *Opt. Lett.,* 2004, **29**, 1626–1628.
117. T. Baehr-Jones, M. Hochberg, C. Walker and A. Scherer, *Appl. Phys. Lett.,* 2005, **86**, 081101.
118. S. Y. Lin, J. J. Hu, L. Kimerling and K. Crozier, *Opt. Lett.,* 2009, **34**, 3451–3453.
119. A. H. J. Yang, S. D. Moore, B. S. Schmidt, M. Klug, M. Lipson and D. Erickson, *Nature,* 2009, **457**, 71–75.
120. A. H. J. Yang, T. Lerdsuchatawanich and D. Erickson, *Nano Lett.,* 2009, **9**, 1182–1188.
121. S. Kita, S. Hachuda, K. Nozaki and T. Babab, *Appl. Phys. Lett.,* 2010, **97**, 161108.
122. X. Serey, S. Mandal, Y. F. Chen and D. Erickson, *Phys. Rev. Lett.,* 2012, **108**, 048102.
123. S. Kita, S. Hachuda, S. Otsuka, T. Endo, Y. Imai, Y. Nishijima, H. Misawa and T. Baba, *Opt. Exp.,* 2011, **19**, 17683–17690.
124. P. T. Lin, S. W. Kwok, H.-Y. G. Lin, V. Singh, L. C. Kimerling, G. M. Whitesides and A. Agarwal, *Nano Lett.,* 2013, **14**, 231–238.
125. Y. Z. Sun and X. D. Fan, *Anal. Bioanal. Chem.,* 2011, **399**, 205–211.
126. H. K. Hunt and A. M. Armani, *Nanoscale,* 2010, **2**, 1544–1559.
127. X. D. Fan, I. M. White, S. I. Shopova, H. Y. Zhu, J. D. Suter and Y. Z. Sun, *Anal. Chim. Acta,* 2008, **620**, 8–26.
128. A. Serpenguzel, S. Arnold and G. Griffel, *Opt. Lett.,* 1995, **20**, 654–656.
129. F. Vollmer, D. Braun, A. Libchaber, M. Khoshsima, I. Teraoka and S. Arnold, *Appl. Phys. Lett.,* 2002, **80**, 4057–4059.
130. F. Vollmer and S. Arnold, *Nat. Methods,* 2008, **5**, 591–596.
131. S. Arnold, M. Khoshsima, I. Teraoka, S. Holler and F. Vollmer, *Opt. Lett.,* 2003, **28**, 272–274.
132. H. C. Ren, F. Vollmer, S. Arnold and A. Libchaber, *Opt. Exp.,* 2007, **15**, 17410–17423.
133. S. Arnold, R. Ramjit, D. Keng, V. Kolchenko and I. Teraoka, *Faraday Discuss.,* 2008, **137**, 65–83.

134. D. K. Armani, T. J. Kippenberg, S. M. Spillane and K. J. Vahala, *Nature,* 2003, **421**, 925–928.

135. M. L. Gorodetsky, A. D. Pryamikov and V. S. Ilchenko, *J. Opt. Soc. Am. B,* 2000, **17**, 1051–1057.

136. J. K. S. Poon, Y. Y. Huang, G. T. Paloczi and A. Yariv, *IEEE Photonics Technol. Lett.,* 2004, **16**, 2496–2498.

137. K. Tada, G. A. Cohoon, K. Kieu, M. Mansuripur and R. A. Norwood, *IEEE Photonics Technol. Lett.,* 2013, **25**, 430–433.

138. X. M. Zhang and A. M. Armani, *Opt. Exp.,* 2013, **21**, 23592–23603.

139. H. Y. Zhu, I. M. White, J. D. Suter, P. S. Dale and X. D. Fan, *Opt. Exp.,* 2007, **15**, 9139–9146.

140. I. M. White, H. Oveys, X. Fan, T. L. Smith and J. Zhang, *Appl. Phys. Lett.,* 2006, **89**, 191106.

141. C. Y. Chao and L. J. Guo, *Appl. Phys. Lett.,* 2003, **83**, 1527–1529.

142. S. Y. Cho and N. M. Jokerst, *IEEE Photonics Technol. Lett.,* 2006, **18**, 2096–2098.

143. T. Lipka, L. Wahn, H. K. Trieu, L. Hilterhaus and J. Müller, *NANOP,* 2013, **7**, 073793.

144. X. M. Zhang, H. S. Choi and A. M. Armani, *Appl. Phys. Lett.,* 2010, **96**, 153304.

145. R. M. Hawk, M. V. Chistiakova and A. M. Armani, *Opt. Lett.,* 2013, **38**, 4690–4693.

146. M. S. Luchansky and R. C. Bailey, *Anal. Chem.,* 2012, **84**, 793–821.

147. S. Arnold, D. Keng, S. I. Shopova, S. Holler, W. Zurawsky and F. Vollmer, *Opt. Exp.,* 2009, **17**, 6230–6238.

148. S. Mandal and D. Erickson, *Opt. Exp.,* 2008, **16**, 1623–1631.

149. S. Pal, P. M. Fauchet and B. L. Miller, *Anal. Chem.,* 2012, **84**, 8900–8908.

150. J. D. Joannopoulos, P. R. Villeneuve and S. H. Fan, *Nature,* 1997, **386**, 143–149.

151. J. S. Foresi, P. R. Villeneuve, J. Ferrera, E. R. Thoen, G. Steinmeyer, S. Fan, J. D. Joannopoulos, L. C. Kimerling, H. I. Smith and E. P. Ippen, *Nature,* 1997, **390**, 143–145.

152. S. Mandal, X. Serey and D. Erickson, *Nano Lett.,* 2010, **10**, 99–104.

153. S. Mandal, J. M. Goddard and D. Erickson, *Lab Chip,* 2009, **9**, 2924–2932.

154. B. Cordovez, R. Hart and D. Erickson, *Proc. SPIE 8594, Nanoscale Imaging, Sensing, and Actuation for Biomedical Applications X,* 2013, 85940Q.

155. A. H. J. Yang and D. Erickson, *Lab Chip,* 2010, **10**, 769–774.

156. H. Cai and A. W. Poon, *Opt. Lett.,* 2010, **35**, 2855–2857.

157. S. Y. Lin, E. Schonbrun and K. Crozier, *Nano Lett.,* 2010, **10**, 2408–2411.

158. S. Y. Lin and K. B. Crozier, *Lab Chip,* 2011, **11**, 4047–4051.

159. S. Lin and K. B. Crozier, *ACS Nano,* 2013, **7**, 1725–1730.

160. A. Kumar, S. J. Williams and S. T. Wereley, *Microfluid. Nanofluid.,* 2009, **6**, 637–646.

161. G. L. Liu, J. Kim, Y. Lu and L. P. Lee, *Nat. Mater.,* 2006, **5**, 27–32.

162. P. Y. Chiou, A. T. Ohta and M. C. Wu, *Nature,* 2005, **436**, 370–372.

163. A. Kumar, S. J. Williams, H. S. Chuang, N. G. Green and S. T. Wereley, *Lab Chip,* 2011, **11**, 2135–2148.

164. G. J. Shah, A. T. Ohta, E. P. Y. Chiou, M. C. Wu and C. J. Kim, *Lab Chip,* 2009, **9**, 1732–1739.

165. S. Glaesener, M. Esseling and C. Denz, *Opt. Lett.,* 2012, **37**, 3744–3746.

166. D. H. Lee, H. Hwang and J. K. Park, *Appl. Phys. Lett.,* 2009, **95**, 164102.

167. M. Esseling, S. Glasener, F. Volonteri and C. Denz, *Appl. Phys. Lett.,* 2012, **100**, 161903.

168. S. J. Lin, S. H. Hung, J. Y. Jeng, T. F. Guo and G. B. Lee, *Opt. Exp.,* 2012, **20**, 583–592.

169. W. Wang, Y. H. Lin, T. C. Wen, T. F. Guo and G. B. Lee, *Appl. Phys. Lett.,* 2010, **96**, 113302.

170. S. M. Yang, T. M. Yu, H. P. Huang, M. Y. Ku, L. Hsu and C. H. Liu, *Opt. Lett.,* 2010, **35**, 1959–1961.

171. S. M. Yang, S. Y. Tseng, H. P. Chen, L. Hsu and C. H. Liu, *Lab Chip,* 2013, **13**, 3893–3902.

172. H. Hwang, H. Chon, J. Choo and J.-K. Park, *Anal. Chem.,* 2010, **82**, 7603–7610.

173. H. Hwang, Y. J. Choi, W. Choi, S. H. Kim, J. Jang and J. K. Park, *Electrophoresis,* 2008, **29**, 1203–1212.

174. H. Hwang and J.-K. Park, *Lab Chip,* 2011, **11**, 33–47.

175. S. S. Ahsan, A. Gumus and D. Erickson, *Lab Chip,* 2013, **13**, 409–414.

176. D. Erickson, D. Sinton and D. Psaltis, *Nat. Photonics,* 2011, **5**, 583–590.

177. Y. F. Chen, L. Jiang, M. Mancuso, A. Jain, V. Oncescu and D. Erickson, *Nanoscale,* 2012, **4**, 4839–4857.

178. C. Song, N.-T. Nguyen, Y. Yap, T.-D. Luong and A. Asundi, *Microfluid. Nanofluid.,* 2011, **10**, 671–678.

179. W. Song and D. Psaltis, *Lab Chip,* 2011, **11**, 2397–2402.

180. P. H. Huang, M. I. Lapsley, D. Ahmed, Y. C. Chen, L. Wang and T. J. Huang, *Appl. Phys. Lett.,* 2012, **101**, 141101.

181. K. S. Lee, H. L. T. Lee and R. J. Ram, *Lab Chip,* 2007, 7, 1539–1545.

182. W. Song, A. E. Vasdekis and D. Psaltis, *Lab Chip,* 2012, **12**, 3590–3597.

183. C. Escobedo, A. G. Brolo, R. Gordon and D. Sinton, *Nano Lett.,* 2012, **12**, 1592–1596.

184. M. Huang, B. C. Galarreta, A. E. Cetin and H. Altug, *Lab Chip,* 2013, **13**, 4841–4847.

185. S. Kumar, N. J. Wittenberg and S.-H. Oh, *Anal. Chem.,* 2012, **85**, 971–977.

186. V. R. Dantham, S. Holler, C. Barbre, D. Keng, V. Kolchenko and S. Arnold, *Nano Lett.,* 2013, **13**, 3347–3351.

187. A. A. Yanik, M. Huang, O. Kamohara, A. Artar, T. W. Geisbert, J. H. Connor and H. Altug, *Nano Lett.,* 2010, **10**, 4962–4969.

188. M. L. Juan, M. Righini and R. Quidant, *Nat. Photonics,* 2011, **5**, 349–356.

CHAPTER 7

Applications of Dielectrophoresis in Microfluidics

BLANCA H. LAPIZCO-ENCINAS

Microscale Bioseparations Laboratory, Biomedical Engineering Department, Rochester Institute of Technology, 160 Lomb Memorial Drive, Rochester, NY 14623-5604, USA
E-mail: bhlbme@rit.edu

7.1 Electrokinetics: An Overview

Electrokinetics (EK) is the general term employed to describe the movement of particles and fluids under the action of electrical fields. EK is the field of science that studies the family of phenomena derived from the electrical double layer (EDL). When an aqueous solution is in contact with a solid surface, it is common for the surface to acquire a charge, due to ionisation and/or adsorption mechanisms.[1,2] To balance this surface charge, a thin layer of highly concentrated ions is formed at the solid/liquid interface. The EDL comprises two layers: (i) the Stern or Helmholtz layer is a compact layer that is just a few Ångströms in thickness and comprises immobilised counterions. (ii) The Gouy–Chapman layer is a diffuse layer that extends from the compact layer to the bulk fluid, where the ions are less affected by the surface charge and are mobile (see Figure 7.1(a)). The electrical potential decays exponentially across the EDL (normal to the interface), until it reaches

RSC Detection Science Series No. 5
Microfluidics in Detection Science: Lab-on-a-chip Technologies
Edited by Fatima H Labeed and Henry O Fatoyinbo
© The Royal Society of Chemistry 2015
Published by the Royal Society of Chemistry, www.rsc.org

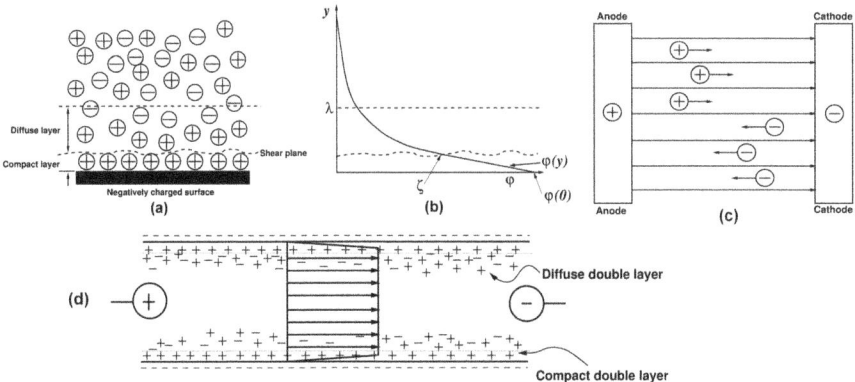

Figure 7.1 (a) Illustration of the EDL in a negatively charged surface, showing the compact and diffuse layers. (b) Representation of the electric potential across the EDL. The potential at the surface is (φ_0) and the zeta potential (ζ) is the potential at the shear plane. (c) Representation of EP particle movement in a uniform electric field. Cations move towards the cathode, while anions move towards the anode. (d) Illustration of EOF in a microchannel with a negative-charged wall, showing the negative charge at the wall, the compact layer of immobile ions, the diffuse layer of mobile ions and the plug like velocity profile.

a nonzero value at a distance slightly beyond the EDL thickness, in the bulk of the fluid (see Figure 7.1(b)). The typical thickness of the EDL is in the range of 1 to 100 nm, depending on medium characteristics. This thickness is known as the Debye length (λ).[3] The EDL is thinner at high electrolyte concentrations and thicker at low electrolyte concentrations, and EDL overlap can occur in nanochannels filled with solutions of low ionic strength. An important parameter used to characterise EK phenomena is the zeta potential (ζ), which is the electric potential at the shear plane between the diffuse and compact layer.[4] The zeta potential groups all the factors related to the surface–liquid chemistry.[5] Important research efforts have been dedicated to the study and measurement of the zeta potential, with two excellent reviews on this topic published by Kirby and Hasselbrink.[6,7]

Electrokinetic phenomena occurs in heterogeneous fluids when an external electric potential is applied. EK represents one of the main pillars within the field of microfluidics; it provides means for manipulating particles and generating liquid flow in microsystems using only an applied electric potential (AC and/or DC). Electrical forces can act on either the particles or the suspending medium with the main EK phenomena used in microfluidic devices being electrophoresis, electro-osmosis, dielectrophoresis, electrorotation, travelling-wave dielectrophoresis and electro-orientation.[2,8] This chapter is written to compliment Chapter 3, with a wider focus on the broader applications of dielectrophoresis technology in microfluidics and bionanoscience.

7.1.1   DC Electrokinetics

DC electrokinetics refers to the "classical" phenomena of electrophoresis (EP) and electro-osmosis (EO), which have a linear behaviour with the electric field and provides effective means for particle and fluid motion. Based on system configuration and design, dielectrophoresis (DEP) can also occur in DC electric fields, and will be discussed in the following sections. EP is the motion of electrically charged particles, relative to a stationary fluid, under the influence of an electric field due to Coulomb force (see Figure 7.1(c)).[2,9,10] Many biological particles (*e.g.*, DNA, proteins and cells) of importance in microfluidic applications possess a net fixed electrical charge (usually negative, due to the presence of acid groups at the surface). When these particles are exposed to an electrical field, they exhibit differences in migration behaviour that allow for effective particle separation and identification. EP has been successfully adapted to many microfluidics applications due to its simplicity, since only electric fields are required. The design of miniaturised EP systems is straightforward, consisting mainly of three elements: (i) sample injection zone, (ii) channel-separation zone, and (iii) detection system.[11] Samples can be introduced electrokinetically, which allows careful measurement of sample size, increasing its potential for analytical applications by decreasing sample dispersion, leading to high-efficiency separations.[12] The current understanding and knowledge on of EP-based microfluidic systems for assessing DNA and proteins molecules is well established. Novel designs are constantly developed, such as the use of micropillars as sieving structures (instead of a gel) for DNA separation with EP.[13] Entropic trapping is another variation of separation of DNA by employing EP, where microchannels with vertical gaps of two different thicknesses are employed and DNA is separated by differences in conformational behaviour.[5] Protein EP separation have also progressed considerably, with the latest advances including free-standing polyacrylamide gel microstructures photo patterned with UV light, allowing for multiple parallel protein separations.[12] EP has also been employed to characterise and separate whole cells, allowing assessment of cell-surface charge and physiological characteristics.[14] The electrophoretic mobility (μ_{EP}) of a particle is proportional to the net charge magnitude (q) and inversely proportional to particle size:[15]

$$\mu_{EP} = \frac{q}{f} = \frac{\vec{v}_{EP}}{\vec{E}} \tag{7.1}$$

where f is the frictional coefficient, $\vec{v}_{EP}$ is the electrophoretic velocity and $\vec{E}$ is the electric field at the location of the particle. For a thin diffuse layer, which is the case for large particles in the micrometre range or under high ionic strength solutions, the electrophoretic velocity is given by the Helmholtz–Smoluchowski equation:

$$\vec{v}_{EP} = \frac{\varepsilon_m \zeta_p \vec{E}}{\eta} = \mu_{EP} \vec{E} \tag{7.2}$$

where ε_m is the permittivity of the electrolyte solution, ζ_p is the zeta potential of the particle, and η is the electrolyte viscosity.[15] Eqn (7.1) and (7.2) do not apply under the nonlinear electrokinetic phenomenon of electrophoresis of the second kind, which occurs under large electric fields.[16]

Electro-osmotic flow (EOF) is one of the leading mechanisms for achieving liquid flow in capillaries and microchannels. This type of fluid motion, referred to as EO, is obtained by applying an electric field that produces electrical forces on the ions in the EDL. A range of common materials used in microfluidics, such as glass, PDMS, *etc.*, acquire charge when in contact with an aqueous solution, and thus, due to their zeta potential, are amenable for EOF.[6,7] EO liquid pumping offers attractive advantages for microfluidic applications, starting with its ease in application, since it is much more straightforward to apply an electric voltage to a microfluidic channel, than handling a miniaturised pump.[1] Additionally, EOF has a plug like velocity profile that significantly decreases sample dispersion (see Figure 7.1(d)). The use of EOF also facilitates integration to other processes within the same device due to lack of moving parts or additional equipment.

EOF originates from the movement of ions in the diffuse layer of the EDL under the action of an external electric field. By assuming a thin EDL, compared to the width of the microchannel, the expression for the EO velocity in the bulk of the fluid (outside the EDL) is defined by:

$$\vec{v}_{EO} = -\frac{\varepsilon_m \zeta_{wall} \vec{E}}{\eta} = \mu_{EO} \vec{E} \tag{7.3}$$

where ζ_{wall} is the zeta potential of the microchannel or capillary wall and μ_{EO} is the electro-osmotic mobility. The zeta potential can be measured experimentally and has been reported for many materials.[1,6,7] Inside the EDL the EO velocity varies as a function of the electric potential $\varphi(y)$ (see Figure 7.1(b)) according to the following relation, which is only valid near the channel wall:[1]

$$\vec{v}_{EO} = -\frac{\varepsilon_m \vec{E}}{\eta} (\phi(y) - \phi_0) \tag{7.4}$$

The EO velocity has a linear dependence on the local electric field, which simplifies its control and application. However, similar to EP, under conditions of strong electric fields, the nonlinear electrokinetic phenomenon of electro-osmosis of the second kind can occur.[17] In many microfluidic applications, the EO velocity of the bulk of the fluid (eqn (7.3)) is the most relevant parameter since it defines the overall velocity of the fluid. DC-EOF is one of the leading liquid-pumping mechanisms due to its high versatility, with important research efforts devoted to the understanding of the fundamentals of EOF.[18-20] Recent microfabrication advances have allowed AC-EOF to become widely popular in a number of micro- and nanoscale applications, this topic will be briefly described in the following subsection.

7.1.2   AC Electrokinetics

AC electrokinetics refers to the movement of particles and fluids under the influence of an AC electric field. The main EK forces achieved with AC electric potentials are: AC electro-osmosis, electrothermal flow, electrorotation, DEP and travelling-wave DEP.[2]

AC voltage-driven microfluidic flow was first reported in 1999 as a result of electrode polarisation. Ramos *et al.*[21] called this mechanism AC electro-osmosis (ACEO). In these experiments, strong fluid flow was observed to vary as function of frequency and had a velocity maxima at intermediate frequencies. At low frequencies (<100 Hz) ACEO is negligible because there is sufficient time for a complete screening of the bulk electric field. At higher frequencies ACEO also decays significantly since there is not enough time for formation of the double layer.[22–24] In planar microelectrode geometries ACEO is generated by the migration of mobile ions under the effect of a tangential component above the electrode surface. In the case of interdigitated planar microelectrodes, the ions from the diffuse layer move from the edge of the electrode to the middle under the action of the AC field, this rotational movement drags the bulk of the fluid producing a net flow along the electrode surface.[23,24] The flow is independent of the sign of the applied voltage as shown in Figure 7.2(a), i.e., reversing the polarity would produce the same flow pattern, and it is most effective at relatively low electrolyte conductivity, which give a thicker EDL.[22] ACEO fluid flow has important applications in microfluidics, from fluid long-range pumping[25] to mixing[26] by employing asymmetrical electrode arrays.

Electrothermal flow (ETF) is a phenomenon that results from Joule heating, which is generated by the presence of an electrical potential gradient and electrical current.[22,27] Joule heating becomes significant at high electric potentials or high ionic concentration, leading to temperature increase and temperature gradients along the length of microchannels. These gradients affect the dielectric properties (conductivity and permittivity) of the electrolyte altering the electric field distribution along the microchannel.[28] The electric field acts on the dielectric properties gradients creating a net body force on the fluid with a velocity that is proportional to the temperature rise, thus, it is also proportional to the conductivity gradient. ETF is much more significant at relative high conductivity conditions, which makes ETF an important phenomenon in applications involving physiological fluids.[22] ETF requires thermally conductive electrode substrates and active cooling in order to generate a sufficient temperature gradient.[24] A schematic representation of the behaviour of ETF as a function of the temperature gradient is presented in Figure 7.2(b).[22]

Electrorotation is a technique that depends on particle polarisability, which is the measure of material's ability to respond to an electric field (polarise) or to produce electric charges at an interface. Electric polarisation is the dipole moment per unit volume, and it reflects the degree to which the bound charge in a medium polarises; with positive components being pulled

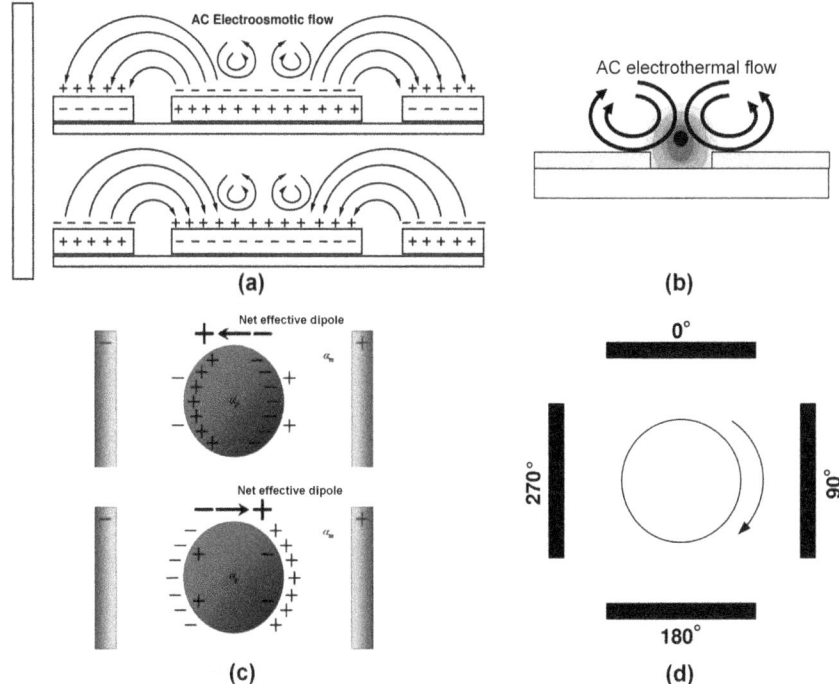

Figure 7.2 (a) Schematic representation of AC electro-osmotic flow above planar electrodes, showing a rotational fluid flow pattern. Adapted from ref. 22 from the original publisher BioMed Central (2011), in accordance with its open-access copyright and license agreement. (b) Representation of AC-ETF generated by a temperature gradient. Reproduced from ref. 22 from the original publisher BioMed Central (2011), in accordance with its open-access copyright and license agreement. (c) Representation of particle polarisation under the influence of an electric field. Top: Particle with higher polarisability than the electrolyte solution. Bottom: Particle with lower polarisability than the electrolyte solution. Reproduced with permission from ref. 29, copyright 2008 Springer-Verlag. (d) Illustration of particle electrorotation in a four-electrode system with 90° advancing phase.

in the same direction of the electric field, while negative components are pulled in the opposite direction of the field.[1,2] Particles can have either a higher or a lower polarisability than that of the electrolyte in which they are suspended. When particle polarisability is higher, more of charges are produced in the interior of the particle/fluid interface, leading to a net dipole in the same direction of the electric field (Figure 7.2(c), top). When particles have a lower polarisability that the suspending electrolyte, more charges are produced on the outer side of the particle/fluid interface, generating a net dipole in the opposite direction of the electric field (Figure 7.2(c), bottom).[9,29] Particle polarisation forms the fundamental physics behind DEP manipulation, and will be discussed in the following subsection.

In electrorotation, particles are exposed to a rotating electric field. If the particle is polarisable, an induced dipole will be formed across the particle, leading to particle rotation in synchrony with the field.[2,30] In a uniform electric field each charge on a dipole experiences an equal and opposite force that aligns the particle dipole to the electric field; however, it takes a finite amount of time for the dipole to become aligned with the field. If the electric field is rotating, this will produce a torque on the polarised particle due to the time delay in the dipole to become realigned with the field. Figure 7.2(d) illustrates the basic system for electrorotation employing four electrodes with 90° advancing phase. The expression for the electrorotational torque induced in a spherical homogeneous particle is described by:

$$\vec{\Gamma} = -4\pi\varepsilon_{\mathrm{m}}r_{\mathrm{p}}^{3}\mathrm{Im}[K(\omega)]|\vec{E}|^{2} \tag{7.5}$$

$$K(\omega) = \frac{\varepsilon_{\mathrm{p}}^{*} - \varepsilon_{\mathrm{m}}^{*}}{\varepsilon_{\mathrm{p}}^{*} + 2\varepsilon_{\mathrm{m}}^{*}} \tag{7.6}$$

where $K(\omega)$ is the Clausius–Mossotti (CM) factor, which is a function of the frequency of the electric field, and $\varepsilon_{\mathrm{p}}^{*}$ and $\varepsilon_{\mathrm{m}}^{*}$ are the complex permittivities of the particle and the electrolyte solution, respectively. The complex permittivity is related to the real permittivity ε by $\varepsilon^{*} = \varepsilon - (j\sigma/\omega)$ where σ and ω are the real conductivity and angular frequency of the applied electric field, respectively, and $j = \sqrt{-1}$.[2] Electrorotation has gained particular importance in applications for biocellular assessments, from bacteria to parasites. One of the major applications is for the study of dielectric properties of cells from an electrorotation spectrum.[31]

7.2 Dielectrophoresis for Nano- and Micromanipulation

Dielectrophoresis (DEP) is an electrokinetic transport mechanism caused by polarisation effects when a dielectric particle is exposed to a nonuniform electric field. As the particle polarises, the density of the electric field at one side of the dipole has a higher strength, producing an imbalance of the forces on the induced dipole.[2,9] DEP was first described by Pohl in 1951,[32] he coined the term DEP in an attempt to describe the force exerted on uncharged dielectric particles as a result of their polarisability. DEP is a leading technique in microfluidics, as evidenced by the growing number of new schemes and applications being developed and the increasing number of publications on this topic.[33–37] DEP offers attractive advantages and higher flexibility than other EK mechanisms. Charged and uncharged particles can be manipulated with DEP, and since it does not depend on the direction of the electric field, DEP can be obtained with AC or DC electric potentials. Additionally, DEP can be used for significant particle enrichment, up to three orders of magnitude, which makes it suitable for front-end applications and sample preparation.[2,38] DEP has been successfully employed for the manipulation of a wide array of particles, from proteins to DNA, to cells and parasites.[39–42] An excellent collection of short articles on the different dielectrophoretic modes can be found on the webpage of the AES Electrophoresis Society.[34]

DEP can be classified as positive or negative. When a particle has a higher polarisability than the suspending electrolyte, the direction of the dipole is with the field, leading to a net particle movement towards the region of higher field intensity, this is called positive DEP (pDEP). On the other hand, negative DEP (nDEP) is observed for particles with a lower polarisability than that of the electrolyte, where the dipole is in the opposite direction of that of the electric field producing particle movement away from the stronger electric field regions. Figure 7.3 illustrates both

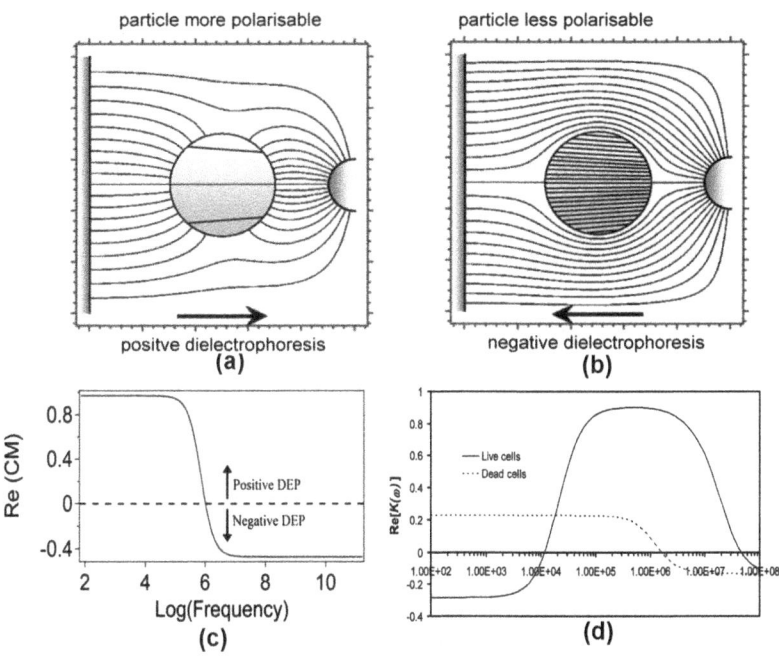

Figure 7.3 Illustration of a particle in a nonuniform electric field. Arrows show the direction of the force and particle movement. (a) pDEP is obtained for a particle that is more polarisable than the medium. (b) nDEP is observed for a particle with lower polarisability than the medium. Reproduced with permission from ref. 29, copyright 2008 Springer-Verlag. Calculated frequency dependence of the real part of the CM factor. (c) Dielectric particle, showing the transition from positive to nDEP as function of electric-field frequency. The parameters used for this estimation are: $\varepsilon_m/\varepsilon_0 = 78$, $\sigma_m = 10^{-5}$ s·m^{-1}, $\varepsilon_m/\varepsilon_0 = 3.0$, $\sigma_p = 10^{-3}$ S·m^{-1}. Reproduced from ref. 46. (d) Live and dead yeast cells, where it can be observed that live cells have two crossover frequencies. Reprinted with permission from ref. 47, copyright 2008 Elsevier.

positive (a) and negative (b) DEP.[29] The dielectrophoretic force exerted on a spherical particle can be expressed as:

$$\vec{F}_{\text{DEP}} = 2\pi\varepsilon_m r_p^3 \text{Re}(K(\omega))\nabla E^2 \qquad (7.7)$$

Where ∇E^2 is the gradient of the squared electric field, and Re $(K(\omega))$ is the real part of the Clausius–Mossotti (CM) factor, which accounts for the polarisation of the particle, can be positive or negative, as defined in eqn (7.6). DEP is a strong function of the properties of the particle and the suspending electrolyte. The CM factor, and thus, the DEP force are frequency dependent, a characteristic that has been successfully employed to separate mixtures of bioparticles by exploiting pDEP and nDEP.[43] This dependence on frequency arises from the nature of the induced dipole on the particle. At low frequencies, the movement of the free charge in a polarised particle can keep up with the changing direction of the electric field, and conductivities dominate. At higher frequencies, the free-charge movement is no longer able to keep up with the changing field and thus, it is not the dominant mechanism; the polarisation of bound charges (permittivity) becomes the leading mechanism. The difference between these two cases is called dielectric dispersion. The CM factor can be written for these limiting cases of low frequency ($\omega \ll \sigma/\varepsilon$) and high frequency ($\omega \gg \sigma/\varepsilon$), by employing the real conductivities and permittivities of the particle and the medium:

$$K = \left[\frac{\sigma_p - \sigma_m}{\sigma_p + 2\sigma_m}\right] \quad \text{conductivity regime,} \ \left(\omega \ll \frac{\sigma}{\varepsilon}\right) \qquad (7.8)$$

$$K = \left[\frac{\varepsilon_m - \varepsilon_m}{\varepsilon_m + 2\varepsilon_m}\right] \quad \text{dielectric regime,} \ \left(\omega \gg \frac{\sigma}{\varepsilon}\right) \qquad (7.9)$$

Usually, below 100 kHz is considered the "conductivity regime" and the CM factor can be approximated in terms of the real conductivities.[44,45] The behaviour of the CM factor for a polymer microparticle suspended in DI water is shown in Figure 7.3(c). It can be observed that pDEP is obtained at lower frequencies, and nDEP is observed at higher frequencies. Details on the properties of the particle and medium are included in the figure caption. The crossover frequency, i.e., the frequency at which CM = 0, lies around 1 MHz.[46] For biological particles, such as cells, estimating the CM factor is a little more complex, since cells contain different structures and layers with distinct dielectric properties. In general cells include a core (cytoplasm), a cell membrane and a cell wall, all of these structures have a specific conductivity and permittivity; these differences have been exploited, for example, to separate live from dead cells.[47] The multishell models (see Section 3.3.1) are commonly used to describe the dielectric properties of cellular compartments. The simplest "smeared out" model is the 2-shell model, giving eqn (7.10), which considers a spherical particle with a core of radius $r_{p,cyto}$ and a total radius $r_{p,mem}$. The complex permittivity ($\varepsilon^*_{p,2shell}$) of this particle, considering that the core is the cytoplasm and the outer shell is the cell membrane, can be described as:[1]

$$\varepsilon^*_{p,2shell} = \varepsilon^*_{mem} \left[\frac{\left(\dfrac{r_{p,mem}}{r_{p,cyto}}\right)^3 + 2\left(\dfrac{\varepsilon^*_{cyto} - \varepsilon^*_{mem}}{\varepsilon^*_{cyto} + 2\varepsilon^*_{mem}}\right)}{\left(\dfrac{r_{p,mem}}{r_{p,cyto}}\right)^3 - \left(\dfrac{\varepsilon^*_{cyto} - \varepsilon^*_{mem}}{\varepsilon^*_{cyto} + 2\varepsilon^*_{mem}}\right)} \right] \qquad (7.10)$$

where $r_{p,cyto}$ and $r_{p,mem}$ are the radius of the cytoplasm core and the cell membrane, respectively; and ε^*_{cyto} and ε^*_{mem} are the complex permittivities of the cytoplasm and membrane, respectively. For particles containing a cell wall, a 3-shell model can be applied, where the new complex permittivity ($\varepsilon^*_{p,3shell}$) is obtained by repeating the application of eqn (7.10), considering $r_{p,wall}$ as the outer radius of the cell wall:[1]

$$\varepsilon^*_{p,3shell} = \varepsilon^*_{wall} \left[\frac{\left(\dfrac{r_{p,wall}}{r_{p,mem}}\right)^3 + 2\left(\dfrac{\varepsilon^*_{p,2shell} - \varepsilon^*_{wall}}{\varepsilon^*_{p,2shell} + 2\varepsilon^*_{wall}}\right)}{\left(\dfrac{r_{p,wall}}{r_{p,mem}}\right)^3 - \left(\dfrac{\varepsilon^*_{p,2shell} - \varepsilon^*_{wall}}{\varepsilon^*_{p,2shell} + 2\varepsilon^*_{wall}}\right)} \right] \qquad (7.11)$$

The behaviour of the CM factor for live and dead yeast cells is illustrated in Figure 7.3(d). As can be seen, the dielectrophoretic spectra differ from that of polystyrene particles. Cells are complex nonhomogeneous particles, which exhibit two crossover frequencies. Also, dead cells behave differently from live cells. When a cell dies, the highly insulating membrane becomes compromised and gets contaminated with the conductive cytoplasm, increasing the conductivity of the membrane. These changes in the dielectric properties of cell membrane have been exploited to separate and identify living from dead cells.[45]

DEP has a strong dependence on particle size, as seen from eqn (7.7), and it is easily applied in the manipulation of particles above 1 μm in diameter, which for biological applications includes bacteria, yeast, microalgae, blood cells, mammalian cells and parasites,[48-52] even worms have been manipulated with DEP.[53] Furthermore, with the aid of microfabrication techniques and novel designs, it has been possible to extend the application of DEP to nanoparticles, ranging from carbon nanotubes, to proteins and DNA.[54-57] Pohl's early DEP set-up employed the use of tinfoil and a piece of tungsten wire as electrodes,[32] but since then, there has been tremendous advances in nano- and microfabrication technologies enabling extremely large field magnitudes, crucial for particle manipulation, to be obtained from low power sources (<20 V_{pk-pk}). The two most popular configurations for carrying out DEP are electrode-based and insulator-based systems.

7.2.1 Electrode-Based Dielectrophoresis

Microelectrode arrays became widely available in the 1990s, driving the development of electrode-based DEP (eDEP). Novel electrode designs are constantly reported, illustrating enhanced and more efficient dielectrophoretic microdevices. The majority of electrode-based systems employ high frequency AC electric potentials that allows the assessment of a particle's behaviour by varying the frequency of the electric field. Traditional eDEP systems consist of a chamber or microchannel with planar electrodes deposited on the bottom. Particles are then introduced and manipulated by applying an AC voltage. Figure 7.4 shows two common electrode configurations employed with eDEP, as well as positive and negative dielectrophoretic particle behaviour as a function of the frequency of the electric field.[58] Negative DEP is observed in Figures 7.4(a) and (b) on polynomial and castellated electrodes, respectively. This behaviour was obtained at high electric field frequencies. In nDEP particles are repelled from regions of stronger electric field magnitude, thus particles cluster away from the electrode edges. Positive DEP is clearly illustrated in Figures 7.4(c) and (d) under a low-frequency electric field, where particles are attracted to the areas of higher field intensity located at the electrode edges.[58]

Greater design flexibility, afforded by rapid prototyping technologies, has enabled a wealth of eDEP microsystems to be created over the past couple of decades. Arrays of microelectrodes can be fabricated from a variety of materials, and due to the small scale, high electric fields can be easily obtained by applying low electrical voltages. Nevertheless, there are some drawbacks: certain microelectrode fabrication processes are expensive and complex, planar microelectrode configurations create fields that only effectively affect particles near the electrode surface and therefore cannot handle high-throughput operations. Fouling, is common when handling bioparticles, and can significantly affect microelectrode performance.[40,59] Many different DEP modes can be developed employing arrays of microelectrodes: 3D cages,[60] DEP field-flow fractionation,[61] travelling-wave DEP,[62] *etc.*; Figure 7.5 shows examples of these applications. Three-dimensional dielectrophoretic field cages can be used to trap and characterise particles and cells in free solution, Figure 7.5(a) shows an octode cage with four-phase rotating fields used by Schnelle *et al.*[60] to trap polystyrene particles by means of nDEP at the centre of the cage. Field-flow fractionation (FFF) is a well-established analytical technique that has been successfully combined with DEP. In FFF-DEP particles inside a microchannel are levitated by means of nDEP by employing electrodes on the bottom of the channel. Particles will be levitated at different heights depending on their dielectric properties and size. Since these systems utilise hydrodynamic flow, a parabolic velocity profile is obtained that allows particles separation. Distinct particles will elute from the system at different times depending on their relative levitation position. Particles closer to the microchannel centre will be eluted first, see figure 7.5(b).[61] FFF-DEP has found important clinical applications, from separation and detection of cancer cells,[63] to assessment of electroporated cells.[64]

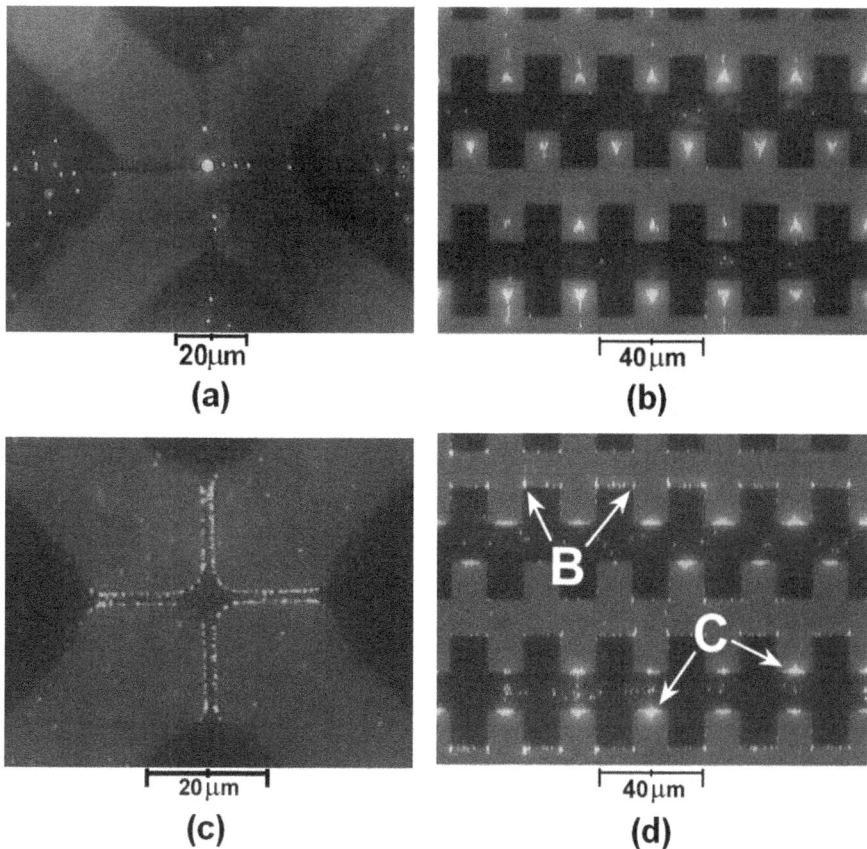

Figure 7.4 Dielectrophoretic behaviour of 557-nm latex particles on polynomial and castellated microelectrodes, DEP response is a function of the electric field frequency. (a) nDEP of the particles on polynomial electrodes obtained at 5 V at 5 MHz. (b) nDEP of the particles on castellated electrodes at 8 V at 8 MHz. (c) pDEP of the particles on polynomial electrodes obtained at 5 V and a low frequency of 500 kHz. (d) pDEP of the particles on castellated electrodes at 8 V and a low frequency of 700 kHz. For pDEP, the particles collect along the edges of the electrodes (marked by points B) and also at the high-field regions at the back of the bays (marked by points C). Reproduced with permission of ©IOP Publishing from ref. 58, copyright 2000 IOP Publishing.

Travelling-wave DEP (twDEP) requires microchannels with a series of electrodes on the bottom that are energised with an AC signal with a 90° phase shift between the electrodes, which produces a nonuniform travelling wave. In these systems the sample is introduced employing pressure-driven flow, and the particles of interest are roughly focused towards one side of the channel by means of DEP (Figure 7.5(c)), selected particles can be separated

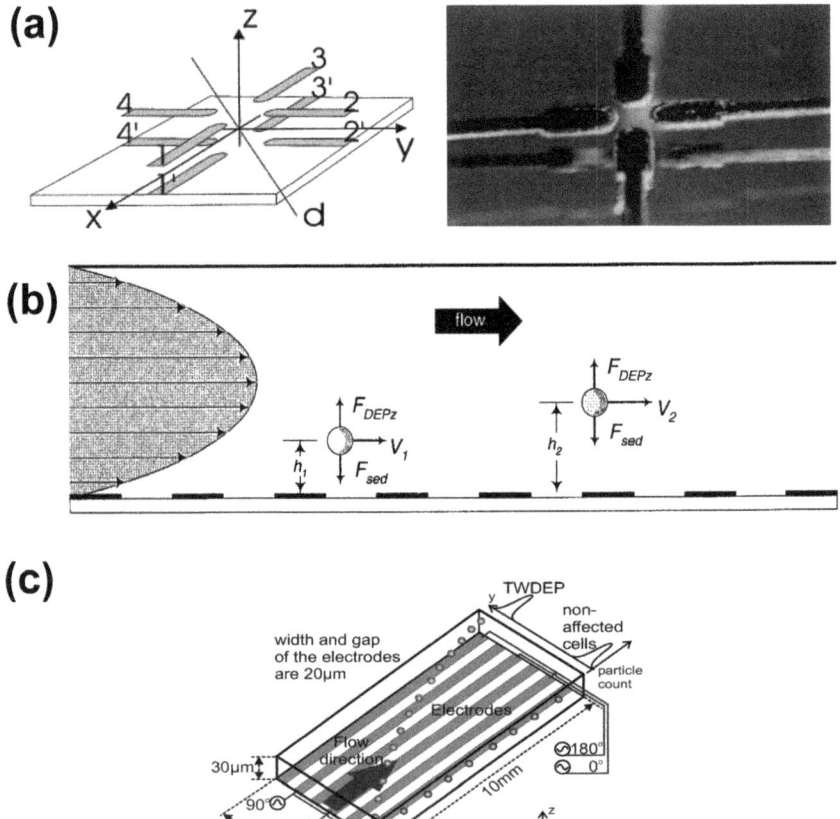

Figure 7.5 (a) 3d octode field cage, image shows sketch and photograph. Liquid with particles flows in the direction of the diagonal d, and particles are trapped by creating AC field cages. Reproduced with permission from ref. 60, copyright 2000 Elsevier. (b) FFF-DEP system, depicting two particles levitated at different height by means of DEP. Reprinted with permission from ref. 61, copyright 2000 American Chemical Society. (c) Schematic representation of travelling-wave DEP system. Reproduced with permission from ref. 62, copyright 2012 Elsevier.

from a mixture depending on their dielectrophoretic response.[62] The twDEP force exerted on a spherical particle is described by:

$$\vec{F}_{\text{twDEP}} = \frac{-4\pi\varepsilon_{\text{m}}r_{\text{p}}^3\text{Im}(K(\omega))\vec{E}^2}{\lambda} \tag{7.12}$$

where λ is the wavelength of the applied signal. This is very similar to electrorotation, since it includes the imaginary part of the CM factor, which dictates the direction of the particle movement with respect to the travelling wave.[65]

7.2.2   Insulator-Based Dielectrophoresis

Insulator-based DEP (iDEP), which was first described by Masuda *et al.* in 1989,[66] is a technique where nonuniform electric fields are created by employing insulating structures instead of electrodes. The presence of these structures distorts the electric-field distribution between two external electrodes. Insulator-based DEP offers attractive advantages over eDEP, since insulators transverse the entire depth of microchannels, high-throughput operations are feasible. Microdevices for iDEP are much simpler, robust and less costly, since they can be made from inexpensive materials such as glass or polymers; and the deposition of metallic electrodes is not required. This opens up the possibility for disposable, single-use microdevices for diagnostic and clinical applications. Fouling effects, which are common when handling bioparticles, do not affect the performance of insulators, an insulator will maintain functionality despite fouling.[67] A majority of iDEP systems have been used with DC electric fields,[45] which enables using EOF to pump the medium and particles through microchannels, avoiding the use of micropumps and making systems simpler and more portable. However, relatively high electric potentials, sometimes in the order of thousands of volts, are required for iDEP systems, producing significant Joule heating, which can damage bioparticles.[28,68] In iDEP, different EK mechanisms are acting on the particles, with the main forces being EP, EOF and DEP. The relative magnitude of these forces can be carefully controlled to achieve precise particle manipulation, sorting and enrichment. At lower applied potentials, linear EK effects (EP and EOF) will dominate particle behaviour, and particles will flow through microchannels. Since DEP is a second-order effect with the electric field, it increases considerably at higher applied voltages and DEP becomes the dominant mechanism, leading to particle "trapping." There are different regimes of DEP that can be exploited to achieve successful particle manipulation. Streaming iDEP is when particles are governed by both DEP and linear EK allowing for continuous and effective particle sorting.[69] Trapping iDEP, obtained when DEP is dominant, captures particles in dielectrophoretic traps (usually under nDEP), and has been used with a wide range of particles, from macromolecules to cells. Some of the most common configurations employed for iDEP particle manipulation are depicted in Figure 7.6. The single and multiple insulating obstacle, serpentine and circular channel configurations are usually used for particle sorting with streaming iDEP. The insulating posts and sawtooth channel configurations are used for trapping iDEP to capture and concentrate particles.[67]

A great majority of iDEP system employ DC fields, where particles are manipulated by linear EK and DEP simultaneously. Successful applications of low-frequency AC electric fields have also been reported. Parameters such as the shape of the applied electric signal and an EOF gradient have been used to achieve highly controlled particle manipulation.[33,70] Under these conditions of DC or extremely low-frequency AC electric fields, the majority of particles exhibit nDEP, which has been observed experimentally for polystyrene particles, bacteria, yeast, microalgae, DNA and proteins.[33,67,69–73] In these

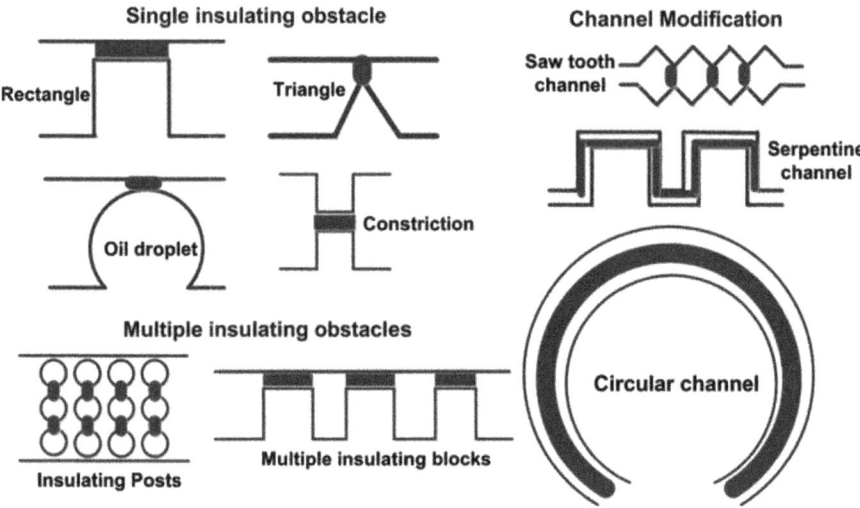

Figure 7.6 Common configurations employed for iDEP, the darker zones indicate the regions where particles experience the maximum dielectrophoretic force. Reprinted with permission from ref. 67, copyright 2010 Springer-Verlag.

systems particles shift from streaming to trapping DEP. The DEP velocity ($\vec{v}_{\text{DEP}}$) of a spherical particle is defined in terms of the DEP mobility (μ_{DEP}) as:

$$\vec{v}_{\text{DEP}} = \mu_{\text{DEP}} \nabla E^2 \tag{7.13}$$

$$\mu_{\text{DEP}} = \frac{r_p^2 \varepsilon_m \text{Re}[K]}{6\eta} \tag{7.14}$$

where the K is given by eqn (7.8). Particle trapping occurs when DEP overcomes linear EK, the EK velocity ($\vec{v}_{\text{EK}}$) of a particle is defined in terms of the EK mobility (μ_{EK}) as:

$$\vec{v}_{\text{EK}} = \mu_{\text{EK}} \vec{E} = (\mu_{\text{EOF}} + \mu_{\text{EP}}) \vec{E} \tag{7.15}$$

It can be assumed that particle trapping is achieved in the regions in the microchannel where DEP overcomes linear EK. It is also possible to define a "trapping condition" by employing the linear EK and DEP mobilities:[71]

$$-\frac{\mu_{\text{DEP}} \nabla E^2}{\mu_{\text{EK}} E^2} \cdot \vec{E} > 1 \tag{7.16}$$

when this condition is satisfied, particles become trapped by means of DEP. There have been significant developments of iDEP systems and novel configurations based on the use of insulators. Techniques such as contactless DEP (cDEP) offer the advantages of both eDEP and iDEP by employing insulating structures and AC fields applied across the microchannel width.[74]

Reservoir-based DEP is another new scheme where particles are manipulated by means of negative dielectrophoresis induced at the reservoir-micro-channel junction.[75] Three dimensional iDEP employs microchannels with constrictions along the width and depth, creating much smaller constrictions that produce higher electric-field gradients, leading to effective particle trapping at much lower applied voltages that with traditional iDEP systems.[76] There is still plenty to be done in iDEP, but the great flexibility of this technique opens up the possibility for applications in many areas, particularly for the manipulation and assessment of biological particles.

7.3 Dielectrophoresis for Nanoanalytical Applications

In the past, DEP was perceived as an efficient technique for handling micrometre-sized particles only. However, the significant advances in microfabrication technologies have made possible more sophisticated microelectrodes and/or insulating structures, which can generate electric-field gradients strong enough to manipulate particles on the nanometre scale.[39,41] DEP has found important applications in the characterisation of the dielectric properties of nanoparticles. Being able to measure dielectric properties supports the development of mathematical models for the design of new systems and devices. An interesting application was reported by Bakewell and Holmes.[77] They employed DEP spectroscopy for the systematic quantification of the properties of 200-nm polystyrene particles using continuously pulsed DEP collection rates. Figure 7.7 shows the experimental setup and an image of the particles trapped at the electrode edges due to pDEP. The crossover frequency (see Section 3.3.2) is a traditional method for probing dielectric properties of particles, but obtaining accurate results is challenging for nanoparticles, since it requires the observation of the transition from pDEP to nDEP.[78] In this study, particle collection by means of pulsed DEP was used to infer dielectric properties, the work also included substantial data fitting, mathematical and statistical models with Monte-Carlo simulations. Pulsed DEP is the application of DEP for short periods of time, in the range of milliseconds to seconds. While the signal is on, particles collect at the electrode edges and are released when the signal is turned off. A novel approach of dual-cycle DEP was employed by switching the frequency while DEP was turned on and off. One of the frequencies was used as a "control" and the second frequency was used as a "probe." The ratio between the probe and control collection rates was used to infer particle dielectric properties. The process included an initial estimation of the particle conductivity, followed by refinement obtained by fitting a nonlinear function of the real part of the CM factor. This novel methodology effectively allowed larger sampling and narrower spread of the estimated values than the traditional crossover method, providing an attractive option for nanoparticles characterisation that could be extended to biological particles.[77]

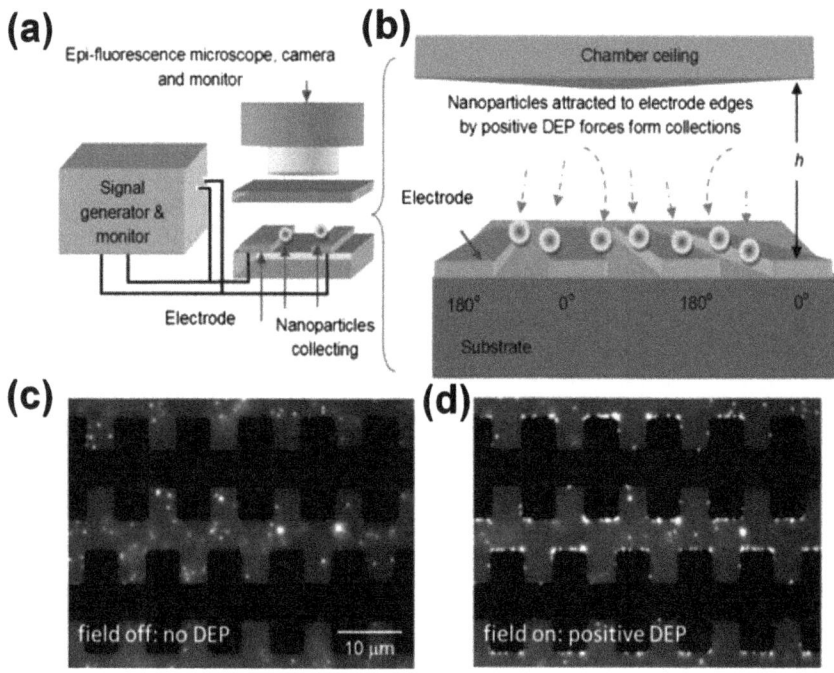

Figure 7.7 Schematic representation of the system for DEP spectroscopy. (a) Experimental setup showing electrodes, signal source and epifluorescence arrangement. (b) Cartoon of the DEP collection of nanoparticles. (c) Image of castellated electrodes without electric field. (d) Particle collection by means of pDEP. Reprinted with permission from ref. 77, copyright 2013 Wiley-VCH Verlag GmbH & Co. KGa.

Manipulation of colloidal (1–1000 nm) particles with DEP has primarily focused on nanowires, nanorods and carbon nanotubes. Only a few groups have focused on the separation and sorting of nanoparticles. Yunus *et al.*,[79] presented an eDEP microdevice for the continuous separation of a mixture of micro- and nanopolystyrene particles (0.5, 1 and 2 μm). They were able to distinguish particles with a diameter difference of only 0.5 μm. A representation of their device is included in Figure 7.8(a). The system had chevron-shaped angled electrodes on the top and bottom of a channel, with three inlets, one for the sample and two more for introducing a strong and a weak sheath flow used for hydrodynamic focusing of the sample to one side of the channel. As the heterogeneous particle sample flowed through the channel, the particles that experienced the highest nDEP response were deflected along the angled electrode defining their exit position. The device had two outlets, where the most deflected (larger) particles went to outlet D (see Figure 7.8(a)). The results reported a 99.9% separation between the micro- and the nanoparticles. Separation and concentration of nanoparticles has also been achieved with iDEP systems. Gencoglu *et al.*,[70] reported the

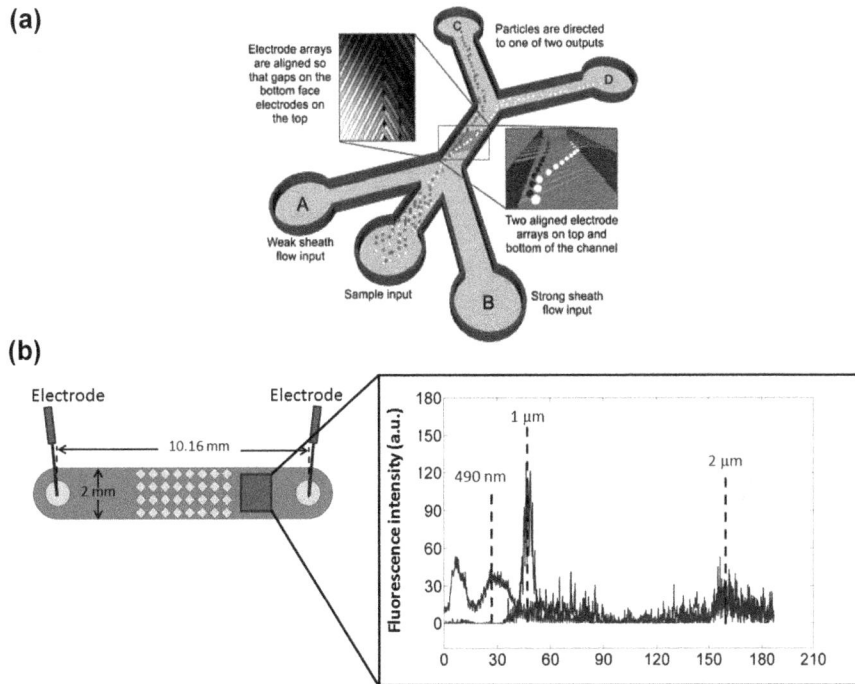

Figure 7.8 Separation of nanoparticles with eDEP and iDEP. (a) Microdevice for continuous eDEP separation of a mixture of 0.5-, 1.0- and 2.0-μm polystyrene particles employing angled electrodes. Reprinted with permission from ref. 79, copyright 2013 Wiley-VCH Verlag GmbH & Co. KGa. (b) Microchannel employed for iDEP separation of a mixture of 0.49-, 1- and 2-μm polystyrene particles, and dielectropherogram of the separation obtained in three minutes, showing a peak for each particle size, from ref. 70.

enrichment and separation of a mixture of 490-nm, 1- and 2 μm polystyrene particles. They employed a microchannel with an array of diamond shaped insulating posts that were 450 μm wide, arranged in a square array of 500 μm centre-to-centre. An asymmetrical low-frequency AC electric potential was employed in order to achieve an EOF gradient. This scheme initially allowed (when EOF was low) for significant particle concentration by means of nDEP inside the insulating post array, followed by selective particle release when the asymmetry of the applied AC signal increased and EOF became more significant. Figure 7.8(b) shows the microchannel as well as the "dielectropherogram" of the separation, which was achieved in three minutes. Particle enrichment was qualitatively assessed by employing an interrogation window for fluorescence detection just at the outlet of the post array (see Figure 7.8(b)). This very simple system allowed separating particles with a diameter difference of only 0.5 μm; demonstrating the great potential of iDEP for effective nanoparticle separation. Further wide-ranging applications of nanoscale DEP, include assembly of colloidal suspensions into crystalline arrays,[80] manipulation of carbon nanotubes,[56] alignment nanofibres,[81] *etc.*, for which dedicated books have been published on this topic.[9] The manipulation of biological nanoscale particles is covered in the following two sections.

7.4   Dielectrophoresis for Bioanalytical Applications

Bioanalysis is a fast-growing field that has benefited significantly from microfluidics, with applications ranging from food and water safety to environmental monitoring. DEP opens up new possibilities for cell assessment, since it uses electrical forces at the microscale for probing the electrical phenotype of bioparticles, in order to characterise or separate different cells.[82] Important advances have also been reported with nanobioparticles, where DNA, proteins, viral particles and cell organelles have been successfully detected and manipulated. Strategically selected examples of DEP bioanalytical applications are summarised and presented below.

DNA plays an important role in molecular biology, where there are significant efforts devoted to the development of DNA analysis, separation and concentration. DEP offers attractive characteristics for DNA manipulation. Traditional methods may have long processing times, while DEP separations require much shorter timescales. Traditional EP separations of DNA are based on the electrical migration, which requires a sieving matrix, in order to obtain a length-dependent migration, since DNA EP mobility is independent on fragment length. DEP exploits another characteristic; the separation of DNA molecules based on their polarisation. Trapping, concentrating, cleaving and stretching of DNA has been successfully achieved in DEP-based microfluidics devices.[39] Electrical characterisation of DNA is essential in order to design DEP-based processes for handling DNA. Numerous groups have focused on studying the electrical properties of DNA.[83]

Washizu's research group demonstrated molecular surgery of DNA in the late 1990s. In this study, DNA molecules where anchored and stretched between electrodes by means of DEP, then the DNA molecules were selectively cleaved employing a restriction enzyme as a cutting tool. The enzyme was immobilised onto a bead, and the bead was manipulated employing optical tweezers (see Figure 7.9(a)). The advantage of this method is that the enzymes cut DNA only at a specific molecular structure, leaving known chemical groups at the cut of the DNA molecule that can be used for downstream chemical processing.[84]

Dielectric characterisation of DNA has also been achieved with eDEP. Zheng *et al.*,[85] employed a microdevice with quadrupole electrodes to trap λ-DNA with pDEP in order to measure the conductance of DNA. In the same study, the properties of bovine serum albumin (BSA) protein and 20-nm latex beads were also assessed. Figure 7.9(b) shows an image of fluorescently labelled DNA particles trapped between the electrodes. Positive DEP of DNA

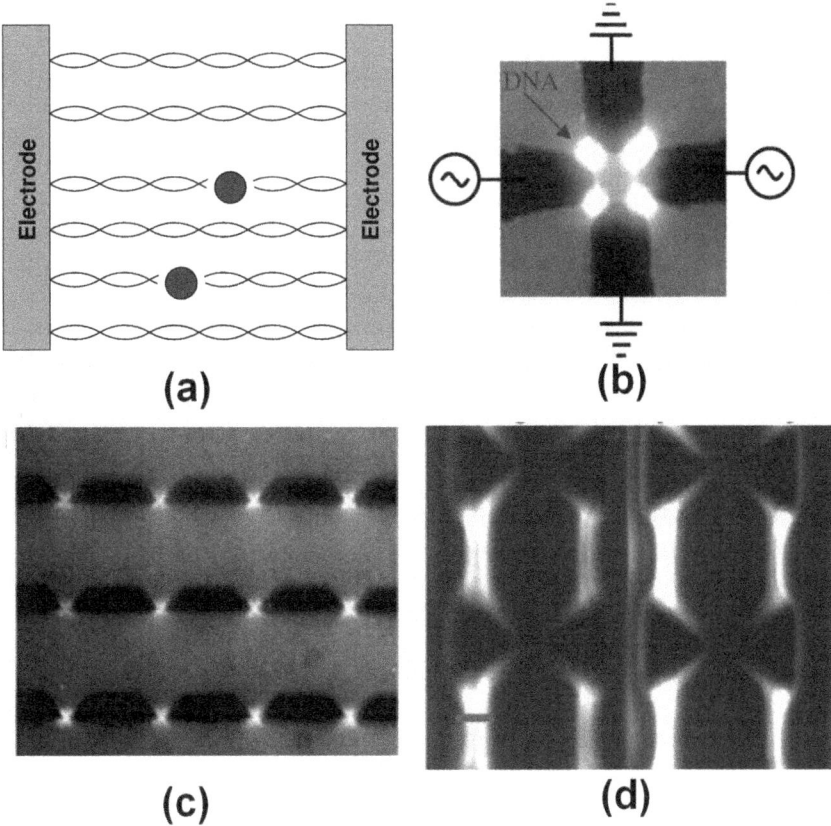

Figure 7.9 DEP of DNA. (a) Representation of molecular surgery of DNA in an eDEP system employing an enzyme immobilised on a bead, from ref. 84. (b) DNA particles trapped in an eDEP system with quadrupole electrode array with 8 V at 1 MHz. Reproduced with permission from ref. 85, copyright 2004 Elsevier. (c) Concentration of DNA particles in an iDEP system with trapezoidal insulating posts at 1 kV at 800 Hz. Reproduced with permission from ref. 86, copyright 2002 Elsevier. (d) Streaming of DNA particles in an iDEP system with triangular insulating posts, at 1 kV at 800 Hz. Reproduced with permission from ref. 54, copyright 2012 Elsevier.

was achieved employing an 8-V at 1-MHz signal. Conductance measurements were performed by monitoring the current *in situ* for 72 h, allowing the solution to dry over and in that way measure only the DNA conductance, which was in the range of 25 nS.

DNA particles have also been successfully manipulated with iDEP systems, Chou *et al.*,[86] reported DEP trapping of single and double stranded DNA employing an array of trapezoidal insulating structures and low frequency AC electric potentials. This is one of the pioneering applications of iDEP for DNA manipulation. Figure 7.9(c) shows an image of 368-bp double stranded fluorescently labelled DNA molecules trapped at the constriction between the trapezoidal posts by pDEP by applying 1 kV at 800 Hz. Another successful demonstration of iDEP manipulation of DNA was reported by Camacho-Alanis *et al.*,[54] where different geometries of insulating structures (micro-posts and nanoposts) were employed to concentrate λ-DNA particles with DC electric potentials. Figure 7.9(d) illustrates streaming iDEP of λ-DNA particles obtained with triangular microposts and a DC potential of 3000 V. The streaming behaviour produces significant concentration close to the centre of the posts by means of pDEP. Negative DEP would have depleted these regions where the DNA particles concentrate.

DEP has also been used for proteins and viruses, though DEP of proteins has far fewer reports than that of DEP of DNA. The small size of protein molecules (*e.g.*, BSA has a diameter of 5 nm)[85] makes the application of DEP more difficult. However, relatively recent reports have demonstrated the use of DEP for the concentration of protein molecules, with an excellent review recently published by Nakano and Ros.[87] Hughes and Morgan[88] reported the trapping of 68-kDa avidin protein particles in a eDEP system, where both positive and negative DEP behaviour were observed by employing polynomial electrodes (Figure 7.10(a)). Positive DEP of avidin was achieved at frequencies below 9 MHz by applying 10 V (see Figure 7.10(a)), while at the same electric potential but for frequencies above 9 MHz, negative DEP was observed (see Figure 7.10(a)). A similar system with four electrodes was employed by Zheng *et al.*,[85] to assess the conductance of BSA. Novel iDEP systems have also been reported for the concentration of protein molecules. Liao *et al.*,[89] employed an iDEP device with a nanoconstriction for the concentration of streptavidin protein particles in physiological media. High electric-field gradients were produced due to the sharp geometry of the triangular posts and the nano-constriction (50–140 nm), as depicted in Figure 7.10(b). The EK force balance was studied in detail, and when only AC potentials were applied the protein particles were repelled from the constriction creating a depleted region. Electrothermal forces, also present, enhanced the nDEP effect (see Figure 7.10(b)). When a DC offset was introduced, EOF and EP forces, which are opposite to each other, came into play and preconcentrated the proteins between the inlet and the constriction (see Figure 7.10(b)). Significant protein concentration was achieved with 200 V and frequencies in the MHz range, demonstrating the potential of iDEP for protein manipulation.[89] Another report on iDEP was the trapping of BSA protein in a microchannel with cylindrical insulating posts employing DC electric potentials, which also combines EOF and nDEP, but required higher potentials than the nano-constriction system.[55] Streaming iDEP, where the particles are only partially dominated by DEP, has great potential for continuous sorting. Abdallah *et al.*,[90] reported the application of a microdevice with one 30-μm constriction used for the sorting of membrane protein nanocrystals. The device was 5 mm long and had a total of five outlets (Figure 7.10(c)). By applying a DC potential, nDEP was obtained that repelled the larger particles and directed them to the central outlet, while the largely unaffected smaller nanocrystals

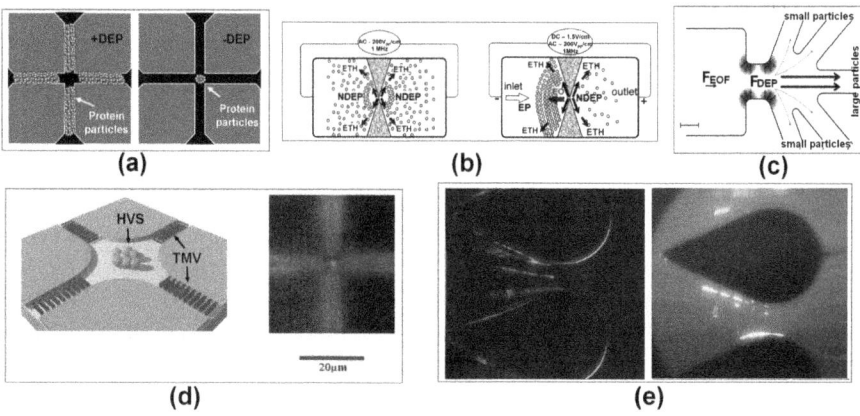

Figure 7.10 DEP of protein and viral particles. (a) Illustration of pDEP and nDEP of avidin protein in an eDEP system with polynomial electrodes from ref. 88. (b) Illustration of an iDEP device with a nanoconstriction employed for the concentration of streptavidin with nDEP; left image shows nDEP obtained with an AC electric potential; right image shows nDEP obtained with an AC potential with a DC offset. Reprinted with permission from ref. 89, copyright 2012 Wiley-VCH Verlag GmbH & Co. KGa. (c) Illustration of streaming iDEP device used for the sorting of membrane protein nanocrystals; darker region at the constriction indicates higher nDEP force. Adapted with permission from ref. 90, copyright 2013 American Chemical Society. (d) Separation of HVS and TMV in an eDEP device with polynomial electrodes; left image shows an illustration, right image shows an actual photograph of the device. Reproduced with permission from ref. 43, copyright 1999 Elsevier. (e) Trapping of TVM in an iDEP microchannel, TVM shows as bright particles trapped at the constriction between the posts. Reproduced with permission from ref. 38, copyright 2005 Elsevier.

flowed into the side outlets. The authors highlighted the possibility of handling a complex mixture obtained from batch crystallisation, without any further treatment, and that nanocrystals of ~100 nm were successfully separated from the bulk solution employing DC potentials ~55 V. These results confirm that streaming DEP can be used for protein sorting and separation.

The detection and characterisation of viral particles is another important area of bioanalysis. DEP has been shown as a valuable tool for the manipulation of viruses. Morgan *et al.*,[43] demonstrated how differences in morphology, which leads to different dielectric properties, can be used to separate and characterise viruses with DEP. They reported the separation of the herpes simplex virus (HSV) and tobacco mosaic virus (TMV). HSV is an enveloped spherical virus (~250 nm in diameter), while TMV is a rod-like (280 nm long) nonenveloped virus. Separation of these viruses were achieved in an eDEP system with a polynomial electrode array with spacing of 2 μm.

TVM exhibited pDEP while HSV exhibited nDEP on application of a 5-V at 6-MHz signal (Figure 7.10(d)). This separation was possible due to their different polarisation behaviours. Viral particles have also been analysed with iDEP, Lapizco-Encinas *et al.*,[38] reported the concentration of TVM in an iDEP microchannel with cylindrical insulating posts that were 200 μm in diameter, with a 50-μm spacing between the posts. TMV were trapped by the effects of nDEP with EOF present. Figure 7.10(e) shows two images of trapped TVM, with the first image (left) depicting TVM as bright particles trapped at the constriction between the posts at an applied DC voltage of 1500 V. The image to the right shows TMV trapped in a suspending medium that also contained 200-nm polystyrene particles at an applied voltage of 1000 V. This last result demonstrated that iDEP has great selectivity, since the polystyrene particles are of the same size range as TMV, but iDEP was able to distinguish between them based on their dielectric properties.

DEP has been successfully employed for the analysis of bacterial cells from water, food and environmental samples. DEP offers the possibility of rapid analysis in portable devices, which is very attractive for this type of applications, where the response time is critical. As an example of the application of DEP in food samples, Koklu *et al.*,[91] reported the capture of bacterial spores in food matrices. The left side of Figure 7.11(a) shows the DEP system with electrodes in a square pattern (60 × 60 μm² inner area) used to analyse samples of apple juice. Spores of *B. subtilis* were trapped at the centre of the electrode squares with nDEP and ETF by applying 10 V at 10 MHz (see Figure 7.11(a)). This report analysed *B. subitlis* and *C. sporogenes* in apple juice and milk, demonstrating the application of DEP for rapid analysis of microbes in food items. DEP has also been successfully employed for environmental applications. In particular, the rapid detection of airborne pathogens, released to the environment in aerosols, is highly challenging. Fatoyinbo *et al.*,[92] reported the isolation of *B. subtilis* spores (a surrogate bacteria for *B. anthracis*) from airborne environmental diesel samples. An eDEP system with slanted interdigitated type electrodes was employed in a microfluidic system. A representation of this system is included in Figure 7.11(b), depicting how diesel particles (white circles) are retained inside the microdevice due to pDEP while the spores (black circles) flow through the device under nDEP. In order to select the appropriate conditions and strategy for this separation, the authors performed a careful crossover frequency study, where they analysed several types of spores: untreated, pasteurised and lyophilised, as well as the diesel particles. There were differences in the DEP response of the spores that were exploited for further separation and characterisation. In general, over the range of frequencies used for experimentation, spores exhibited nDEP while diesel particles experienced pDEP. This difference in DEP behaviour was used to successfully separate the spores from the diesel particles by applying 10 V at 1 MHz, obtaining separation efficiencies ∼99%. These encouraging results demonstrate the high potential of DEP for biosensing environmental applications. Cell viability is another parameter that can be evaluated with

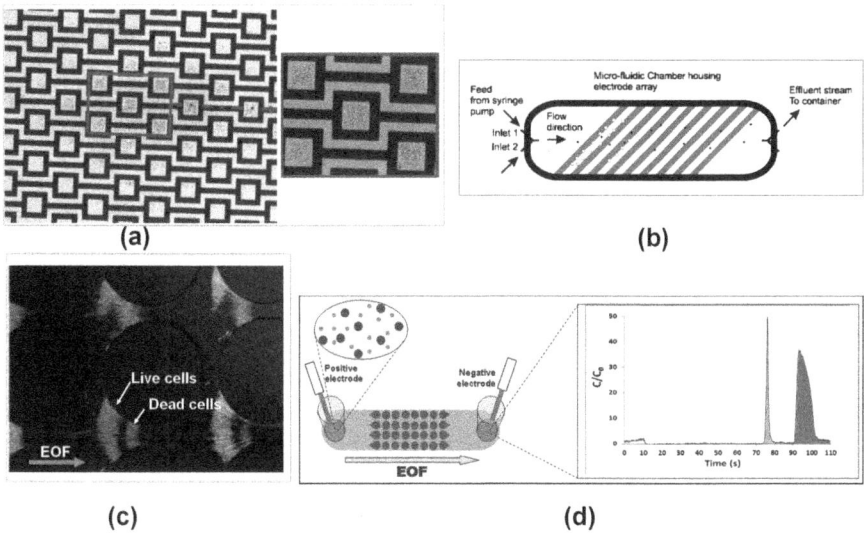

Figure 7.11 DEP for the manipulation of microbes. (a) Square electrode eDEP system employed for the trapping of *B. subtilis* spores. Reprinted with permission from ref. 91, copyright 2010, AIP Publishing LLC. (b) Representation of an eDEP system employed for the separation of *B. subitilis* spores from an environmental diesel particles. White particles represent diesel, and black particles represent the spores. Reprinted from ref. 92. (c) Trapping of live and dead *E. coli* cells with nDEP in an iDEP system. Reprinted with permission from ref. 45, copyright 2004 American Chemical Society. (d) Illustration of the enrichment and separation of *E. coli* and *S. cerevisiae* cells in an iDEP system. Dielectropherogram shows a peak for each cell type. Reproduced with permission from ref. 48, copyright 2010 Springer-Verlag.

DEP, providing a rapid alternative for cell assessment. Lapizco-Encinas *et al.*,[45] reported the identification and trapping of live and dead *E. coli* cells with nDEP in an iDEP microchannel with cylindrical insulating posts (see Figure 7.11(c)). The posts were 200 μm in diameter, with a gap of 50 μm. Only DC potentials were used and EOF was present, producing "bands" of concentrated cells, since EOF is in the opposite direction of nDEP. Dead cells have a weaker nDEP response and are trapped closer to the constriction, while live cells have a stronger nDEP response and are *"pushed back"* farther from the constriction. The differences in DEP response of live and dead cells are due to changes in membrane properties. Fast separation and concentration of microbes has also been reported with iDEP systems.[48] Samples of water containing *E. coli* and *S. cerevisiae* cells were analysed in microchannels with cylindrical posts (480 μm diameter, 80 μm gap).[48] Simultaneous enrichment and selective release was obtained by employing a DC field gradient. A high potential of 1500 V was applied first to trap all the cells in the sample, then the potential was lowered to 500 V, releasing concentrated *E. coli* cells at ~75 s. Then, the potential was lowered again to 200 V releasing *S. cerevisiae* cells at ~95 s. These results are plotted as "peaks" of concentrated cells in the "dielectropherogram" shown in Figure 7.11(d), where it can be seen that the entire process of concentration and separation took less than two minutes.

7.5   Dielectrophoresis for Biomedical and Clinical Applications

The isolation and characterisation of rare cells from complex biological samples is the core challenge of microfluidic devices in biomedical and clinical applications.[93] DEP is one the main techniques used in miniaturised systems for rare cell and blood analysis. Sonnenberg *et al.*,[94] reported the isolation of cell-free circulating (cfc) DNA, mitochondria and viruses from blood samples. They employed an eDEP system with circular electrodes with an 80 μm diameter (Figure 7.12(a)). They achieved the isolation of cfc DNA from lymphocytic leukemia patients employing an AC signal of 20 V at 10 kHz (DNA shows white in the image). This type of test has great potential for rapid diagnostics. Bacteriophage viruses were also isolated from whole blood, while mitochondria were isolated from buffer samples. Srivastava *et al.*,[95] studied the DEP response of RBC as a function of the blood type with an eDEP system. RBCs were exposed to AC potentials of 0.025 V at 1 MHz delivered by one platinum electrode. Figure 7.12(b) shows the response for A+ (left) and AB+ (right) RBC, where differences in orientation can be observed. These differences can be exploited to develop a DEP-based technique for blood typing. The advantage of this method is that untreated diluted samples can be analysed, making for a fast and simple process.

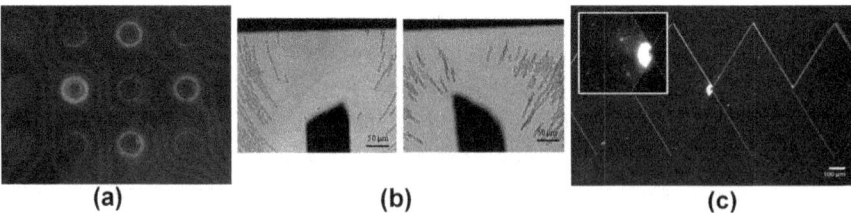

(a)   **(b)**   **(c)**

Figure 7.12   DEP of blood samples. (a) Trapping of cfc DNA from a blood sample in an eDEP device with circular electrodes; DNA is shown white. Reprinted with permission from ref. 94, copyright 2013 Wiley-VCH Verlag GmbH & Co. KGa. (b) DEP response of A+ (left) and AB+ (right) RBC obtained using one platinum electrode. Reprinted with permission from ref. 95, copyright 2008 Wiley-VCH Verlag GmbH & Co. KGa. (c) Concentration of RBC in an iDEP sawtooth channel, cells are shown white. Reproduced with permission from ref. 96, copyright 2011 Springer-Verlag.

Regarding iDEP systems, Jones *et al.*,[96] used an iDEP sawtooth-patterned microchannel to test diluted blood samples. This channel configuration creates an electric field gradient along the microchannel length, which allows for spatial separation of particles. Figure 7.12(c) shows an image of isolated and concentrated RBCs, shown as a white bright band, trapped due to nDEP. The authors reported that high reproducibility was achieved with DC electric potentials in the 200–700 V range. The results demonstrate that eDEP and iDEP systems have the capabilities for rapid RBC characterisation, isolation and concentration.

Clinical and biomedical applications require fast and low-cost methods for the isolation and detection of cancer cells. EK methods, such as DEP are particularly attractive; since subtle changes in cell morphology and physiological state can be probed.[82,97] However, obtaining sufficient data for assessing DEP results can be challenging and a single spectrum is not enough. Fatoyinbo *et al.*,[97] developed a programmable eDEP multichannel dot-electrode system for continuous monitoring of cells. The system had the capability for testing up to eight channels; allowing for the measurement of 16 spectra points in 90 s. The response of human leukaemic K562 cells was monitored after exposure to staurosporine and valinomycin. Figure 7.13(a) shows the initial and final cells response obtained by applying 10 V in a range of frequencies (1 kHz–1 MHz), each dot shows a different frequency, which allows obtaining a whole spectra. It can be observed from the figure how the cell behaviour changes from nDEP at low frequencies, to the crossover frequency, to pDEP. The authors were able to assess the viability of the cells after exposure to the drugs, where changes in the cells electrophysiological state were identified from the spectra data. These encouraging results illustrate that DEP offers a great alternative over traditional methods (*e.g.*, patch clamp) for rapid cell assessment in clinical and biomedical applications. The study of drug efficacy on cells faces significant challenges, since studies are carried out employing 2d monolayers of cells, which do not mimic complex 3D cell–cell interactions, leading to unreliable results. To this end, Abdallat *et al.*,[98] proposed a novel methodology based on DEP for the creation of 3d cell aggregates with the potential to be used for drug testing. An eDEP dot electrode microarray was employed to manipulate human leukemic (K562) and HeLa cell by means of nDEP, followed by encapsulation with a synthetic hydrogel. Negative DEP was strategically used to avoid cell damage, which resulted in impressive cell viabilities, above 90% for over seven days. Samples of the cells, suspended in polyethylene glycol diacrylate (PEG-DA), were introduced into the device where they positioned onto the electrodes with nDEP to form clusters. After nDEP was achieved, the device was exposed to UV light for ~30 s, in order to photopolymerise the PEG-DA, creating the encapsulated cell aggregates. Figure 7.13(b)[98] shows a large array of encapsulated evenly spaced HeLa cells aggregates. The application of this novel system can be extended to other types of cells, with special focus on mammalian cells for the study of drug effects.

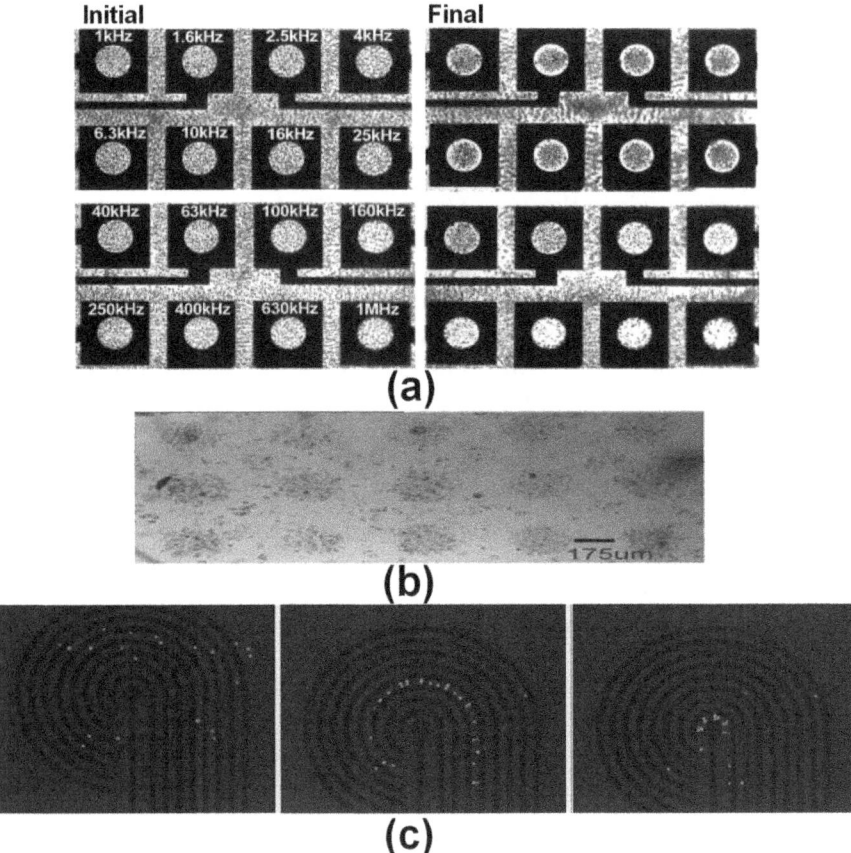

Figure 7.13   DEP of cancer cells. (a) DEP behaviour of human leukaemic K562 cells
obtained with a dot-electrode eDEP systems; image shows the initial
and final results of 16 measurements at different frequencies.
Reprinted with permission from ref. 97, copyright 2011 Wiley-VCH
Verlag GmbH & Co. KGa. (b) Large array of HeLa cells clusters
obtained with nDEP. Reprinted with permission from ref. 98,
copyright 2013 Wiley-VCH Verlag GmbH & Co. KGa. (c) Concentration
of HeLa cells with pDEP in an eDEP system with circular electrodes,
the images show the progress of the cell enrichment, finalising with
the concentrated cells at the electrode centre. Reprinted with
permission from ref. 99, copyright 2011, AIP Publishing LLC.

Another important challenge in medical diagnostics is the ability to
concentrate cells, such as circulating tumour cells, in order to perform
further analyses. An eDEP system with circular electrodes was reported by Jen
and Chang[99] for the enrichment of HeLa cells by pDEP by applying 16 V in the
MHz range. The electrode chamber had a radius of 600 μm and electrodes
were 30 μm wide with a gap of 30 μm. Figure 7.13(c) shows the progress of

this process, where initially (left image) the cells were randomly distributed, followed by partial patterning of the cells (central image), and finally with strong cell concentration by pDEP at the centre of the circular electrode (right image). The entire preconcentration process lasted 160 s, and 76% efficiency was achieved. This system was integrated into a handheld device, which could be applied in point-of care applications.

7.6   Summary

Dielectrophoresis (DEP) has become one of the main pillars in microfluidics, allowing for a multitude of processes from particle concentration to continuous sorting operations; while offering high flexibility and ease of integration with other technologies. The growth in DEP-based studies has been impressive, as demonstrated by more than 300 papers published yearly since 2008. DEP offers the potential for a wide array of applications, from nanoanalysis, to bioanalytical and clinical assessments. We expect DEP to keep growing at a fast pace, with new designs and configurations becoming available as highly integrated systems, enabling more sophisticated analysis and manipulation of nanoparticles and biological samples.

References

1. B. J. Kirby, *Micro- and Nanoscale Fluid Mechanics. Transport in Microfluidic Devices*, Cambridge University Press, New York, 2010.
2. T. B. Jones, *Electromechanics of Particles*, Cambridge University Press, USA, 1995.
3. S. Chakraborty, in *Encyclopedia of Microfluidics and Nanofluidics*, ed. D. Li, Springer, US, 2008, pp. 444–453.
4. K. Dorfman, in *Encyclopedia of Microfluidics and Nanofluidics*, ed. D. Li, Springer, US, 2008, pp. 580–588.
5. J. Han, S. W. Turner and H. G. Craighead, *Phys. Rev. Lett.,* 1999, **83**, 1688–1691.
6. B. J. Kirby and E. F. Hasselbrink, *Electrophoresis,* 2004, **25**, 187–202.
7. B. J. Kirby and E. F. Hasselbrink, *Electrophoresis,* 2004, **25**, 203–213.
8. M. P. Hughes, *Nanoelectromechanics in Engineering and Biology*, CRC Press, Boca Raton, FL, 2002.
9. H. Morgan and N. G. Green, *AC Electrokinetics: Colloids and Nanoparticles*, Research Studies Press LTD, Hertfordshire, England, 2003.
10. S. Roy, H. Kumar and R. Agarwal, in *Encyclopedia of Microfluidics and Nanofluidics*, ed. D. Li, Springer, US, 2008, pp. 588–592.
11. V. M. Ugaz and J. L. Christensen, in *Microfluidic Technologies for Miniaturized Analysis Systems*, ed. S. Hardt and F. Schonfeld, Springer, New York, 2007, pp. 393–438.
12. T. A. Duncombe and A. E. Herr, *Lab Chip,* 2013, **13**, 2115–2123.

13. K.-i. Inatomi, S.-i. Izuo, S.-s. Lee, H. Ohji and S. Shiono, *Microelectron. Eng.*, 2003, **70**, 13–18.
14. J. N. Mehrishi and J. Bauer, *Electrophoresis*, 2002, **23**, 1984–1994.
15. Y. Kang and D. Li, in *Encyclopedia of Microfluidics and Nanofluidics*, ed. D. Li, Springer, US, 2008, pp. 517–518.
16. E. N. Kalaïdin, E. A. Demekhin and A. S. Korovyakovskiï, *Dokl. Phys.*, 2009, **54**, 210–214.
17. N. A. Mishchuk and P. V. Takhistov, *Colloids Surf. A Physicochem. Eng. Aspects*, 1995, **95**, 119–131.
18. A. Sze, D. Erickson, L. Ren and D. Li, *J. Colloid Interface Sci.*, 2003, **261**, 402–410.
19. A.-M. Spehar, S. Koster, V. Linder, S. Kulmala, N. F. de Rooij, E. Verpoorte, H. Sigrist and W. Thormann, *Electrophoresis*, 2003, **24**, 3674–3678.
20. D. Yan, C. Yang, N.-T. Nguyen and X. Huang, *Electrophoresis*, 2006, **27**, 620–627.
21. A. Ramos, H. Morgan, N. G. Green and A. Castellanos, *J. Colloid Interface Sci.*, 1999, **217**, 420–422.
22. M. Sin, J. Gao, J. Liao and P. Wong, *J. Biol. Eng.*, 2011, **5**, 6.
23. M. Bazant, in *Encyclopedia of Microfluidics and Nanofluidics*, ed. D. Li, Springer, US, 2008, pp. 8–14.
24. D.-H. Lee, C. Yu, E. Papazoglou, B. Farouk and H. M. Noh, *Electrophoresis*, 2011, **32**, 2298–2306.
25. A. Ajdari, *Phys. Rev. E*, 2000, **61**, R45–R48.
26. O. P. Bi and S. Simon, *J. Micromech. Microeng.*, 2012, **22**, 115034.
27. B. Cetin and D. Li, *Electrophoresis*, 2008, **29**, 994–1005.
28. R. C. Gallo-Villanueva, M. B. Sano, B. H. Lapizco-Encinas and R. Davalos, *Electrophoresis*, 2014, **35**, 352–361.
29. H. Morgan and N. Green, in *Encyclopedia of Microfluidics and Nanofluidics*, ed. D. Li, Springer, US, 2008, pp. 350–357.
30. T. B. Jones, *IEEE Eng. Med. Biol. Mag.*, 2003, **22**, 33–42.
31. D. Voyer, M. Frenea-Robin, F. Buret and L. Nicolas, *Bioelectrochemistry*, 2010, **79**, 25–30.
32. H. A. Pohl, *J. Appl. Phys.*, 1951, **22**, 869–871.
33. J. L. Baylon-Cardiel, N. M. Jesús-Pérez, A. V. Chávez-Santoscoy and B. H. Lapizco-Encinas, *Lab Chip*, 2010, **10**, 3235–3242.
34. A. R. Minerick and B. H. Lapizco-Encinas, *Dielectrophoresis Resource Page*, AES Electrophoresis Society, November 2013, http://www.aesociety.org/areas/dielectrophoresis.php.
35. B. H. Lapizco-Encinas, *Electrophoresis*, 2013, **34**, 951–951.
36. B. H. Lapizco-Encinas and F. Foret, *Electrophoresis*, 2011, **32**, 2231–2231.
37. B. H. Lapizco-Encinas and F. Foret, *Electrophoresis*, 2011, **32**, 2401–2401.
38. B. H. Lapizco-Encinas, R. Davalos, B. A. Simmons, E. B. Cummings and Y. Fintschenko, *J. Microbiol. Methods*, 2005, **62**, 317–326.
39. B. H. Lapizco-Encinas and M. Rito-Palomares, *Electrophoresis*, 2007, **28**, 4521–4538.

40. K. Khoshmanesh, S. Nahavandi, S. Baratchi, A. Mitchell and K. Kalantar-zadeh, *Biosens. Bioelectron.,* 2011, **26**, 1800–1814.
41. A. Kuzyk, *Electrophoresis,* 2011, **32**, 2307–2313.
42. B. Çetin and D. Li, *Electrophoresis,* 2011, **32**, 2410–2427.
43. H. Morgan, M. P. Hughes and N. G. Green, *Biophys. J.,* 1999, **77**, 516–525.
44. L. Benguigui and I. J. Lin, *J. Appl. Phys.,* 1984, **56**, 3294–3297.
45. B. H. Lapizco-Encinas, B. A. Simmons, E. B. Cummings and Y. Fintschenko, *Anal. Chem.,* 2004, **76**, 1571–1579.
46. P. Cheng, M. J. Barrett, P. M. Oliver, D. Cetin and D. Vezenov, *Lab Chip,* 2011, **11**, 4248–4259.
47. P. Patel and G. H. Markx, *Enzyme Microb. Technol.,* 2008, **43**, 463–470.
48. H. Moncada-Hernández and B. H. Lapizco-Encinas, *Anal. Bioanal. Chem.,* 2010, **396**, 1805–1816.
49. M. B. Sano, J. L. Caldwell and R. V. Davalos, *Biosens. Bioelectron.,* 2011, **30**, 13–20.
50. B. Sankaran, M. Racic, A. Tona, M. V. Rao, M. Gaitan and S. P. Forry, *Electrophoresis,* 2008, **29**, 5047–5054.
51. K. M. Leonard and A. R. Minerick, *Electrophoresis,* 2011, **32**, 2512–2522.
52. Y.-H. Su, M. Tsegaye, W. Varhue, K.-T. Liao, L. S. Abebe, J. A. Smith, R. L. Guerrant and N. S. Swami, *Analyst,* 2014, **139**, 66–73.
53. H.-S. Chuang, D. M. Raizen, A. Lamb, N. Dabbish and H. H. Bau, *Lab Chip,* 2011, **11**, 599–604.
54. F. Camacho-Alanis, L. Gan and A. Ros, *Sens. Actuators B,* 2012, **173**, 668–675.
55. B. H. Lapizco-Encinas, S. Ozuna-Chacón and M. Rito-Palomares, *J. Chromatogr. A,* 2008, **1206**, 45–51.
56. M. Duchamp, K. Lee, B. Dwir, J. W. Seo, E. Kapon, L. Forró and A. Magrez, *ACS Nano,* 2010, **4**, 279–284.
57. J. Regtmeier, R. Eichhorn, L. Bogunovic, A. Ros and D. Anselmetti, *Anal. Chem.,* 2010, **82**, 7141–7149.
58. N. G. Green, A. Ramos and H. Morgan, *J. Phys. D: Appl. Phys.,* 2000, **33**, 632–641.
59. N. M. Jesús-Pérez and B. H. Lapizco-Encinas, *Electrophoresis,* 2011, **32**, 2331–2357.
60. T. Schnelle, T. Muller and G. Fuhr, *J. Electrostatics,* 2000, **50**, 17–29.
61. X. B. Wang, J. Yang, Y. Huang, J. Vykoukal, F. F. Becker and P. R. C. Gascoyne, *Anal. Chem.,* 2000, **72**, 832–839.
62. S. van den Driesche, V. Rao, D. Puchberger-Enengl, W. Witarski and M. J. Vellekoop, *Sens. Actuators B,* 2012, **170**, 207–214.
63. P. R. C. Gascoyne, J. Noshari, T. J. Anderson and F. F. Becker, *Electrophoresis,* 2009, **30**, 1388–1398.
64. J. Čemažar and T. Kotnik, *Electrophoresis,* 2012, **33**, 2867–2874.
65. S. Williams, in *Encyclopedia of Microfluidics and Nanofluidics*, ed. D. Li, Springer, US, 2008, pp. 357–364.
66. S. Masuda, M. Washizu and T. Nanba, *IEEE Trans. Ind. Appl.,* 1989, **25**, 732–737.

67. S. Srivastava, A. Gencoglu and A. Minerick, *Anal. Bioanal. Chem.,* 2010, **399**, 301–321.

68. J. Zhu, S. Sridharan, G. Hu and X. Xuan, *J. Micromech. Microeng.,* 2012, **22**, 075011.

69. E. B. Cummings, *IEEE Eng. Med. Biol. Mag.,* 2003, **22**, 75–84.

70. A. Gencoglu, D. N. Olney, A. LaLonde, K. S. Koppula and B. H. Lapizco-Encinas, *Electrophoresis,* 2014, **35**, 363–373.

71. J. L. Baylon-Cardiel, B. H. Lapizco-Encinas, C. Reyes-Betanzo, A. V. Chávez-Santoscoy and S. O. Martínez Chapa, *Lab Chip,* 2009, **9**, 2896–2901.

72. R. C. Gallo-Villanueva, C. E. Rodríguez-López, R. I. Díaz-de-la-Garza, C. Reyes-Betanzo and B. H. Lapizco-Encinas, *Electrophoresis,* 2009, **30**, 4195–4205.

73. B. H. Lapizco-Encinas, B. A. Simmons, E. B. Cummings and Y. Fintschenko, *Electrophoresis,* 2004, **25**, 1695–1704.

74. H. Shafiee, J. Caldwell, M. Sano and R. Davalos, *Biomed. Microdev.,* 2009, **11**, 997–1006.

75. S. Patel, D. Showers, P. Vedantam, T.-R. Tzeng, S. Qian and X. Xuan, *Biomicrofluidics,* 2012, **6**, 034102–034112.

76. W. A. Braff, A. Pignier and C. R. Buie, *Lab Chip,* 2012, **12**, 1327–1331.

77. D. J. Bakewell and D. Holmes, *Electrophoresis,* 2013, **34**, 987–999.

78. I. Ermolina and H. Morgan, *J. Colloid Interface Sci.,* 2005, **285**, 419–428.

79. N. A. M. Yunus, H. Nili and N. G. Green, *Electrophoresis,* 2013, **34**, 969–978.

80. P. J. Beltramo and E. M. Furst, *Electrophoresis,* 2013, **34**, 1000–1007.

81. M. Sano, A. Rojas, P. Gatenholm and R. Davalos, *Ann. Biomed. Eng.,* 2010, **38**, 2475–2484.

82. J. Voldman, *Ann. Rev. Biomed. Eng.,* 2006, **8**, 425–454.

83. J. Regtmeier, T. T. Duong, R. Eichhorn, D. Anselmetti and A. Ros, *Anal. Chem.,* 2007, **79**, 3925–3932.

84. T. Yamamoto, O. Kurosawa, H. Kabata, N. Shimamoto and M. Washizu, *IEEE Trans. Ind. Appl.,* 2000, **36**, 1010–1017.

85. L. F. Zheng, J. P. Brody and P. J. Burke, *Biosens. Bioelectron.,* 2004, **20**, 606–619.

86. C. F. Chou, J. O. Tegenfeldt, O. Bakajin, S. S. Chan, E. C. Cox, N. Darnton, T. Duke and R. H. Austin, *Biophys. J.,* 2002, **83**, 2170–2179.

87. A. Nakano and A. Ros, *Electrophoresis,* 2013, **34**, 1085–1096.

88. M. P. Hughes and H. Morgan, *1st European Workshop on Electrokinetics and Electrohydrodynamics*, Glasgow, UK, 2001.

89. K.-T. Liao, M. Tsegaye, V. Chaurey, C.-F. Chou and N. S. Swami, *Electrophoresis,* 2012, **33**, 1958–1966.

90. B. G. Abdallah, T.-C. Chao, C. Kupitz, P. Fromme and A. Ros, *ACS Nano,* 2013, **7**, 9129–9137.

91. M. Koklu, S. Park, S. D. Pillai and A. Beskok, *Biomicrofluidics,* 2010, **4**, 034107–034115.

92. H. O. Fatoyinbo, M. P. Hughes, S. P. Martin, P. Pashby and F. H. Labeed, *J. Environ. Mon.,* 2007, **9**, 87–90.

93. E. D. Pratt, C. Huang, B. G. Hawkins, J. P. Gleghorn and B. J. Kirby, *Chem. Eng. Sci.,* 2011, **66**, 1508–1522.
94. A. Sonnenberg, J. Y. Marciniak, J. McCanna, R. Krishnan, L. Rassenti, T. J. Kipps and M. J. Heller, *Electrophoresis,* 2013, **34**, 1076–1084.
95. S. K. Srivastava, P. R. Daggolu, S. C. Burgess and A. R. Minerick, *Electrophoresis,* 2008, **29**, 5033–5046.
96. P. Jones, S. Staton and M. Hayes, *Anal. Bioanal. Chem.,* 2011, **401**, 2103–2111.
97. H. O. Fatoyinbo, N. A. Kadri, D. H. Gould, K. F. Hoettges and F. H. Labeed, *Electrophoresis,* 2011, **32**, 2541–2549.
98. R. G. Abdallat, A. S. Ahmad Tajuddin, D. H. Gould, M. P. Hughes, H. O. Fatoyinbo and F. H. Labeed, *Electrophoresis,* 2013, **34**, 1059–1067.
99. C.-P. Jen and H.-H. Chang, *Biomicrofluidics,* 2011, **5**, 034101–034110.

CHAPTER 8

Novel Lab-on-a-Chip Sensing Systems: Applications of Optical, Electrochemical, and Piezoelectric Transduction in Bioanalysis

ANTHONY J. TAVARES, SAMER DOUGHAN,
M. OMAIR NOOR, MATTHEW V. DACOSTA,
PAUL A. E. PIUNNO, AND ULRICH J. KRULL*

Chemical Sensors Group, University of Toronto Mississauga, 3359
Mississauga Rd., Mississauga, Ontario, L5L 1C6, Canada
*E-mail: ulrich.krull@utoronto.ca

8.1   Introduction

8.1.1   Overview and Classifications of Microfluidics

Microfluidics is often defined as the manipulation of small volumes of fluids (10^{-9} to 10^{-18} L) in droplet form or in channels that are on the order of tens to hundreds of micrometres in size in at least one dimension.[1] These miniaturised platforms have found numerous applications in analysis in chemistry, biology and medicine as a result of the advantages they offer; reduced sample volumes, decreased processing time, low reagent consumption,

RSC Detection Science Series No. 5
Microfluidics in Detection Science: Lab-on-a-chip Technologies
Edited by Fatima H Labeed and Henry O Fatoyinbo
© The Royal Society of Chemistry 2015
Published by the Royal Society of Chemistry, www.rsc.org

without compromising specificity and sensitivity of analysis. This form of fluid handling offers a platform that is amenable to automation, often at a fraction of the cost of many other analytical formats. Furthermore, laminar flow, the capability for parallel analysis, and versatility of design and portability has made this technology popular for in-field point-of-care (POC) diagnostics.[2,3] Microfluidic technologies can be classified under three designations: channel based, droplet, and digital microfluidics.[4] Both channel and droplet fluidics require fluid processing in microchannels; the former entails continuous streams of fluid and the latter uses a sequential stream of droplets that are constrained within a channel. In digital microfluidics (see Chapter 4), droplets of fluid are transported on a Teflon surface by an underlying array of patterned electrodes. Control of voltages that are applied at numerous electrode pads systematically drive droplet movement across the surface of the device based on capacitive charge, providing a system that can manipulate very small volumes in analogy to macroscale wet-solution handling by volumetric pipetting.

Microfluidic devices can be fabricated using different technologies (refer to Chapter 1) such as photolithography, soft lithography, electron-beam lithography, wax printing and nanoimprinting. Of these, soft lithography has been the most popular since it is cost effective and offers high-throughput production. Specifically, the use of polydimethylsiloxane (PDMS), poly(methyl methacrylate) (PMMA) and polycarbonate (PC) is popular due to ease of use, low cost and excellent optical transparency from the UV to the near-IR region of the spectrum, which allows for implementation of optical detection strategies.[2] Electrochemical and mass-sensitive techniques are also used in addition to optical methods for the detection of cells, biomolecules and metabolites.[5]

Recent advances in polymer materials and microfabrication technology have made it possible to incorporate micropumps, micromixers, microvalves, microfilters, microreactors and microseparators onto microfluidic chips.[5] This permits on-chip sample preparation, amplification and purification that enables the fabrication of self-contained microfluidic devices for POC applications. Fluid manipulation in microfluidic devices can be mechanical involving piezoelectric, pneumatic, electrostatic or electromagnetic pumps. Alternatively, nonmechanical techniques can be implemented and these include electro-osmotic flow (EOF), electrophoretic flow, electrowetting, optoelectric manipulation and magnetohydrodynamic pumping.[5]

Recently, much interest has been directed toward microfluidic paper-based analytical devices (μ-PADs), which are based on the concepts of dipstick and lateral flow assays. By the use of wax printing, photolithography, cutting or plasma etching, for example, fluid can be constrained in micro-channels in porous cellulose or nitrocellulose paper, where enclosed fluid is transported *via* capillary action. Microfluidic paper-based electrochemical devices (μ-PEDs) are a subclass of μ-PADs that can be prepared from conducting materials like carbon or graphite, the latter being implemented by printing or pencil drawing on paper or plastic. The cost of each μ-PAD "device" is minimal, and fluid manipulation requires no external pump or power supply. Moreover, μ-PADs can be easily manufactured in bulk, handled and stored for high-throughput analyses. Assay development in microfluidics and μ-PADs can offer the ASSURED criteria: affordable, sensitive, specific, user friendly, rapid, equipment free, deliverable to end users. Currently, challenges associated with μ-PADs include variable specificity and sensitivity, poor reproducibility, and difficulties with sample pretreatment on paper.[6]

8.1.2   Chapter Outline

Microfluidic platforms offer an attractive approach for the design of biosensors and bioassays allowing for the realisation of lab-on-a-chip (LOC) and POC systems for clinical analyses.[7] In tandem with the prevalent interest in the development of functional nanomaterials, the two have provided the bioanalytical community with new opportunities for sensing and assay development. This widespread interest in nanomaterials has accelerated numerous advances in analytical capability and performance, with many of these derived from the intrinsic physical properties of the nanoparticle and the selection of diverse surface coatings that enable routine coupling of biomolecules.[8] Moreover, the size scale of these materials allows for ease of integration into microfluidic and LOC-type sensors and assays, where the unique properties of the nanoparticle can be exploited to provide a diverse array of transduction strategies for a given analyte. Recent reviews have highlighted the impact of nanomaterials on biosensing and for bioassays in LOC formats.[4,9–13]

The scope of this chapter is aimed at providing the reader with an overview of the many transduction strategies that have been integrated into LOC based biosensors and bioassays. Focus is directed toward optical, electrochemical and piezoelectric transduction technologies. Optical based detection techniques include: fluorescence and fluorescence resonance energy transfer (FRET), surface plasmon resonance (SPR) and localized surface plasmon resonance (LSPR), in addition to Raman and surface enhanced Raman spectroscopy (SERS). Voltammetric and field-effect transistors (FET), and piezoelectric techniques, are based on electrochemical and mass/viscosity responses, respectively. Emphasis is provided on discussion of bioassays and biosensors that integrate nanoparticles (NPs) into transduction strategies. The chapter concludes with a prospective outlook on the current direction of biosensing with novel LOC systems.

8.2   Optical Detection

Optical methods of transduction are generally desired to have high sensitivity and low detection limits. In reference to fluorescence, plasmonic, and Raman techniques, single-molecule detection is the ultimate detection limit.[14-17] However, for LOC systems discussed herein detection limits are often between femtomole to picomole quantities of target. Integration of optical detection strategies in LOC systems tends to require offline analysis due to the sophisticated instrumentation required for sorting of optical signals. Much of the earlier work that combined optical detection with fluidic technologies has made use of laser-induced fluorescence (LIF) for detection,[18,19] and this methodology will not be discussed as it has been widely used and reviewed elsewhere.[20] More recent advances have made use of the near-field phenomena associated with plasmonics and resonance energy transfer, as well as engineering designs that have integrated optical waveguides into LOC systems, improving the potential for optical transduction in field portable and POC detection.[21] In recent work, integrated waveguides have shown successful application in simple plasmonic aggregation assays.[22] Moreover, complementary metal oxide semiconductor (CMOS) detectors with integrated filters have been used in fluorescence based assays.[23,24] Herein, we provide the reader with an overview of the theoretical concepts and some applications of FRET, LSPR and SERS in combination with LOC systems.

8.2.1 Transduction using Fluorescence-Based Methods

8.2.1.1 Fluorescence Resonance Energy Transfer

The process of FRET is defined as a nonradiative, through-space coupling of transition dipoles. The mechanism considers an excited state donor D and a ground state acceptor A in close proximity, where the difference in energy between the excited state and ground state of D match resonant transitions in the molecular orbitals of A. The accepted FRET range for energy transfer is 1–10 nm.[25] The D–A separation distance (r) determines the degree of energy transfer, where the FRET efficiency is a function of the inverse sixth power of the distance between the D–A as governed by the Förster formalism. The rate of energy transfer (k_{ET}) is given in eqn (8.1),[26]

$$k_{ET} = \frac{9(\ln 10)\kappa^2 \Phi_D J}{128\pi^5 n^4 N \tau_D r^6} = \left(\frac{1}{\tau_D}\right)\left(\frac{R_0}{r}\right)^6 \tag{8.1}$$

where κ^2 is an orientation factor that describes the angle between the two transient dipoles, the refractive index (n) of the medium, the quantum yield (Φ_D) and fluorescence lifetime (τ_D) of the donor in absence of the acceptor, Avogadro's number (N) and J, the spectral overlap between the donor emission band and the absorption spectrum of the acceptor. Often the rate of energy transfer is expressed as a function of the characteristic Förster distance (R_0), which is defined as the distance between the D–A when the efficiency of energy transfer is 50%. The Förster distance can be determined experimentally as given in eqn (8.2) where the spectral overlap as a function of the wavelength is shown in eqn (8.3).[26]

$$R_0^6 = (8.79 \times 10^{-28}) n^{-4} \Phi_D \kappa^2 J(\lambda) \tag{8.2}$$

$$J(\lambda) = \frac{\int F_D(\lambda)\varepsilon_A(\lambda)\lambda^4 d\lambda}{\int F_D(\lambda)} \tag{8.3}$$

The fluorescence emission of the donor, $F_D(\lambda)$ and the acceptor absorption coefficient, $\varepsilon_A(\lambda)$ in units of M^{-1} cm^{-1} are integrated across the resonant wavelength range of characteristic overlap. The energy transfer efficiency, E at a given D–A separation distance is given in eqn (8.4). In the case for NP applications in FRET, eqn (8.4) takes on another form to account for the acceptor valence, a around a given donor since it is possible to conjugate multiple biomolecules to the surface of a NP.[13]

$$E = \left(\frac{R_0}{R_0 + r}\right)^6 = \left(\frac{aR_0}{aR_0 + r}\right)^6 \tag{8.4}$$

The extent of acceptor valency ultimately provides for a higher FRET efficiency since there are more pathways available for energy transfer. Furthermore, in solid-phase assays consisting of a 2-dimensional array of donors with proximal acceptors, the efficiency is enhanced as energy-transfer pathways extend beyond single-donor multiple-acceptor formulations and one must consider a higher order of donor–acceptor pathways. Transduction of analytical targets of interest by FRET offers the opportunity to integrate ratiometric detection by considering the magnitude of both the donor and acceptor emission. This reduces the impact of inconsistency in assay preparation and other sources of error that affect analytical precision.[13]

8.2.1.2   Applications of Fluorescence and FRET Transduction in LOC Platforms

A fluorescent LOC immunoassay for the recognition of cancer biomarkers was developed using immobilised zinc oxide nanorods (ZnO NRs) in a glass capillary. Portable analysis was accomplished using an in-house fabricated fluorescence analyser with an integrated flow pump.[27] ZnO NRs were synthesised on the inner walls of the capillary followed by overcoating with epoxy terminated silane to facilitate immobilisation of recognition antibodies for prostate specific antigen (PSA), α-fetoprotein (AFP) and carcinoembryonic antigen (CEA). Bovine serum albumin (BSA) was then coated within the capillary interior to prevent nonspecific adsorption of subsequent sample components since all analytes were detected in 10% human serum. A recognition antibody mixture was then dispensed and contained a Cy3-tagged IgG from rabbit. Enhancement of Cy3 fluorescence from the immobilised ZnO NRs was reported to improve the LOD for PSA, CEA, and AFP, and these values were reported to be 1, 5, and 5 ng ml^{-1}, respectively. Transduction of all markers was possible over a dynamic range spanning more than two orders of magnitude.[27]

Optical barcoding technologies have shown success in nucleic acid transduction and are also amenable to LOC-based sensing.[28,29] Moreover, optical barcodes using semiconductor quantum dots (QDs) are highly advantageous in comparison to counterparts that use molecular fluorophores. This can be explained by the unique photophysical properties of QDs that include: high quantum yield, long photoluminescence (PL) lifetimes, resistance to photobleaching, broad adsorption spectra, and size-tuneable narrow PL emission profiles.[30,31] The latter two photophysical properties offer ease of multiplexing since a single excitation source can excite multiple colours of QDs. This has important implications for barcoding technologies, allowing for a large number of permutations by variation of both the composition of mixtures of QD with different PLs and the respective concentrations of these QDs to control relative emission intensities. Multiplexed detection of up to four oligonucleotide markers characteristic of HIV, hepatitis B, and syphilis has been demonstrated on-chip with QD optical barcodes comprised of magnetic polystyrene microbeads.[32] In this report, the feasibility of full automation of QD barcodes was investigated by implementation of the assay into a LOC platform. In this format, sample-processing methods including purification steps that would typically require an advanced laboratory setting were avoided. This enabled a POC diagnostic assay that could be suitable for use in undeveloped countries given that sample processing was conducted on-chip without the need for highly trained personnel.[32] The chip design is shown in Figure 8.1 and used three channel inlet ports, where two enabled mixing of reagents followed by subsequent flow into a reaction chamber. An additional inlet port for washing and removal of excess reagents was implemented prior to bead alignment in the detection zone. Three magnets were affixed along the chip to deflect the QD magnetic barcodes within the channel and to enable alignment for detection. Four different QD barcodes were created by varying the concentration of QDs with PL maxima at 570 and 650 nm, with the QDs

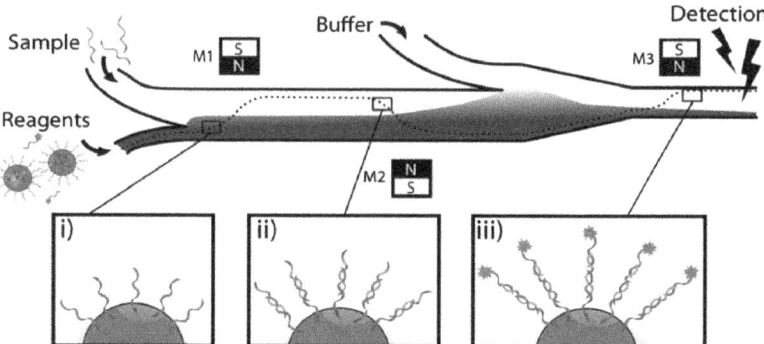

Figure 8.1 LOC channel design where three magnets positioned adjacent to the microchannel enabled mixing of magnetic QD barcodes with target sequences (i) followed by attraction towards the second magnet in order to hybridise with reporter probe (ii) before aligning the barcodes in the detection zone (iii). Reprinted with permission from ref. 32. Copyright 2013 American Chemical Society.

being integrated with magnetic iron NPs within polystyrene microbeads. Oligonucleotide probes were then conjugated to the barcoded beads using standard carbodiimide chemistry, followed by challenging the beads with target oligonucleotides. Detection was based on a sandwich assay format where an additional oligonucleotide reporter tagged with Alexa Fluor 488 (A488) hybridised to a remaining unhybridised portion of the previously hybridised target sequence (see Figure 8.1). The reporter sequence was designed to be identical for all four targets. Detection of both QD and A488 PLs provided for self-calibration and reduced the variance in PL from bead-to-bead. In single-target analysis, a dynamic range spanning *ca.* two orders of magnitude and a LOD of 1.2 nM was reported for the on-chip analysis. This was comparable to the 1.0 nM LOD reported for off-chip interrogation using flow cytometry. In multiplexed analyses, transduction of four different biomarkers was possible within 20 min with no crosstalk, regardless of whether single or multiple oligonucleotide targets were present.[32]

The distance sensitivity of FRET has attracted much attention in the bio-analytical community as a detection strategy for bioassays. Examples of NP and organic dye FRET assays have been developed for the detection of small-molecule toxins, proteins, and oligonucleotides. Detection of the highly potent botulinum neurotoxin a (BoNT-A) using a FRET-based LOC assay has been reported by Sun *et al.*[33] The assay utilised a peptide substrate tagged with fluorescein isothiocyanate (FITC) and 4-(dimethylaminoazo) benzene-4-carboxylic acid (DABCYL) to form a FRET pair. In the presence of BoNT-A, peptide cleavage prevents energy transfer from FITC to DABCYL and regen-eration of FITC fluorescence is used to quantify the amount of BoNT-A. Concentrations as low as 50 nM of BoNT-A could be detected, which is well below the lethal dose. The on-chip assay also permitted the simultaneous quantitative analysis of eight 10-μl samples.[33] A LOC FRET biosensor has also been reported for the detection of T-cell acute lymphoblastic leukemia cells.[34] A schematic of the multichannel microfluidic chip is shown in Figure 8.2(a) and represented pictorially in Figure 8.2(b). The operating principles of the biosensor design are displayed in Figure 8.2(c) that utilised graphene oxide (GO) sheets and a recognition aptamer (sgc8) labelled with carboxy-fluorescein (FAM). Adsorption of the FAM–sgc8 to the GO sheets resulted in quenching of FAM fluorescence upon excitation. In the presence of cancer cells, the aptamer adopted a hairpin-loop structure and preferentially bound to the cell with high selectivity ($K_d = 10^{-10}$ M) causing the FAM–sgc8 aptamer to dislodge from the GO sheets. FAM fluorescence was restored and was used to quantify the number of cells that were present, where as few as 25 cells ml^{-1} could be detected.[34] The biosensor maintained high selectivity and displayed no analytical response when tested against four additional cancer cell lines.

Our team has been investigating solid-phase QD-FRET assays for the transduction of oligonucleotides that are characteristic of pathogens and genetic disorders.[35] The foundation of this work began in solution-based analyses[36,37] and was then transferred to solid-phase formats[38,39] by means of

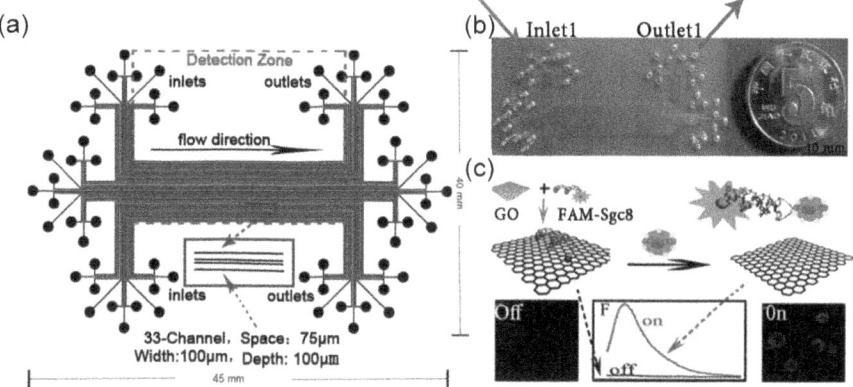

Figure 8.2 (a) General schematic of the multichannel GO-FRET aptasensing chip, pictorally represented in (b) the scale bar in the image is 10 mm. (c) Transduction method for cancer cell detection where GO/FAM–sgc8 interactions are destabilised in the presence of cells as a result of higher aptamer affinity. Binding of sgc8 to the cell restored the fluorescence of FAM, which was initially quenched by energy transfer with GO sheets in channel. Emission spectra and epifluorescence images are also shown for "on/off" operating modes. Adapted from ref. 34 with permission from the Royal Society of Chemistry.

immobilisation of QDs.[40] Initially the sensors were designed to operate on either glass beads[41] or optical fibres and offered concurrent detection of up to four targets using a combination of fluorescence and FRET transduction.[42] Current efforts in our group have aimed at miniaturisation of solid-phase QD-FRET assays in electrokinetically controlled biochips,[43–45] and more recently on PADs.[46,47] The fundamental design of all these assays utilised QDs as donors in FRET and dye-labelled oligonucleotide targets as acceptors, albeit the interfacial chemistry and form of quantitative analysis differed significantly. The assay designs for multiplexed oligonucleotide transduction on-chip and on PADs are shown in Figures 8.3 and 8.4, respectively. In duplexed analysis, green- and red-emitting QDs were paired with Cy3 and Alexa 647 (A647) tagged targets, respectively. In the microfluidic assay, streptavidin (Sav)-coated QDs with PL maxima at 525 nm and 605 nm were coimmobilised on a biotinylated glass coverslip in a PDMS–glass chip and were subsequently conjugated with biotinylated oligonucleotides. Assay assembly was done in-channel within a few minutes, and required only 5–7 µl of reagent at each step. EOF and electrophoresis enabled efficient dispensing of QDs and nucleic acids, respectively. Injection of Cy3- and A647-labelled target oligonucleotides resulted in hybridisation with immobilised probe sequences, where hybridisation provided the necessary proximity for FRET from the interfacial QD donors. Cy3 and A647 FRET sensitised PL served as the analytical signal for quantitative transduction of nucleic acid targets. Electrophoretic delivery of oligonucleotide targets proceeded, and

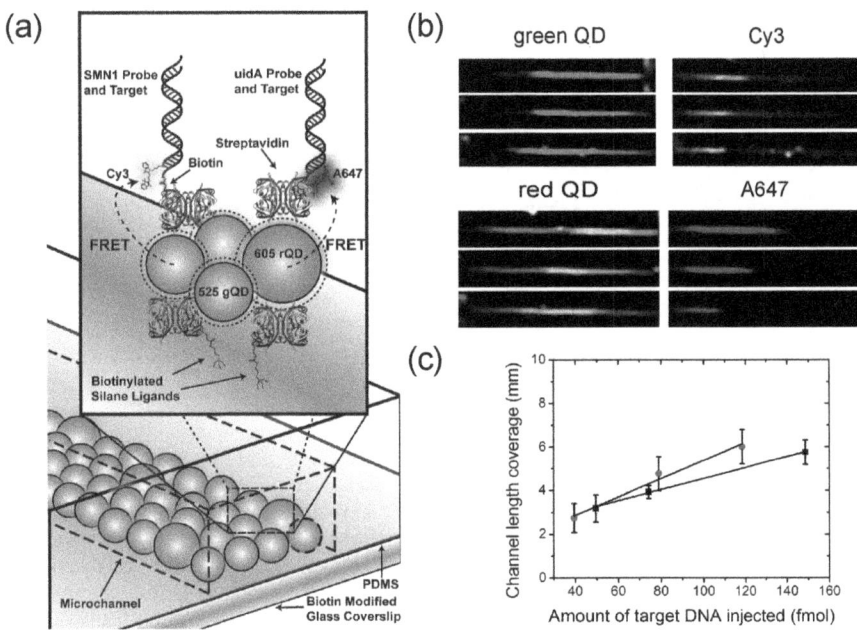

Figure 8.3 (a) Cross-sectional schematic of the on-chip multiplexed solid-phase QD-FRET assay. Sav-coated red and green (605 QD) and (525 QD) QDs were immobilised on a biotinylated glass coverslip in a PDMS–glass microfluidic chip and subsequently conjugated with biotinylated oligonucleotide probes that were oligonucleotides characteristic of human spinal muscular atrophy (SMN1) and *E. coli* (uidA). Hybridisation with Cy3 and A647 tagged targets provided the necessary proximity for FRET from the QDs. (b) epi-fluorescence microscope images displaying the spatial PL and FRET-sensitised PL profiles from immobilised QDs and QD–oligonucleotide hybrids, respectively. The amount of the Cy3 SMN1 target was fixed at 50 fmol in all channels while the amount of A647 uidA target was 118 fmol (top), 78.9 fmol (middle), and 39.4 fmol (bottom). (c) Response curve for quantitative nucleic acid transduction by measuring the length of a FRET-sensitised spatial PL profile against the quantity of oligonucleotide injected. Adapted with permission from ref. 45. Copyright 2012 Elsevier.

hybridisation was observed to vary along the length of the assay interface in correlation with the quantity of target that was introduced. Once probe sites became saturated, any remaining target was delivered to probes downstream until the sample had in effect been extracted onto the channel surface. Hybridisation was transduced within 5 min by determining the spatial FRET profiles along the length of channels. Quantification of the target (see Figure 8.3) in a sample was achieved by measurement of the length of the channel from the injection site to the point where the FRET-sensitised acceptor and QD donor PLs were at equal intensity. This crossover point was determined to be a reliable location to measure the length of a spatial profile. In multiplexed analysis, improved sensitivity was reported in contrary to

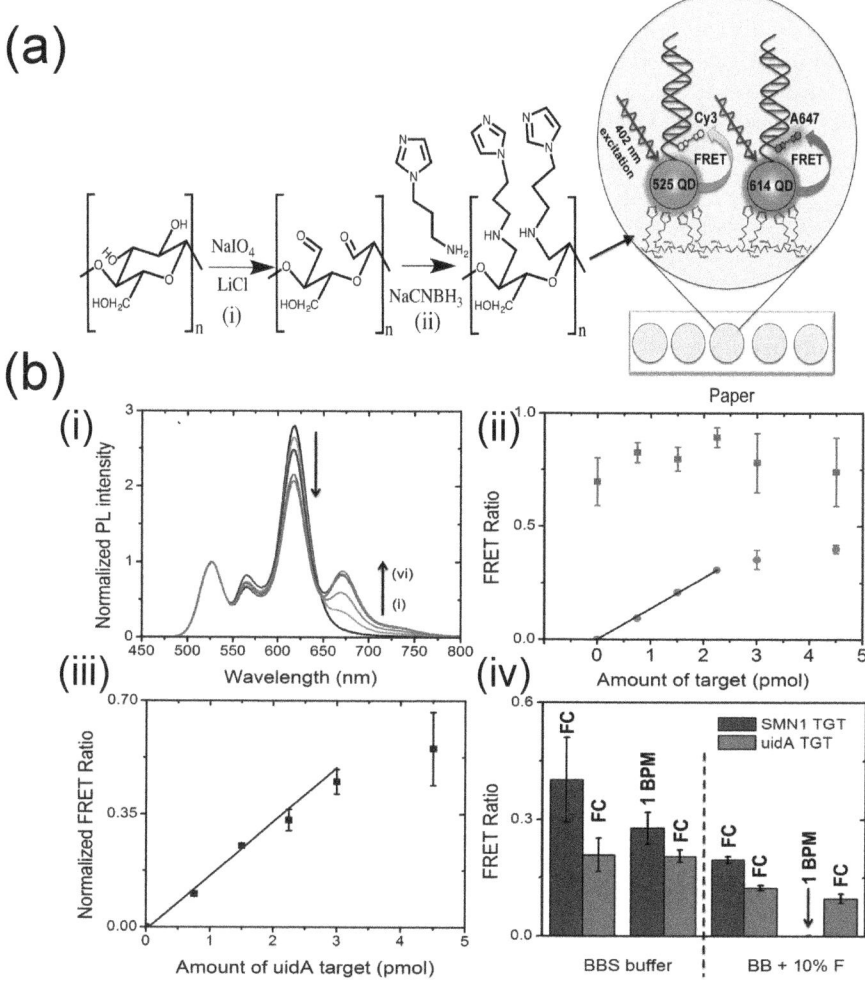

Figure 8.4 (a) Solid-phase QD-FRET mutiplexed nucleic acid hybridisation on a PAD. Glutathione-capped green- and red-emitting QDs are immobilised on cellulose paper derivatised with monodentated imidazole ligands; the two-step paper modification protocol involved (i) periodate oxidation followed by reductive amination (ii) with sodium cyanoborohydride as highlighted. Dithiol-tagged oligonucleotide probes are self-assembled on immobilised QDs, where hybridisation with Cy3 and A647 oligonucleotides result in energy transfer from the QDs and FRET-sensitised acceptor emission serves as the analytical signal from quantitative transduction. (b) (i) PL spectra after subsequent hybridisation with both targets where the Cy3 target served as an internal standard and was fixed at 3 pmol. The A647 target represented the transduced analyte and varied from 0–4.5 pmol. The response curve for both targets (squares – Cy3, and circles – A647) is shown in (ii). (iii) The normalised FRET ratio response for the A647-labelled target response shown in (ii). (iv) Discrimination against a 1 BPM target using 10% v/v formamide. Adapted with permission from ref. 46. Copyright 2013 American Chemical Society.

traditional ratiometric QD-FRET assays where probe dilution decreased sensitivity due to reduction in the FRET ratio. In this transduction method, addition of two probes of the same concentration resulted in less target capture per unit area to achieve duplex saturation. While the FRET ratio was also reduced, the length of a spatial FRET PL profile increased for a given target concentration relative to the same quantity of oligonucleotide in single target assays. Selective discrimination against a target containing a three base-pair mismatch (3 BPM) was possible and as little as 3 fmol of 20 mer target could be detected as determined by extrapolation of the linear response range.[45]

In QD-FRET assays on PADs, glutathione-capped alloyed QDs were immobilised on cellulose paper that had been derivatised with mono-dentate imidazole ligands as shown in Figure 8.4(a). Given the success of hexahistidine[48] and multidentate imidazole ligands[49,50] for QD conjugation and immobilisation, the collective abundance of imidazole ligands in the cellulose matrix displayed excellent avidity toward the Zn-rich shell of the QDs. Tethering of oligonucleotide probes was done prior to QD-bioconjugate immobilisation on paper, and was facilitated using a dithiol tag on the probe sequence. Hybridisation of Cy3 and A647 tagged targets resulted in FRET sensitised emission from the defined assays spots. The matrix and interfacial chemistry also displayed excellent resistance to adsorption of oligonucleotides. Discrimination against a single base-pair mismatch (1 BPM) with a contrast ratio of 50 : 1 was reported and multiple cycles of assay regeneration were possible. In multiplexed analysis, control over respective donor QD concentration ratios enabled a "tuneable" dynamic range spanning over an order of magnitude and a LOD of 90 fmol. Each assay spot integrated both a control and test zone, where the green QD-Cy3 (control) and red QD-A647 (test) FRET pairs worked in tandem to offer internal assay standardisation.[46]

8.2.2   LSPR and SERS-Based Detection

8.2.2.1   LSPR Theory

Incident electromagnetic radiation on a thin metal film of *ca.* 50 nm can, under certain conditions, produce a coherent oscillation of electrons in the conduction band at the surface of the metal in a process termed surface plasmon resonance (SPR). Commonly used metals include gold and silver, where plasmon resonance can translate hundreds of micrometres across the metal film and has characteristic decay lengths (1/*e*) of electric-field strength of up to a few hundreds of nanometres into the surrounding environment.[51] In localised surface plasmon resonance (LSPR), an incident photon causes a distortion in the electron density of a metal nanoparticle. The size of the NP is typically about 10-fold smaller than the wavelength of light. The distortion of the electron cloud about the nuclei composing the metal initiates a restoring force due to Coulombic attraction. This results in a charge-density oscillation localised about the NP, where the frequency of oscillation is governed by the effective electron mass, the electron density, and the size and shape of the charge localisation. LSPR is often called a dipole plasmon resonance to distinguish this resonance from those on planar metal films or in the bulk material.[52] Both SPR and LSPR are sensitive to the interfacial dielectric environment, where changes induced by selective biomolecule binding and or nonspecific adsorption can be transduced. This inherent sensitivity of the LSPR has been implemented in many bioassays and biosensors for the detection of a wide variety of analytes.[11]

To understand the extinction profile of a spherical metal nanoparticle of varying size, Mie presented a solution to Maxwell's equation in 1908 that describes both absorption and scattering components of the spectrum.[52] This model considers a spherical particle of radius a interacting with a photon of wavelength λ, where a/λ is <0.1, in this notation the electric field of z-polarised light can be considered constant, where this assumption is known as the quasistatic approximation.[51,52] The LSPR of the particle is given by eqn (8.5),

$$E_{out}(x, y, z) = E_0\hat{\mathbf{z}} - \left[\frac{\varepsilon_{in} - \varepsilon_{out}}{\varepsilon_{in} + 2\varepsilon_{out}}\right] a^3 E_0 \left[\frac{\hat{\mathbf{z}}}{r^3} - \frac{3z}{r^5}(x\hat{\mathbf{x}} + y\hat{\mathbf{y}} + z\hat{\mathbf{z}})\right] \tag{8.5}$$

where the dielectric constant of the interfacial environment and the particle are denoted by ε_{out} and ε_{in}, respectively. The resulting plasmon frequency outside of the particle is a function of the radius (a) and the dielectric constant of the outside matrix, while the resonance condition is determined by the term $[(\varepsilon_{in} - \varepsilon_{out})/(\varepsilon_{in} + 2\varepsilon_{out})]$ in eqn (8.5) since the dielectric constant of the particle is strongly dependent on wavelength. The electromagnetic field outside the particle is described as a function of the Cartesian coordinates x, y, and z, the corresponding vectors: $\hat{\mathbf{x}}$, $\hat{\mathbf{y}}$, and $\hat{\mathbf{z}}$ radial distance (r). The extinction spectrum of the particle is given by eqn (8.6)

$$E(\lambda) = \frac{24\pi^2 N \alpha^3 \varepsilon_{out}^{3/2}}{\lambda \ln(10)} \left[\frac{\varepsilon_i(\lambda)}{(\varepsilon_r(\lambda) + \chi\varepsilon_{out})^2 + \varepsilon_i(\lambda)^2}\right] \tag{8.6}$$

where the dielectric of the metal particle is described as a function of both real (ε_r) and imaginary (ε_i) components. The geometrical shape of the particle also has implications on the resulting extinction spectrum, where χ assumes a value ranging from 2–20 depending on the particle aspect ratio; the initial value of 2 denotes a sphere.[51] Control over shape, in addition to particle size, has enabled a myriad of particle ensembles that have LSPR bands spanning from the ultraviolet to the infrared region of the electromagnetic spectrum. For a detailed discussion of the correlation between NP shape and the LSPR, the interested reader can refer to a comprehensive review.[53] The discrete dipole approximation and the finite-difference time-domain methods are used to estimate χ for particles of distinctive shapes since χ can only be calculated for spheres and spheroids.[51]

For quantitative analysis it is necessary to define a relationship that describes the resultant shift of the peak maximum (λ_{max}) upon biomolecule binding. This transduction method can be described using eqn (8.7),[51]

$$\Delta\lambda_{max} = m\Delta n\left[1 - e^{2d/l_d}\right] \tag{8.7}$$

which considers the change in refractive index (Δn) caused by an interfacial binding or adsorption event, the bulk refractive-index response of the NP(s) (m), the thickness (d) of the resulting biorecognition assembly, and (l_d) the characteristic decay length of the electromagnetic field. In addition to wavelength-shift measurements, proximal plasmon coupling, angularly resolved and imaging detection modalities have also been demonstrated using LSPR.[51,54,55]

8.2.2.2 Extraordinary Optical Transmission in Nanohole Arrays

Metal nanohole arrays have also found application in SPR biosensing and are amenable to the development of LOC-based sensing formats.[56] The utility of the nanohole substrate arises from the enhanced extraordinary optical transmission (EOT) phenomenon that was first identified in 1998 by Ebbesen *et al.*[57] when incident incoherent light on a nanohole array displayed zero-order transmission spectra where the photons detected were colinear. Intensities of the transmitted bands were also higher than those of the incident photons of corresponding wavelength, contrary to what was predicted by aperture theory. Nanohole arrays were fabricated through deposition of a *ca.* 200-nm gold film on a quartz substrate and cylindrical holes were then created through focused-ion-beam milling. Regardless of the metals, the hole diameters, and film thicknesses investigated, it was found that the position of the bands in the spectrum changed only as a function of the array periodicity. Furthermore, this phenomenon was also observed for incident wavelengths that were longer than the array period where diffraction does not occur.[57] Moreover, the width of the bands in the spectrum were a function of the aspect ratio between film depth and hole diameter, and transmission changed as a result of the angle of incidence. It was hypothesised that the EOT was due to plasmonic excitation followed by resonant coupling of adjacent plasmons within the array.[58]

$$\lambda_{spr}(i,j) = \frac{P\sqrt{\dfrac{\varepsilon_d\varepsilon_m}{\varepsilon_d + \varepsilon_m}}}{\sqrt{i^2 - j^2}} \tag{8.8}$$

The wavelength of SPR resonance from an array (λ_{spr}) is described by eqn (8.8), where P is the periodicity of the array (space between adjacent holes), ε_d and ε_m are the dielectric constants of the medium and metal, respectively, and the scattering orders of the array are given by the integers i and j.[58]

8.2.2.3   Applications of LSPR in LOC Sensing Formats

In recent work, Au NPs have been integrated into a LOC sensing format for the detection of bovine growth hormone (BGH) found in milk.[59] In this system a PDMS–PDMS microfluidic chip was used and operated under pressure-induced flow. Herein, the chip served both to synthesise the Au NPs from gold precursor and for assembly of the biorecognition interface and subsequent target detection. In-channel synthesis of the Au NPs was mediated by the diffusion of the prepolymer from the oxidised PDMS to the channel interface, which caused reduction of the gold precursor. The embedded Au NPs were grown over a 24-h period since the kinetics of the reaction were limited by monomer diffusion from within the PDMS to the channel interface. Coating the Au NPs with mercaptoundecanoic acid followed by standard carbodiimide activation of the terminal carboxylic acid enabled immobilisation of anti-BGH. Assembly of the assay recognition chemistry was monitored by a shift in the LSPR band from its initial position of 525 nm to 536 nm after antibody coupling. Transduction of BGH was also facilitated by monitoring a shift in the LSPR band. As little as 3.7 ng ml^{-1} (185 pM) of BGH could be detected *via* interpolation of the dynamic range, which spanned *ca.* 1.5 orders of magnitude.[59]

An LSPR biochip has also been developed for the enantioselective detection of the anticoagulant therapeutic melagatran.[60] Gold nanorods (Au NRs) with an aspect ratio of 2.6 were immobilised in a glass microfluidic chip and further decorated with α-thrombin, which is highly selective for RS-melagatran. The immobilised Au NRs exhibited a bimodal extinction spectrum that is characteristic of particles with high aspect ratio due to the presence of longitudinal and transverse plasmon resonances. It was noted that the longitudinal band (λ_{max}: *ca.* 725 nm) displayed a greater sensitivity to a local change in the dielectric in comparison to the transverse plasmon (λ_{max}: *ca.* 560 nm) where the shifts in λ_{max} were 3.5 nm and 1.2 nm, respectively. Monitoring the peak shift of the 725 nm plasmon resonance as a function of RS-melagatran binding offered a S/N of 5.5 and a linear response ranging from 0.9–25 nM. No false-positive signals were seen in the presence of a 10^4-fold excess of the SR enantiomer and across the entire dynamic range. Analysis was also possible in a complex matrix composed of human serum. Transduction of as low as 0.9 nM RS-melagatran was demonstrated where analyte recovery was statistically indifferent to what was seen for a pure sample.[60]

Gold nanohole arrays have also found application in a LOC system that was reported by Escobedo *et al.*[61] for the quantitative transduction of r-PAX8, an upregulated transcription factor found in ovarian cancer. A gold nanohole array was coated with dithiobissuccinimidylundecanoate *via* self-assembly followed by conjugation of r-PAX8 capture antibody. A 6-channel microfluidic chip with an integrated concentration-gradient generator was overlaid on the functionalised gold nanohole array. The use of a concentration gradient generator offered the advantage of constructing an analyte calibration curve upon detection of an unknown quantity of r-PAX8. Detection of analyte binding was facilitated by an SPR imaging modality where the light intensity from a 632.8-nm laser transmitted through the nanohole array by EOT was monitored as a function of r-PAX8 concentration. Binding of r-PAX8 to the immobilised capture antibody caused a change in the local refractive index and decreased transmittance through the nanohole array. A dynamic range greater than an order of magnitude was realised (0.25–9 μg ml^{-1}) and as low as 5 nM of r-PAX8 could be detected.[61]

Monitoring of the SPR angle shift has been used in an on-chip immuno-assay of the trace cardiac marker B-type natriuretic peptide (BNP), which is indicative in the diagnosis of heart failure.[62] In this LOC system, a T-shaped microfluidic device with two patterned gold films facilitated the indirect detection of BNP. To the first gold film, BNP was self-assembled by derivati-sation with cystamine where the thiol group facilitated immobilisation. Actual detection of analyte BNP was facilitated using anti-BNP antibody that was modified with acetylcholinesterase (BNP-Ab-AChE) in a two-step capture – enzymatic reaction protocol. Sample BNP was first mixed with BNP-Ab-AChE to enable immunoreaction followed by flowing the reaction contents over the initial gold film that had been decorated with BNP. This allowed for unreacted Ab-AChE to be captured in the chip. Subsequent delivery of ace-tylthiocholine afforded enzymatic cleavage and production of thiocholine. This product was delivered to the second gold film downstream, where self-assembly of thiocholine caused a change in the angle of the plasmon reso-nance. Transduction of BNP was done within 30 min of sample delivery on-chip and over a wide concentration range from 5 pg·mL^{-1} – 100 ng·mL^{-1}. For comparison, BNP in the blood of a patient would normally be found at levels between 20 pg·mL^{-1} – 2 ng·mL^{-1}. Transduction of BNP was also possible in spiked human serum samples. However, it was necessary to first heat the samples to 67 °C in order to denature endogenous pseudocholin-esterase, which adsorbed in the microchannels and further catalysed acetylthiocholine, thus compromising assay accuracy.[62]

Detection of nucleic acid hybridisation has been demonstrated on-chip using imaging SPR (SPRi) through electrowetting-on-dielectric (EWOD) in a DMF device.[63] The EWOD device represented in Figure 8.5 offered both assay assembly and sample detection, where immobilisation of 20- and 24-mer oligonucleotide sequences was facilitated *via* thiol coordination on 300-μm detection spots. Delivery of samples ranging from 1–2 μl to all four detection spots was accomplished through droplet splitting. The top trans-parent ITO plate allowed SPRi by use of an 800-nm LED and a CCD camera to capture the percentage of reflected light (%R) from the gold detection spots. Immobilised oligonucleotide probes retained high specificity, where hybridisation with only complementary target displayed a difference in %R compared to background. Regeneration was also possible and subsequent hybridisation cycles did not display signal reduction. Hybridisation kinetics were on the order of minutes and displayed a quantitative response that correlated with probe density. Differences in oligonucleotide density were

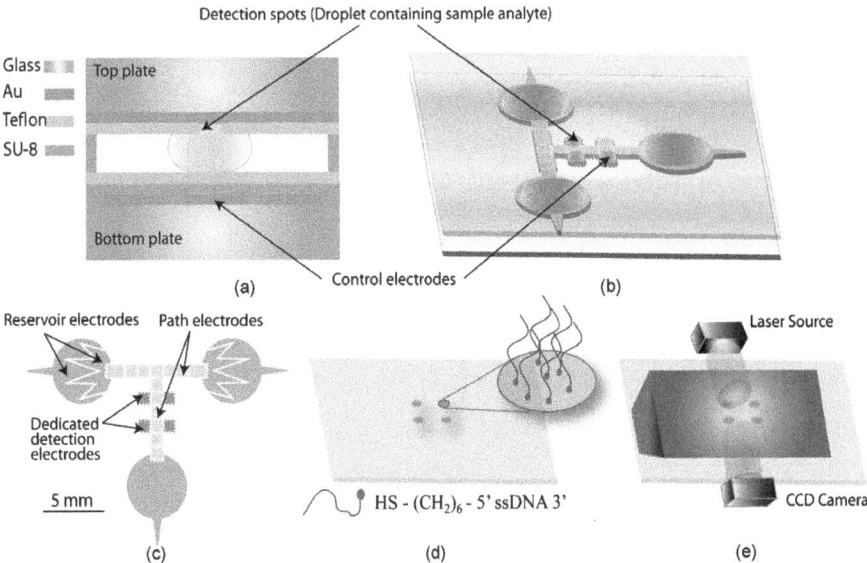

Figure 8.5 Cross-sectional representation of the DMF-EWOD device displaying all lithographic fabrication layers in (a) and an overview of the entire EWOD chip design in (b). The patterned step zones, reagent reservoirs and detection pads are shown in (c). (d) Each detection zone contained a biorecognition spot composed of self-assembled thiol terminated probe oligonucleotides. The detection strategy for SPRi is given in (e) that integrated a laser source and CCD detector. Reprinted with permission from ref. 63. Copyright 2009 Elsevier.

visible when immobilisation was done in the presence of a +90 V or −90 V potential in addition to passive immobilisation (0 V) due to attractive and repulsive electrostatic forces, respectively. In the presence of the −90 V potential, an increase in hybridisation efficiency was noted and a two-fold increase in the SPRi signal magnitude was measured. The authors attribute this to an improved probe orientation which enabled transduction of as low as 500 pM of target, that is an order of magnitude lower than the typical LOD for SPRi.[63]

Sensing within LOC systems offers the additional advantage of access to fluid-handling methods, which can significantly improve the analytical figures of merit of an assay. An on-chip SPR assay for the transduction of femtomole amounts of short (20-base) oligonucleotide targets characteristic of *E. coli* has been reported by Springer *et al.*,[64] using a dispersion-less sample-injection method. The fluidic design is shown in Figure 8.6 where the detection region is flanked by two inlet and outlet ports. The former are connected to multichannel peristaltic pumps, while the latter are connected to microelectric valves. Opening and closing of adjacent output ports (state A to state B) provided solution exchange over the interfacial sensing region and prevented fluid intermixing and dispersion – a prominent issue in

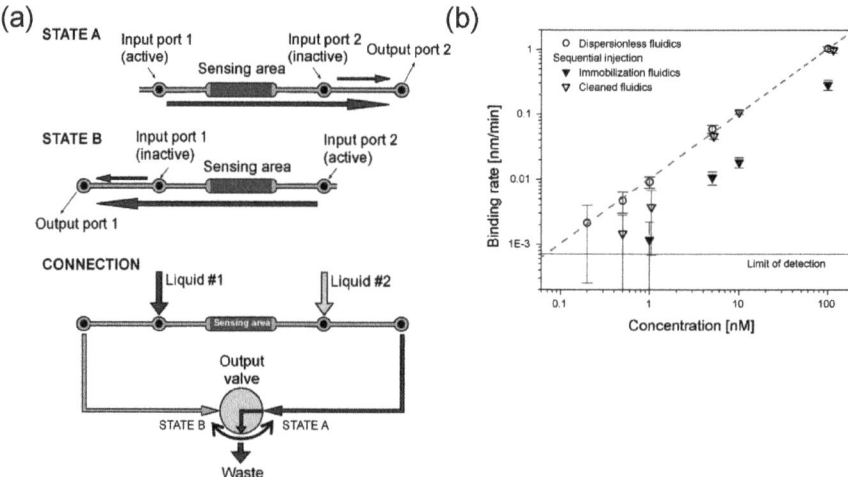

Figure 8.6 (a) Dispersionless sample dispensing technique where the control of flow by switching input/output ports enabled rapid sample mixing adjacent to the sensing area. (b) Response curve for the transduction of oligonucleotide targets within 4 min at a flow rate of 10 μl min^{-1} for three different sample dispensing methods as shown by the inset legend. Adapted with permission from ref. 64. Copyright 2010 Elsevier.

a microchannel with high surface to volume ratio. In-channel immobilisation of biotinylated probe oligonucleotides on a covalently immobilised Sav layer comprised the interfacial biorecognition element. SPR transduction of oligonucleotide hybridisation was accomplished through a 50-nm gold layer detection region in each of the four microchannels, where the reflected light from the sensing zone was coupled into a four-channel spectrometer by use of four in-line optical fibres. Transduction of complementary oligonucleotide target was accomplished within minutes and with a dynamic range spanning three orders of magnitude. In comparison to sequential flow, the dispersionless injection method offered higher sensitivity and a *ca.* 3.5-fold lower detection limit (70 pM opposed to 250 pM); which corresponded to 3 fmol of target oligonucleotide.[64]

8.2.2.4 Surface-Enhanced Raman Spectroscopy (SERS)

The intrinsic Raman scattering fingerprint of many molecules and biomolecules has provided a basis for a powerful spectroscopic method for identification and quantification.[65] The characteristically narrow Stokes and anti-Stokes scattering bands of analytical targets of interest offer an order of magnitude, if not better multiplexing capacity than traditional fluorescence methods. Moreover, SERS that is based on the acquisition of Raman spectra of molecules on a rough metal film has provided for a greatly improved Raman signal.[65] The improved sensitivity can be described as a product of two enhancement factors: an electromagnetic mechanism (EM) and a chemical mechanism. Given that the magnitude of Raman scattering is proportional to the square of the induced dipole moment, the presence of an adjacent LSPR site provides for an attenuated electromagnetic field, enhancing the probability of Raman scattering. The EM enhancement can be understood by first considering eqn (8.5) that describes the LSPR for a spherical particle. In eqn (8.5) the maximum electromagnetic field enhancement outside the NP occurs when $\varepsilon_{in} \approx -2\varepsilon_{out}$. The term $[(\varepsilon_{in} - \varepsilon_{out})/(\varepsilon_{in} + 2\varepsilon_{out})]a^3$ is often abbreviated as ga^3 and is referred to as the polarisability of the metal (α).[65] Since LSPR can provide a field enhancement up to a factor of 10, and the Raman scattering is proportional to the fourth power of the incident electromagnetic field, an EM enhancement factor of 10^4 can be realised. Given the attenuated field (E_{out}) of LSPR, and that the intensity of the Raman signal is linear with respect to the square of the incident field (E_0^2), the resultant Raman intensity is proportional to $|E_{out}|^2$. A first-order approximation of the enhancement factor (EF) at the Raman Stokes-shifted frequency is shown in eqn (8.9).

$$\text{EF} = \frac{(|E_{out}|^2 |E'_{out}|^2)}{|E_0|^4} = 4|g|^2|g'|^2 \qquad (8.9)$$

In eqn (8.9), the terms with the prime notation describe the field at the scattered wavelength. For a small Stokes shift an enhancement of E^4 is predicted.[65] For gold and silver nanostructures, the dielectric resonance condition resides in the visible region of the spectra and thus, no modifications to Raman instrumentation is required. The chemical enhancement mechanism is often viewed as a result of an adsorbed analyte that forms a charge-transfer state at the interface.[11] This is referred to as surface-enhanced resonance Raman scattering (SERRS) and allows for enhancements of up to 10^2.[65] In cases where both enhancement mechanisms are acting concurrently, EFs on the order of 10^9–10^{10} are realised as a result of target molecules residing in the vicinity of "hot spots"; narrow gaps where the fields of two or more NPs overlap, which is not uncommon due to NP aggregation.[11] The distance dependence of SERS is shown by eqn (8.10), where the Raman mode (I_{SERS}) scales with a r^{-10} distance dependence from the average size of a surface-enhancing feature (a).

$$I_{SERS} = \left(\frac{a + r}{a}\right)^{-10} \qquad (8.10)$$

The r^{-10} distance dependence arises from considering the r^{-3} decay of the electromagnetic field enhancement from the surface of the NP, the fourth-power dependency of the EM enhancement and a r^2 increased surface area as a result of the Debye lengths of molecules in the vicinity of the NP surface.[65]

8.2.2.5 *Applications of SERS in LOC Sensors*

Integrating SERS detection in LOC-sensing formats has allowed for the analysis of targets ranging from small molecule toxins to an entire bacterium. A common theme in LOC-based sensing with SERS is the advantage of continuous flow, enabling removal of adsorbed NP aggregates that create memory artifacts and offering rapid and homogenous mixing of biomolecules and or Raman labels with colloidal NPs.[66] One of the earliest examples of SERS-based detection in a LOC format was reported by Monaghan *et al.*[67] for nucleic acid targets that are indicative of *Chlamydia trachomatis*. The detection strategy used a Rhodamine (Rh) tagged oligo-nucleotide that served as a Raman-active probe for on-chip SERS measurements. Initially, target nucleic acids were amplified using 30 cycles of PCR with a biotinylated forward primer. The biotin-tagged amplicons are then hybridised with the Rh-probe and extracted from the mixture by the use of 10-μm Sav-coated beads. The beads were affixed in the channel by integrating a bead trap feature into the PDMS layer (see Figure 8.7). This solid-phase extraction capability enabled opportunity for copious washing to remove excess probe and the complex PCR matrix. The Rh-probe was then released from the beads through melting of the duplex by use of a Peltier element. The two inlet ports enabled mixing of the released Rh-probe with silver NPs downstream for SERS analysis. Monitoring of the 1365 cm^{-1} band of the Rh label as a function of probe concentration enabled quantitative detection of nucleic acid sequences with a dynamic range spanning *ca.* an order of magnitude. Examination of 1 pl volume using a 20 × objective lens indicated a LOD of 20.6 nM or 20 zmol of nucleic acid target.[67]

On-chip SERS detection of Hg(ɪɪ) ion in water has also been demonstrated using a Rhodamine B (Rh-B) Raman label.[68] The assay utilised Rh-B electrostatically adsorbed to Au NPs in a droplet-based continuous-flow microfluidic system. Addition of Hg^{2+} results in a competitive displacement of the Rh-B from the Au NP surface resulting in a decrease in the 1647 cm^{-1} SERS signal as a function of mercury ion concentration. Equilibration of Hg^{2+}/Rh-B exchange was possible within 5 min in each droplet of fluid. Integrating the 1647 cm^{-1} signal as a function of the ion concentration afforded a dynamic range from 0.1–2 ppb of Hg^{2+}. The LOD as defined by three standard deviations of the background noise was estimated to be between 100 and 500 ppt, which is an order of magnitude lower than fluorescence methods.[68] The SERS detection of the herbicide paraquat has also been reported using a droplet-based microfluidic approach. Adsorption of paraquat on Ag NPs in-channel, followed by addition of NaCl to induce aggregation and the formation of hotspots, enabled quantitative transduction of paraquat by monitoring the Raman shift at 1651 cm^{-1}. A dynamic range spanning three orders of magnitude was realised with a LOD of 2 nM of paraquat.[66]

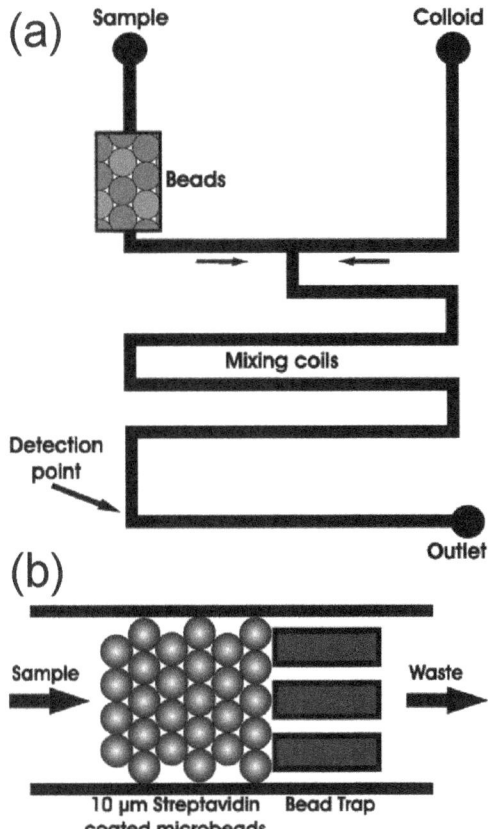

Figure 8.7   (a) Chip design incorporating two inlet ports and a serpentine mixing chamber. (b) Schematic representation of the bead trap used to fixate the Sav microbeads. The positive relief structures are composed of PDMS. Adapted with permission from ref. 67. Copyright 2007 American Chemical Society.

The capability of multiplexing through combination of SERS with LOC was highlighted in a study by Walter *et al.*[69] for the detection and classification of 9 different strains of *E. coli*. Cells were lysed by ultrasonication and the contents were mixed on-chip with Ag NPs that were then further aggregated with KCl. The intensities of a multitude of SERS bands ranging from 661–2954 cm^{-1} that were characteristic of cell wall, cytoplasmic components and other biomolecules were measured. The SERS signals that were characteristic of strain-specific signatures of cell walls were used in classification. A total of 11 200 spectra were acquired using a 1-s integration time. These data were split into two sets, where principle components analysis (PCA) was done on the first set and the data were classified using a support vector machine (SVM). The SVM model for the first set was then trained for accuracy and the subsequent set was classified against this reference library in the same PCA space. Validation accuracies between 91 and 92.6% were realised using the trained library as a sorting technology. In comparison to traditional SERS and Raman measurements where acquisition times are 10 s and 60 s, respectively, the on-chip SERS measurements reduced the total analysis time from lysing to sorting to within 5 h.[69] The same group also demonstrated that this SERS LOC sensor was amenable for POC diagnostics since it could be used to evaluate the toxicity and monitor the efficacy of the immunosuppressive therapeutic 6-mercaptopurine. Metabolism of 6-mercaptopurine is catalysed by the enzyme thiopurine-*s*-methyltransferase, and the activity of this enzyme was monitored by observing SERS signals associated with the evolution of the metabolite 6-methylmercaptopurine. Using the same PCA-SVM chemometric analysis, the enzyme activity could be identified in lysed red blood cells.[70] In more recent work, an on-chip SERS biosensor in combination with data processing by a chemometric dendrogram and cluster analysis has been used to identify the presence of methicillin-sensitive *Staphylococcus aureus* (MSSA) and the methicillin-resistant form (MRSA). Isolates from both China and the United States were detected and each strain was classified with a 95% recognition rate. Again, the LOC-SERS biosensor significantly reduced analysis times and offered the ability to track the origin and course of an outbreak.[71]

8.3   Electrochemical Detection

8.3.1   Voltammetric Transduction

Electrochemical methods have enjoyed widespread use in chemical analysis, spanning application scales from production reactors with macroscale electrodes to microfluidic systems employing nanometre to micrometre-sized electrode systems for the analysis of samples of picolitre to nanolitre volumes. Voltammetry offers several advantages with respect to other analysis techniques including: ease of electrode fabrication, facile interfacing to control and data acquisition electronics, insensitivity to electromagnetic radiation and the optical characteristics of the sample (*e.g.* solution turbidity), use of relatively inexpensive components, and the capability of achieving low (subattomolar) detection limits.[72] Voltammetric detection systems have found widespread use as in-line detection systems incorporated into microfluidic devices.[72–99] The discussion of voltammetric theory is beyond the scope of this chapter. A theoretical description of transduction using field-effect transistors is provided in Section 8.3.3 and the interested reader is referred elsewhere.[100]

8.3.2   Applications of Voltammetric Sensors in LOC Systems

Voltammetric electrode systems may be incorporated into microfluidic channels by standard photolithographic methods or by *in situ* synthesis. With regards to the latter, microfluidic synthesis methods take advantage of laminar flow within microchannels, which permits assembly of features that can be as small as a few micrometres in size with control of feature placement to within 5 μm. An elegant process for *in situ* fabrication of a three-electrode in-line voltammetric detector has been reported by Kenis *et al.*[101] The process was initiated with the deposition of a gold strip (~200 μm wide) laterally across a glass microscope slide. A PDMS casting was then bonded to the glass substrate with the main microchannel feature in the PDMS oriented orthogonally to the gold strip. At the point of intersection with the gold, the microchannel existed as a single channel of 200 μm width. Ahead of the intersection, the microchannel branched into three reagent inlet channels. Beyond the intersection, the microchannel intersected a smaller (~100 μm wide) side channel. The gold strip was divided into two independent gold electrodes by introduction of a gold etchant in the central reagent channel and water in the left and right reagent channels. The etchant stream remained centred in the main channel between the two water streams due to laminar flow conditions that maintained separation of the three reagent streams. The divided gold strip then served as the working and counter-electrodes of the voltammetry system. A silver/silver chloride (Ag/AgCl) reference electrode was then synthesised *via* the two-phase delivery of solutions containing silver ion and a reductant, respectively, along the main microchannel. This was achieved by delivery of reagents *via* the left and right reagent channels (and not the central reagent channel). Silver metal was deposited at the interface between the two reagent streams, at the centre of the main channel and between the gold working and counterelectrodes. The reagents used for silver deposition exited the side channel beyond the gold strip, which continued the formation of conductive silver metal that extended to a fluid port containing a silver contact pad. Subsequent introduction of dilute hydrochloric acid in the microfluidic system resulted in the formation of AgCl on the surface of the silver wire that had been synthesised, thereby completing the reference electrode. Cyclic voltammograms of $Ru(NH_3)_6Cl_3$ were successfully recorded using the fabricated electrode system, which interrogated 5-nl solution volumes (the volume of solution in the microchannel above the electrodes) and yielded voltammetric signals from as little as 10 pmol of $Ru(NH_3)_6Cl_3$ in the detection region.

Incorporation of voltammetric detection in DMF platforms has recently been described.[93,102,103] DMF is a microfluidic technique that is capable of manipulating nanolitre to microlitre volumes of liquids by translating discrete fluid droplets over an array of dielectric coated electrode pads. Individual droplets are manipulated by application of voltage to the independently addressable electrode pads. DMF provides a platform for microfluidic investigations that is not limited to diffusion-based mixing, that can manipulate multiple droplets at a time, and that is amenable to automation and multiplexing.[104] One-plate configurations have been described in which the top of the droplets are exposed to air and can be delivered to microelectrodes suspended above the plate for voltammetric analysis.[93,102,103] Use of a two-plate configuration, in which the fluid(s) to be manipulated are sandwiched between two plates coated with a hydrophobic dielectric material (with underlying electrodes), provides the capability for more sophisticated fluid manipulations. This includes droplet splitting and droplet dispensing from reservoirs, in a configuration that minimises issues associated with evaporation. Such an approach necessitates incorporation of the electrodes within the plate structure and therefore the intrinsic capability for in-line voltammetric analysis. An in-line voltammetric analyser can be built directly into one of the standard gold electrode pads used for fluid manipulation through standard photolithography and etching techniques (as used to prepare the array of electrode pads). It is possible to introduce three subelectrodes within the larger electrode pad element, as described by Dryden *et al.*[93] Additional photolithographic steps may be employed to create a Teflon-AF dielectric overlayer above the electrode pads, with the exception of the voltammetric detection electrodes. While the working and counterelectrodes are gold, a silver reference electrode may be fabricated by treatment of one of the electrodes with a solution of $AgNO_3$ followed by electroplating of metallic silver onto the gold electrode surface by application of a -0.8 V potential between the reference microelectrode and an external counterelectrode. Use of a silver reference electrode (as opposed to a gold pseudoreference electrode) provided significantly improved repeatability, with not more than 2 mV drift in potential observed between scans.[93] The in-line voltammetric detection element provided response characteristics similar to that for a commercially available electrochemical analysis system based on the use of disposable screen-printed electrodes. Patterned nanostructured microelectrodes (NMEs) have also been integrated into LOC devices for bioanalysis.[105-107] Bacterial detection proposed for POC analysis without the need for PCR was reported in a study by Lam *et al.*[107] using a system that offered results within 30 min of dispensing of a sample. The chip design integrated a bacterial cell lysis zone and NME transduction using $Ru(NH_3)_6^{2+}$ and $Fe(CN)_6^{3}$ electrochemical reporters. Gold working, reference and auxiliary electrodes were patterned on a silicon chip and 100-µM NMEs were electroplated into 5 µM apertures in the SiO_2 overcoat layer as shown in Figure 8.8. Peptide nucleic acid (PNA) probes were then immobilised onto 20 working NMEs. The electrochemical lysis chamber was comprised of Au electrodes with a 500-µM spacer. The target was intracellular β-mRNA characteristic of RNA polymerase, which hybridised to the immobilised PNA probes after lysis of cells. Addition of $Ru(NH_3)_6^{2+}$ resulted in accumulation on the polyanionic mRNA backbone. Using differential pulse voltammetry, the NME surface was scanned over a potential window where the reduction of $Ru(NH_3)_6^{2+}$ to $Ru(NH_3)_6^{3+}$ took place. The $Ru(NH_3)_6^{3+}$ was then oxidised back to the 2+ state using $Fe(CN)_6^{3}$ to create an electrocatalytic current. The clinically relevant detection limit of 1 cfu mL^{-1} was met in samples of urine spiked with bacteria within 30 min of cell lysis.[107]

In addition to devices prepared from PDMS on glass and for DMF, electrochemical detection systems have also been integrated into µPADs.[79,97,108–111] Screen printing of electrodes is conventionally done on

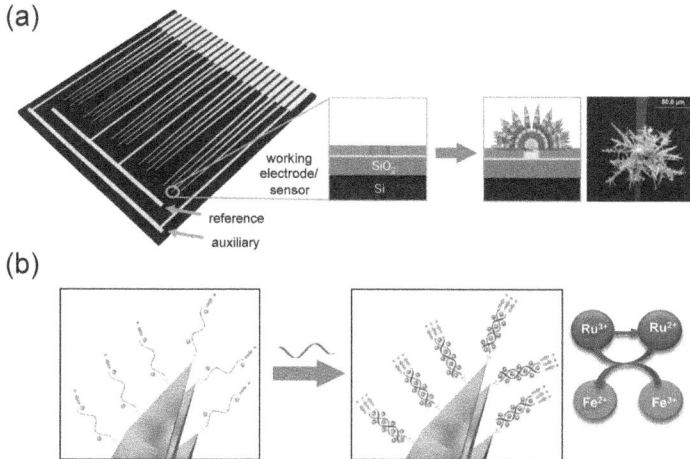

Figure 8.8 (a) Schematic diagram of the NME patterned electrochemical sensors. Gold working, auxiliary and reference electrodes are patterned on a Si/ SiO₂ chip. Working electrodes are then overcoated with an additional SiO₂ to which 5-µm wells are etched in. Gold NMEs are then electroplated in apertures and are on the order of 100 µm. The scale bar in (a) is 50 µm. (b) Au NMEs are derivatised with PNA oligonucleotide probes, where hybridisation with bacterial mRNA is transduced using an electrocatalytic current. This was accomplished using differential pulse voltammetry and the electrochemical Ru(NH₃)₆²⁺/Fe(CN)₆³ reporter system where Ru(NH₃)₆²⁺ is initially associated to the polyanionic backbone of the PNA-mRNA duplex. Adapted with permission from ref. 107. Copyright 2012 American Chemical Society.

a plastic substrate, typically polyethylene terephthalate (PET). This is later connected to the µPAD that contains embedded diagnostic reagents. Using such an approach, Wang *et al.*[97] devised a µPAD device for conducting a multiplexed immunoassay of tumour markers based on a sandwich assay. Exposure of *o*-phenylenediamine to a horseradish peroxidase labelled antibody (in the presence of H₂O₂) was used to detect the sandwich complexes that were formed at the detection zones of the µPAD. Voltammetric analysis was based on detection of 2,2′-diaminoazobenzene that was produced by the enzyme reaction. Enhanced electrical coupling of the redox active product generated in the reaction pads was achieved by introduction of multiwalled carbon nanotubes in a chitosan matrix to the paper reaction pads before immobilisation of the antibody. The device showed a dynamic range spanning up to four orders of magnitude, with detection limits of 1 pg ml⁻¹ and 50 pg ml⁻¹ for cancer antigen 125 and carcinoembryonic antigen, respectively. It was stated that progress will continue to be made toward the use of paper as a substrate for printed electronics[112] so that inexpensive, recyclable and biodegradable devices may be produced. In clinical diagnostic applications, the construction of devices using only paper as a platform is desired for both lower cost and as a matrix that is suitable for facile disposal of bio-hazardous materials by incineration. Preliminary progress toward an all-paper system has been reported by Dungchai *et al.*[79] Screen printing was done to fabricate electrodes directly onto paper substrates that had been prefunctionalised with hydrophilic fluid transport and reaction zones bounded by barriers comprised of hydrophobic polymer embedded in the paper substrate. The presence of glucose, lactic acid and uric acid in bio-logical samples was detected by use of associated oxidase enzyme reactions for each of the test species. Each enzyme was placed in discrete reaction zones and voltammetric detection of H_2O_2 produced from the enzyme–substrate reactions provided the basis for electrochemical transduction. Millimolar detection limits for each of the target compounds were observed, with good agreement shown between the µ-PAD results and those provided by established testing protocols.

8.3.3   Field-Effect Transistors

Electrochemical sensors based on field-effect transistors (FETs) exhibit a number of features that are attractive for biosensing applications. FETs offer direct, label-free and real-time detection of target analytes.[113] FETs that are based on one-dimensional (1D) nanomaterials (NMs) offer an additional advantage of improved sensitivity as compared to much larger planar FET devices owing to large surface area to volume ratio.[114] In the case of nanowire (NW)-based FETs (NW-FETs), the NW dimensions are on the order of the thickness of the conduction channel region in the FET. As a result, the entire cross section of a NW experiences the field effect upon selective interaction. In the case of much larger planar FETs (*e.g.* metal oxide semiconductor FETs), the conduction channel is buried inside the bulk material and as a result, the conduction channel only experiences a partial field effect.[114] The resulting improvement in sensitivity that is achieved by reducing the size of the FET device is sufficient to allow detection of single virus particles.[115] Numerous 1D NMs have been described as nanoscale FETs for biosensing applications. These NMs include carbon nanotubes (CNTs),[116,117] silicon nanowires (SiNWs),[118,119] and metal-oxide nanowires.[120–122] Among these 1D NMs, examples of use of SiNWs as FET transducers are the most common due to a combination of unique physiochemical and electronic properties. The conduction mecha-nism in SiNW-FETs is solely attributed to the field effect[123] (*cf.* a combina-tion of charge transfer, field effect and Schottky barrier in CNTs[114]). SiNWs can be fabricated either by top-down or bottom-up approaches, and their physiochemical properties such as doping density, dopant types and NW dimensions can be controlled during the synthesis process.[124] In the case of a top-down approach, a bulk silicon wafer (such as silicon-on-insulator) is subjected to microfabrication processing steps to reduce its dimensionality to a nm-size scale. Bottom-up synthesis makes use of molecular precursors to fabricate SiNWs by implementation of methods such as chemical vapour deposition.[125] Details of the top-down and bottom-up approaches for synthesis of SiNWs have been reviewed elsewhere.[125,126] A further advantage derived from materials based on SiNWs is that a thin layer of silicon oxide develops spontaneously when the surface is subjected to atmospheric conditions. This sheath serves as a dielectric but also facilitates introduction of various functional groups (thiols, epoxides and amines) on the surface of SiNWs through silane coupling agents.[126] These functional groups can subsequently be used for immobilisation of selective agents on the surface of SiNWs. Fabrication methods of CMOS technology can be used for the development of SiNW-FET devices, which offers a significant practical advantage in terms of infrastructure that is already well established for manufacturing.

8.3.3.1 Operating Principles of SiNW-FETs

In a FET configuration, a semiconductor such as a p-doped or n-doped silicon is sandwiched between the source and drain metal electrodes.[127] A voltage is applied between the source and drain electrodes to allow a current flow through the channel region of a sandwiched semiconductor. The conductance of the sandwiched semiconductor is modulated by a third gate electrode that through capacitance modulates the density of charge carriers in the channel region of the sandwiched semiconductor. FETs can either exhibit p-type or n-type conductivity.[123] In the case of p-type conductivity, an application of a positive (negative) gate voltage decreases (increases) the density of charge carriers in the channel region, resulting in a decrease (increase) in conductivity. For n-type conductivity, an application of positive (negative) gate voltage increases (decreases) the density of charge carriers in the channel region, resulting in an increase (decrease) in conductivity. FETs are configured as chemical and biological sensors by modifying the dielectric interface with a selective chemistry. Upon selective interaction, the electric field of the analyte that is accumulated at the interface is analogous to applying a new gate voltage, which in turn modulates the conductivity of the channel region and shifts the subthreshold voltage. The change in conductivity and resulting shifts in subthreshold voltage upon selective interaction serves as an analytical signal.

8.3.3.2 Examples of Assays using FETs

Implementation of various types of assays using SiNW-FETs include detection of ions,[118,128,129] nucleic acids,[130–132] protein–protein interactions,[71,133,134] small molecule–protein interactions,[135] and immunodetection.[136,137] These assays have been extensively reviewed (readers are referred to review articles by Patolsky and Lieber,[127] Chen *et al.*,[114] Wanekaya *et al.*,[117] and Curreli *et al.*[138]). The following section provides an account of some selected examples of assays that are based on SiNW-FETs integrated with LOC for bioanalysis.

The negative charge associated with the backbone of nucleic acids provides SiNW-FET technology an attractive opportunity to serve as the basis for a sensor for detection of nucleic acid markers. Bunimovich *et al.* reported real-time detection of nucleic acid hybridisation using SiNW-FETs that were fabricated using a top-down approach.[139] The surface of SiNWs was modified with amine functionality to electrostatically adsorb oligonucleotide probes. The delivery of sample solutions containing oligonucleotide targets (16 mer) into PDMS-based microfluidic channels was done using a syringe pump at a flow rate of 2.0 µl min^{-1} (see Figure 8.9(a)). Additionally, a multilayer PDMS device was fabricated to automate fluid injection and sample exchanges. The authors compared the FET performance of SiNWs that were modified with oligonucleotide probes in the presence and absence of a silicon oxide layer. The absence of a silicon oxide layer offered two orders of magnitude lower

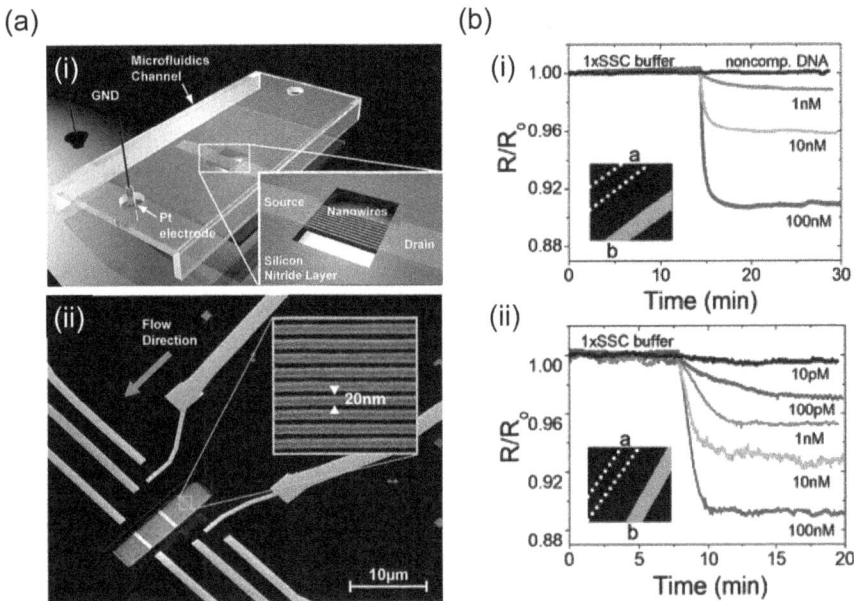

Figure 8.9   (a) (i) Schematic diagram and (ii) scanning electron microscope (SEM) image of SiNW-FET device with an integrated microfluidic channel. The FET device consisted of three groups of *ca.* 10 SiNWs. High-resolution image of one group of SiNWs is shown in a zoomed-in image in (ii). (b) Real-time responses of SiNW-FET device modified with oligonucleotide probes to increasing concentrations of target DNA in the (i) presence and (ii) absence of a silicon oxide layer. The traces in black correspond to SiNW-FETs responses in the presence of 100 nM of noncomplementary target. (Inset) Fluorescence images of microfluidic channels in the presence of labelled (a) noncomplementary and (b) fully complementary targets. Reprinted with permission from ref. 139. Copyright 2006 American Chemical Society.

LOD and two orders of magnitude higher dynamic range as compared to SiNWs that were modified with oligonucleotide probes in the presence of a silicon oxide layer. The LODs in the absence and presence of silicon oxide layer were reported to be 10 pM and 1 nM, respectively (Figure 8.9(b)). The inferior FET performance of SiNWs in the presence of a silicon oxide layer is attributed to the presence of interfacial trap states.[140] Additionally, a silicon oxide layer serves as a dielectric and reduces the extent of the field effect that is experienced by a FET upon selective interaction.[141]

From the standpoint of nucleic acid detection with FETs, analogs of DNA probes that are electrostatically neutral offer improved signal-to-noise ratio as compared to negatively charged DNA probes. Neutral analogs of DNA such as peptide nucleic acid (PNA) probes offer reduced background signal that originates from conductivity. A PNA–DNA interaction offers a proportionally greater signal based on the change in charge than would a DNA–DNA hybridisation of the same target. For example, Zhang *et al.* reported a 5-fold improvement in assay sensitivity for a FET-based transduction of nucleic acid hybridisation when SiNWs were modified with PNA probes as compared to DNA probes.[132] The strength of a dipole moment that develops on the surface of SiNW-FETs also determines the extent of the shift in the subthreshold voltage upon selective interaction, which is a function of degree of alignment of interfacial chemistry. The alignment of interfacial chemistry can be achieved by an application of electric field as was reported by Chu and coworkers.[142] The authors reported a LOD of 0.1 fM for a 15-base-pair oligonucleotide hybridisation assay with SiNW-FETs, where the sample delivery to the surface of SiNW-FETs was done using a microfluidic channel made in PDMS. Gao *et al.* reported top-down-fabricated SiNW-FETs with a triangularly shaped cross section for an oligonucleotide hybridisation assay that made use of a 24-base-pair.[143] The LOD of the assay was reported to be 0.1 fM. The assay conditions were optimised in terms of concentration of probe (0.5 μM), ionic strength (0.01 × PBS) of the solution and the backgate voltage that was in the subthreshold region. In another study by Gao *et al.*, the authors showed that it is possible to lower the LOD of an oligonucleotide hybridisation assay to 50 aM by integration of a signal enhancement step that was achieved by means of rolling circle amplification following target hybridisation.[144] The sensitivity of a response exhibited by FETs is a function of ionic strength of the solution. An increase in the ionic strength of a solution results in increased screening of analyte charge by counterions. This ultimately decreases the Debye length and lowers the extent of the field effect that is experienced by FETs. From the standpoint of immune detection, this consideration adversely affects the sensitivity of FET response due to the dimensions of capture antibodies ($\geq$nm), which places the charges on an antigen that is bound with a capture antibody well beyond the Debye length. As a result of this limitation, low ionic strength of a solution (1 μM–1 mM) is desired for conducting immunoassays. At these low ionic strength conditions, the Debye length is typically greater than the size of capture antibody.[115] However, the ionic strength of most physiologically relevant solutions exists in a range of 100–150 mM (Debye length <1 nm). Physiological samples require desalting, filtering and buffer exchange to allow sensitive detection of low concentrations of biomarkers by SiNW-FETs. To address such limitations, Stern *et al.* developed a microfluidic purification chip (MPC) to selectively "extract" biomarkers from physiological fluid sample (*e.g.* blood) and subsequently release the captured biomarkers into a solution that is more suitable for detection using SiNW-FETs.[137] The MPC consisted of two chambers, a purification chamber and a SiNW-FET sensing chamber. The purification chamber was used to selectively capture biomarkers from blood samples using primary antibodies as capture agents that were immobilised *via* a photocleavable crosslinker. Following selective capture, the MPC was rinsed with a buffer to wash away the unwanted components of blood. The MPC was subsequently irradiated with UV light to cleave the photolabile crosslinker and release the captured biomarkers into solution for delivery into a SiNW-FET sensing chamber. In the sensing chamber, SiNWs were modified with secondary antibody for selective detection of the primary antibody–biomarker complex. The authors employed MPC for selective detection of two cancer biomarkers, prostate specific antigen (PSA) and carbohydrate antigen 15.3 (CA15.3) in whole blood at concentrations of 2.5 ng ml^{-1} and 30 U ml^{-1}, respectively.

An alternative to sample processing for detection of biomarkers in physiological fluids is to use fragments of antibody that are smaller in dimensions than an intact antibody. Elnathan *et al.* showed that the sensitivity of immunoassay conducted with SiNW-FETs can be improved by size fragmentation of capture antibody.[145] Two types of fragments from an intact capture antibody were generated by sequential treatments with Pepsin and 2-mercaptoethylamine (2-MEA). Treatment of an intact antibody (size 9–10 nm) with Pepsin-generated F(Ab)$_2$ fragments (size: 4–5 nm), while a sequential treatment of the F(Ab)$_2$ fragments with 2-MEA generated FAb fragments (size: 2–3 nm). Studies of SiNW-FETs that were modified with an intact antibody (anti-cTnT) and associated F(Ab)$_2$ and FAb fragments showed that the sensitivity of the response improved as the size of the selective agent decreased, especially at conditions of high ionic strength (0.01 × 0.1 × and 1 × PBS; associated Debye lengths are 7.3 nm, 2.3 nm and 0.7 nm, respectively). No difference in the assay response between an intact anti-cTnT and the associated fragments was seen under low ionic strength conditions (0.001 × PBS; Debye length of 20 nm). Selective detection of cTnT in undiluted serum samples was also demonstrated using SiNW-FETs modified with an anti-cTnT F(Ab)$_2$ fragment with a LOD of 27 pM. Hideshima *et al.* also reported an improvement in assay sensitivity upon reducing the size of a selective agent.[146] The authors modified the surface of SiNWs with two different selective agents, sialic acid containing oligosaccharide (size: *ca.* 2 nm) and an antibody (size: 4–14 nm) for selective detection of hemagglutinins associated with avian and human influenza virus. FETs that were modified with an oligosaccharide provided 6 orders of magnitude lower LOD (50 aM) as compared to an antibody modified FETs (LOD 50 pM).

Zheng *et al.* reported a multiplexed detection of three cancers biomarkers that included prostate-specific antigen-α1-antichymotrypsin (PSAα1), carcinoembryonic antigen (CEA) and mucin-1 using SiNW-FETs.[119] Multiplexing was achieved by modifying different regions of a device that consisted of arrays of SiNWs with different monoclonal antibodies (mAbs). The device was bonded to a PDMS cover to form a microfluidic channel that was used for sample delivery; a schematic of the chip is shown in Figure 8.10(a). SiNW-FETs that were modified with mAbs for PSAα1 exhibited a change in conductance response that was proportional to the concentration of PSAα1 from 5 to 90 ng ml^{-1} (Figure 8.10(b)). Subsequent injection of a buffer solution into the microfluidic channels caused the conductance signal to

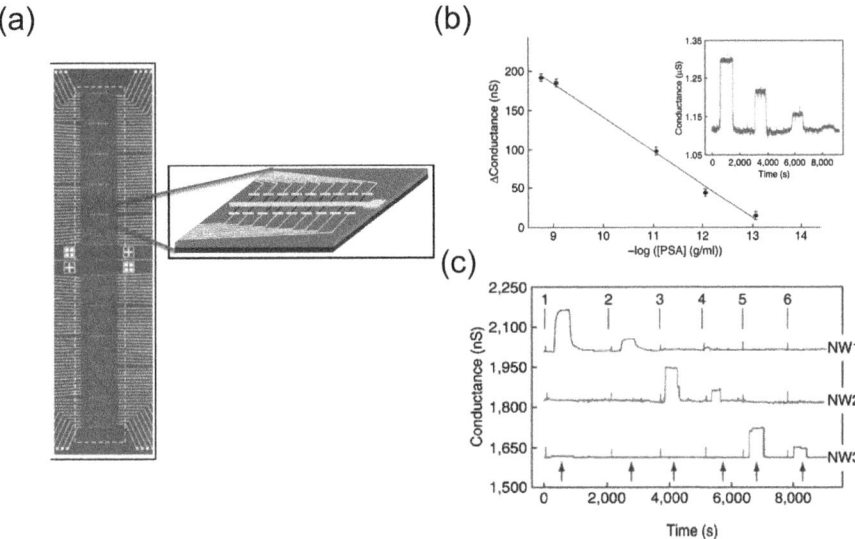

Figure 8.10 (a) Image of a SiNW-FET device with the NWs arranged in an array format. The metal electrodes are shown in white lines. The area of the device that was covered with a microfluidic channel is shown in white-dashed rectangular box. The arrangements of source and drain electrodes in the red rectangular box are shown in the zoomed in image. (b) Concentration-dependent response of SiNW-FET modified with PSAα1 mAb to increasing concentration of PSAα1. (Inset) Real-time response of the assay to decreasing concentrations of PSAα1. (c) Real-time multiplexed detection of three cancer biomarkers, where NWs 1, 2 and 3 were modified with mAbs for PSAα1, CEA and mucin-1, respectively. The antigens were sequentially delivered into a microfluidic channel. (1) 0.9 ng ml^{-1} PSAα1, (2) 1.4 pg ml^{-1} PSAα1, (3) 0.2 ng ml^{-1} CEA, (4) 2 pg ml^{-1} CEA, (5) 0.5 ng ml^{-1} mucin-1 and (6) 5 pg ml^{-1} mucin-1. After the injection of each of the aforementioned solutions, a buffer solution was injected into a microfluidic channel at point indicated by the upright arrows. Figure reprinted with permission from ref. 119. Copyright 2005 Nature Publishing Group.

drop to the background level, suggesting that the binding was reversible. The LOD of PSAα1, CEA and mucin-1 were reported to be 75 fg ml^{-1} (2 fM), 100 fg ml^{-1} (0.55 fM) and 75 fg ml^{-1} (0.49 fM), respectively. Within the same assembly, NWs designated as 1, 2 and 3 were modified with mAbs for PSAα1, CEA and mucin-1, respectively. Real-time conductance response recorded from each NW showed that the change in conductance response was only observed in the presence of the corresponding relevant target antigen in the sample solution (Figure 8.10(c)).

8.4 Piezoelectric Sensors

Piezoelectricity is a phenomenon exhibited in a number of crystals and synthetic ceramics. It is defined as the electric polarisation in a substance resulting from the application of mechanical stress. By corollary, the application of an electric field imparts mechanical stress to the substance and can manifest as a mechanical wave. Bulk acoustic waves (BAW) propagate into the material perpendicular to the surface, whereas surface acoustic waves (SAW) propagate on the material interface.[147] The change in the resonant frequency of the generated wave due to chemical interactions at the surface of the material has been used as a transduction mechanism in biosensors. By immobilising a selective biorecognition element on the surface of the piezoelectric material, both adsorption and selective binding of species generate a signal by changing the viscoelastic coupling of the surface to the solution that ultimately alters the resonant frequency. Mathematical equations such as the Sauerbrey equation and its derivatives are used in quantitative analysis.[147] Piezoelectric sensors offer a label-free transduction mechanism that eliminates labelling interferences, reduces costs and analysis time, and allows real-time detection.

8.4.1 Theoretical Aspects of Piezoelectricity

Piezoelectric materials are generally classified into two groups: crystals and ceramics. Crystals are ordered structures with repeating identical unit cells and common materials include quartz (SiO_2), tourmaline and sodium potassium tartrate.[148] The application of mechanical stress to a crystal results in a net polarisation as a result of the displacement of dipoles or a change in the separation between positive- and negative-charged sites in each element. Electric potential that is induced by mechanical displacement can be measured *via* electrodes placed above and below the crystal.[148] Ceramics form another group of piezoelectric material with lead zirconate titanate Pb $[Zr_{(x)}Ti_{(1-x)}]O_3$, PZT, being the most common. PZT is made from identical ferroelectric perovskite unit cells, which have an electric dipole that can be selected to lie in specific directions.[149] Other examples include ZnO aluminium nitride (AIN), and polyvinylidene fluoride (PVDF) and its copolymers.[150]

The resonators are usually employed in either SAW or BAW modes. In the SAW configuration, two interdigitated (interleaved metal) electrodes, or IDT electrodes, are placed on the surface of the piezoelectric material. An electrical signal from the input electrode generates a mechanical wave *via* the piezoelectric effect. The output electrode transduces the mechanical wave back into an electrical signal. Given the potential for variation in IDT geometry, a wide variety of SAW devices can be built. Rayleigh, shear wave, Lamb wave and Love wave modes are commonly employed in the operation of SAW devices.[147] Changes in the frequency, amplitude, phase or time delay between the input and the output electrical signal can be used to interrogate the surface.[151] SAW applications are limited in frequency range and power-handling capability. The relatively low surface wave velocity limits the working frequencies to the MHz range, whereas the tight localisation of the surface wave reduces the power-handling capabilities. Since the displacement wave travels along the surface of the piezoelectric material, it is easily dampened in solution, making SAW sensors less than ideal for liquid applications. BAW sensors on the other hand can be operated with one interface exposed to a solution, at GHz frequencies and are more widely used due to their availability, robustness, and affordable electronics.[152] BAW devices involve a piezoelectric material sandwiched between two electrodes, in a configuration that is also known as the quartz crystal microbalance (QCM). An AC electrical waveform is applied to the crystal resulting in a displacement wave that propagates through the material in alignment with the applied electric field. The wave is reflected at boundaries and the resultant mechanical stress generates an electrical field in the external circuit. This current in turn generates the next mechanical wave. For all subsequent currents to have the same magnitude, more energy is added to the system to account for frictional loss. The process repeats in a cycle generating a resonant wave that is usually on the order of MHz to GHz.[148]

In general, for a thin film of mass, m, on the surface of the crystal, the change in resonance frequency change, Δf, for BAW can be predicted by the Sauerbrey equation (eqn (8.11)),[147]

$$\Delta f = \frac{-2f_0^2 m}{A\sqrt{\mu_s \rho_s}} \qquad (8.11)$$

where f_0 is the resonance frequency of the bare crystal, A is the area of the film deposited on the crystal, μ_s is the shear modulus of the piezoelectric crystal, and ρ_s is the density of the piezoelectric crystal. From a practical perspective this equation only considers mass changes, and therefore is applicable to gas-phase detection. Once the crystal is exposed to a solution, the change in frequency is no longer a function of only mass. The first layer of solvent, usually water, is strongly adsorbed to the surface of the crystal and moves in unison. Cohesive forces hold a second layer of solvent to the first. Since there is some slippage between these layers as the crystal moves, the oscillation of the second layer is considered to be damped. Furthermore, the third layer exhibits more slippage and this continues farther into solution until there is no connectivity between the piezoelectric material and the solvent. The binding of a molecule onto the surface alters the coupling of the crystal to the external circuit, which in turn affects the resonant frequency and the resulting signal is measured as the change in resonant frequency. Quantitative measure is often obtained using eqn (8.12),[153]

$$\Delta f = -f_0^{3/2} \left(\frac{\eta_1 \rho_1}{\pi \rho_q \mu_q} \right)^{1/2}$$ (8.12)

where η_1 is the viscosity of the liquid, ρ_1 is the density of the liquid, μ_q is the shear modulus of quartz and ρ_q is the density of quartz. Surface guided shear-horizontal waves (SHWs) are an alternative to traditional SAW devices. They originate from shear BAWs but are physically guided along the surface by loading a continuous solid layer onto the surface. Guided SHWs have a higher velocity than traditional SAWs making them usable at higher frequencies. Field confinement to the surface is weaker and allows for the decrease of power flow density and reduces nonlinear effects.[154] Thin-film bulk acoustic wave (TFBAR) devices are also commonly employed BAW-based sensors.[155] A typical TFBAR consists of a thin piezoelectric film sandwiched between two metal layers. Its dimensions can be as small as $10 \times 10 \times 1 \ \mu m^3$. This technology is typically an order of magnitude more sensitive than typical quartz crystals.[155,156]

8.4.2 Applications of Piezoelectric LOC Systems in Bioanalyses

The earliest example of piezoelectric transducers in LOC systems was reported in 2003, where a QCM was mechanically fixed in a microchannel plate for the detection of muscle cell contractions.[157] In 2005, a PDMS flow cell was attached to a QCM substrate for the detection of antiprotein A using protein A as the recognition element.[158] These devices depended on passive detection modes and therefore cannot be used to detect real samples for POC testing since they lack suitable manipulation components. Furthermore, regeneration of surfaces in microfluidics can be time consuming and it requires specific surface modifications to bond biochemical species to prevent loss and deactivation.[159] Recent work has focused on more versatile, regenerable, and integrated devices.

Li *et al.*[160] reported a piezoelectric polymer microdiaphragm array in a microfluidic chip for protein analysis. A polyvinylidene fluoride trifluoro-ethylene (PVDF-TrFE) piezoelectric diaphragm was fabricated by high-throughput and reproducible mould-transfer and hot-embossing techniques. The PVDF-TrFE exhibited a hydrophobic surface that bound proteins *via* hydrophobic interactions. All fluids were manipulated by syringe pumps. BSA was used as a model protein to generate a response curve for the sensor

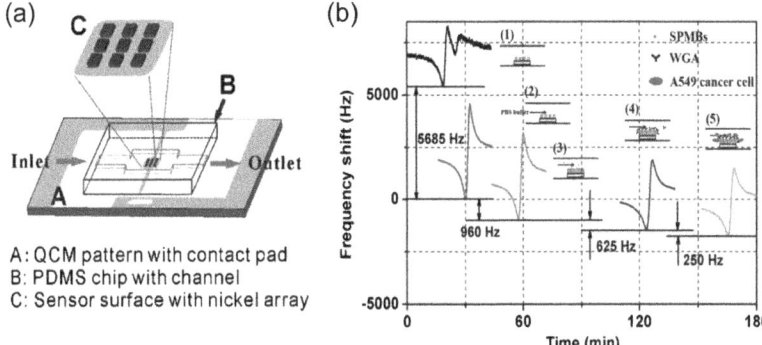

Figure 8.11 (a) Schematic of the nickel array/QCM sensor showing its different components. (b) Monitoring frequency shifts of QCM in real time, (1) Microfluidic system in air, (2) Rushing PBS buffer solution in chip, (3) Trapping SPMBs on the nickel array, (4) Immobilising WGA to EDAC and NHS activated surface-functionalised SPMBs, (5) Capturing of A549 cancer cells by WGA protein attached SPMBs. Reprinted with permission from ref. 159. Copyright 2010 Elsevier.

as a function of protein concentration between 1 and 1000 µg ml^{-1}. This approach is compatible with array technology, which allows for the simultaneous detection of multiple targets by immobilising recognition elements on the PVDF-TrFE film that capture targets of interest.

Zhang *et al.*[159] have reported a versatile piezoelectric sensor based on a nickel array patterned on the surface of a QCM as shown in Figure 8.11(a). The nickel array served to enhance the trapping ability of super paramagnetic microbeads (SPMBs) on the surface using an external magnet. The trapped beads, which increased the active sensing area, were further modified with probe molecules to capture targets. The system was sealed with PDMS to form microfluidic channels and samples were dispensed using pressure-driven flow. The QCM was operated in shear wave BAW mode and was connected to an impedance analyser for readout. In this work, wheat germ agglutinin (WGA) proteins immobilised on SPMBs were used for the detection of A549 cancer cells in tens of microlitres of sample. SPMBs were first trapped on the QCM and decorated with WGA protein. Cancer cells were incubated over the beads for 30 min followed by subsequent washing for 40 min. Experimental results showed that an estimated 90.6 ng of A549 cancer cells were captured on the surface with a sensitivity of 5 Hz mm^2 ng^{-1}. The result of real-time monitoring of frequency shifts at each step of the experiment is shown in Figure 8.11(b).

Zhang *et al.*[161] incorporated a lead magnesium niobate-lead titanate (PMN-PT) single crystal-based piezoelectric resonator in a microfluidic system for application in cell detection. PMN-PT ceramics, which are typically operated in shear wave BAW, are suitable dielectric materials because of their large dielectric constants. The device was fabricated using soft lithography, where a 1 mm diameter transducer was flanked by a glass slide and PDMS to form a microchannel. To test the applicability of the device, a 50 µl solution of *E. coli* suspension was injected at 100 µl h^{-1} and monitored by the acoustic sensor system. Impedance spectra of the resonator before and after the injection of the *E. coli* sample showed a resonant frequency shift of 2639 Hz, which indicated the presence of 1.94 µg of *E. coli* on the sensor interface.

Hsu and Tang[162] proposed a novel device that has the potential to be used for the detection of cells that respond mechanically to diseases such as cancer and malaria. Cancer cells can exhibit contractile forces that can be twice as large as normal cells, and malaria-infected red blood cells are stiff and sticky. Thus, it is important to ensure that there are minimal influences of mechanical forces of the fluid flow on cellular detection. A perfusion-based microbioreactor integrated with a TFBAR array in the culture chamber has been used to ensure that fluid reached the chamber predominantly by diffusion. This minimised shear stress on living cells and avoided mechanical perturbation of the sensor by the environment. Piezoelectric transducers based on SiO$_2$ were suspended on top of the perfusion channel, as shown in Figure 8.12. The transducers were equally spaced about 50 µm apart. The spacing was small enough to suppress convective fluid flow in the culture chamber but wide enough to allow for effective mass transport and media exchange *via* diffusion. The height of the PDMS chamber did not exceed 100 µm, which further controlled diffusion length and mimicked a capillary vessel environment surrounded by a thin layer of tissue. The piezoelectric sensor responded to traction exerted by cells, and this was determined through impedance measurements. Details about the relevance of transducer size and analysis of cell traction have been previously reported.[162,163] In a typical experiment, the sensors were treated with a 1% collagen culture medium for 10 min, followed by 1×10^6 cells per ml A549 lung cancer cells with incubation for 2 h. Excess cells were flushed out and cellular adhesion activities were then monitored. Preliminary results have shown that the cells remained viable for up to two days. This device shows a novel design that can be used to minimise stress from the environment on piezoelectric transducers in microfluidic chips and enables greater versatility for use of piezoelectric sensors.

Piezoelectric systems are not only used as transducers in microfluidic channels, but can also be used for fluid manipulation. García-Gancedo *et al.*[152] reported a lab-on-a-chip system with SAW-based fluid manipulation and BAW-based sensing. SAW-based devices can induce acoustic streaming and motion of microdroplets with high sensitivity. Work made use of TFBAR devices that were fabricated by coating silicon wafers with ZnO followed by Cr/Au deposition to form top and bottom electrodes. The physical adsorption of BSA from solutions at concentrations over 100 µg ml^{-1} yielded shifts in the sensor output on the order of kHz.

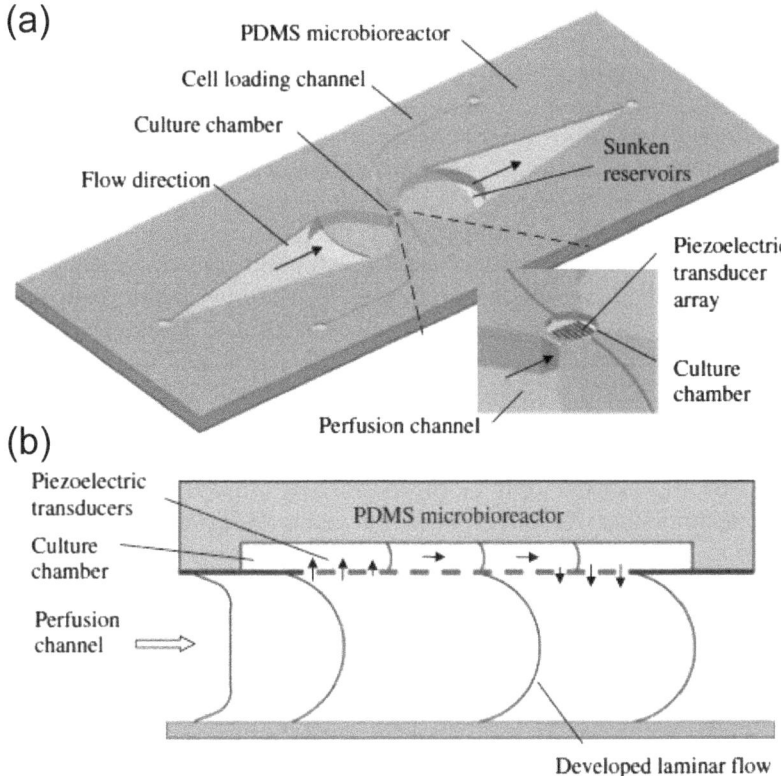

Figure 8.12 (a) Schematic of perfusion-based microbioreactor integrated with TFBAR and (b) illustration of velocity profiles within perfusion channel and culture chamber. Reprinted with permission from ref. 162. Copyright 2011 Springer-Verlag.

8.5 Summary and Future Potential

The adoption of microfluidic methods as tools for analytical chemistry has evolved to fill three primary roles; sampling and sample manipulation, separation of sample components, and assembly of complete integrated systems that offer sample handling and detection as pioneered by Manz[164–166] and others[167–169] in the area of miniaturised total analysis systems, μ-TAS. In the recent literature, there has been significant interest and advancement of methods for sampling and sample preparation as exemplified by work in the area of digital microfluidics. The miniaturisation of traditional wet-bench sample-processing methods offers substantial advantages in reduction of volume of sample, which can be an important consideration for applications such as tissue biopsies. From the perspective of practical implementation, the small volumes used in microfluidic sample preparation can reduce reagent costs dramatically and often reduces the time required for processes to equilibrate or complete. It is therefore a natural evolution to take the final step in analysis, the detection step, into/onto the platform that deals with sample preparation. While in some cases the final step may require instrumentation that is not readily mounted onto a chip such as mass spectrometry or nuclear magnetic resonance, there are many cases where selective detection strategies in combination with transducers of small size are appropriate for addressing specific problems in analysis. The combination of digital microfluidics with on-chip detection technology will become significant in the future.

It is apparent that the technology that has been proposed for biosensor development finds strong alignment with the opportunity to handle samples in a single device assembly of small size. In the biosensors field, optical methods based on fluorescence, and electrochemical methods based on voltammetry and impedance, have traditionally dominated the technologies that are best suited for selective detection of very small quantities of targets. With advances in nanotechnology that have appeared, it is now possible to add to the repertoire of detection strategies to make use of physical phenomena that appear only at the nanoscale. Implementation of nano-particle technologies for target detection using field strength is a theme common to plasmonics, FRET methods and nanowire-FET technology, with potential in each case to approach capacity for the measurement of the presence of one or a few molecules as a detection limit. The advent of nanotechnology in the world of fluidics offers other interesting opportuni-ties, such as linearisation of polyelectrolytes like DNA when transported by EOF channels of nanometre dimensions. This offers potential for concepts such as detection by sequencing in fast parallel analysis.

In some cases the amalgamation of traditional biosensor technologies with microfluidics is only beginning to manifest. For example, the use of piezoelectric sensors in microfluidic chips is still at an early stage. Much of the work done to date has focused on understanding how piezoelectric transducers behave and respond to fluid flow when located inside micro-channels. However, it is possible to appreciate the opportunities presented by use of high-frequency cantilever oscillators in fluidic streams for multi-plexed high-throughput analysis that can avoid use of labelling of targets as a requisite step for signal generation.

The practical implementation of microfluidics may take an unexpected turn by moving away in some applications from carefully designed and controlled microchannel systems. The renewed interest in paper substrates that provide a version of fluid transport in microfluidic dimensions, coupled with printing methods such as those to deposit wax on paper to define fluid-transport pathways, may lead to introduction of semiquantitative, fast, sensitive and inexpensive platforms for a wide variety of applications in clinical, environ-mental, agricultural and forensic analysis. Electrochemical detection using screen-printed electrodes would be relatively easy to integrate with paper substrates. However, it is noteworthy that optical detectors would not neces-sarily be mounted on the paper device. For example, the use of a cell phone camera with software to isolate the RGB[170] colour palette has already been demonstrated in numerous applications[171,172] to work well as a means of spectroscopic detection for analysis done using paper-based fluidic systems.

List of Abbreviations

μ-PADs	Microfluidic paper-based analytical devices
μ-PEDs	Microfluidic paper-based electrochemical devices
μ-TAS	Miniaturised total analysis systems
2-MEA	2-Mercaptoethylamine
AC	Alternating current
AChE	Acetylcholinesterase
AF	Amorphous fluoroplastic
AFP	α-Fetoprotein
Au NRs	Gold nanorods
BAW	Bulk acoustic wave
BGH	Bovine growth hormone
BNP	B-type natriuretic peptide
BoNT-A	Botulinum neurotoxin A
BP	Base-pair
BPM	Base-pair mismatch
BSA	Bovine serum albumin
CA15.3	Carbohydrate antigen 15.3
CCD	Charge-coupled device
CEA	Carcinoembryonic antigen
CMOS	Complementary metal oxide semiconductor
CNT	Carbon nanotubes
cTnT	Cardiac troponin T
DABCYL	4-(Dimethylaminoazo)benzene-4-carboxylic acid
DMF	Digital microfluidics
DNA	Deoxyribonucleic acid
E. coli	*Escherichia coli*
EF	Enhancement factor
ELISA	Enzyme-linked immunosorbent assay
EM	Electromagnetic mechanism
EOT	Extraordinary optical transmission
EWOD	Electrowetting-on-dielectric
FAM	Carboxyfluorescein
FET	Field-effect transistors
FITC	Fluorescein isothiocyanate
FRET	Fluorescence resonance energy transfer
GO	Graphene oxide
HIV	Human immunodeficiency virus
ITO	Indium tin oxide
LED	Light-emitting diode
LIF	Laser-induced fluorescence
LOC	Lab-on-a-chip
LOD	Limit of detection
LSPR	Localised surface plasmon resonance
mAbs	Monoclonal antibodies
MPC	Microfluidic purification chip

MRSA	Methicillin-resistant *Staphylococcus aureus*
MSSA	Methicillin-sensitive *Staphylococcus aureus*
NME	Nanostructured microelectrodes
NP	Nanoparticle
NW	Nanowire
PBS	Phosphate-buffered saline
PC	Polycarbonate
PCA	Principle components analysis
PCR	Polymerase chain reaction
PDMS	Polydimethylsiloxane
PET	Polyethylene terephthalate
PL	Photoluminescence
PMMA	Poly(methyl methacrylate)
PMN-PT	Lead magnesium niobate-lead titanate
PNA	Peptide nucleic acid
POC	Point of care
PSA	Prostate specific antigen
PSAα1	Prostate-specific antigen-α1-antichymotrypsin
PVDF	Polyvinylidene fluoride
PZT	Lead zirconate titanate
QCM	Quartz crystal microbalance
QD	Quantum dot
Rh-B	Rhodamine B
RNA	Ribonucleic acid
Sav	Streptavidin
SAW	Surface acoustic wave
SERRS	Surface-enhanced resonance Raman scattering
SERS	Surface-enhanced Raman spectroscopy
SPMB	Super paramagnetic microbead
SPR	Surface plasmon resonance
SPRi	Imaging SPR
SVM	Support vector machine
TFBAR	Thin-film bulk acoustic wave
TrFE	Trifluoroethylene
ZnO NRs	Zinc oxide nanorods

References

1. G. M. Whitesides, *Nature,* 2006, **442**, 368–373.
2. S. Kumar, M. A. Ali, P. Anand, V. V. Agrawal, R. John, S. Maji and B. D. Malhotra, *Biotechnol. J.,* 2013, **8**, 1267–1279.
3. A. Rios, M. Zougagh and M. Avila, *Anal. Chim. Acta,* 2012, **740**, 1–11.
4. C. H. Vannoy, A. J. Tavares, M. O. Noor, U. Uddayasankar and U. J. Krull, *Sensors,* 2011, **11**, 9732–9763.
5. K. F. Lei, *JALA,* 2012, **17**, 330–347.

6. J. Hu, S. Wang, L. Wang, F. Li, B. Pingguan-Murphy, T. J. Lu and F. Xu, *Biosens. Bioelectron.,* 2014, **54**, 585–597.

7. A. M. Foudeh, T. F. Didar, T. Veres and M. Tabrizian, *Lab Chip,* 2012, **12**, 3249–3266.

8. W. R. Algar, D. E. Prasuhn, M. H. Stewart, T. L. Jennings, J. B. Blanco-Canosa, P. E. Dawson and I. L. Medintz, *Bioconjugate Chem.,* 2011, **22**, 825–858.

9. B. Perez-Lopez and A. Merkoci, *Anal. Bioanal. Chem.,* 2011, **399**, 1577–1590.

10. E. Petryayeva, W. R. Algar and I. L. Medintz, *Appl. Spectrosc.,* 2013, **67**, 215–252.

11. E. Petryayeva and U. J. Krull, *Anal. Chim. Acta,* 2011, **706**, 8–24.

12. K. E. Sapsford, W. R. Algar, L. Berti, K. B. Gemmill, B. J. Casey, E. Oh, M. H. Stewart and I. L. Medintz, *Chem. Rev.,* 2013, **113**, 1904–2074.

13. W. R. Algar, A. J. Tavares and U. J. Krull, *Anal. Chim. Acta,* 2010, **673**, 1–25.

14. A. M. Michaels, M. Nirmal and L. E. Brus, *J. Am. Chem. Soc.,* 1999, **121**, 9932–9939.

15. K. Kneipp, Y. Wang, H. Kneipp, L. T. Perelman, I. Itzkan, R. Dasari and M. S. Feld, *Phys. Rev. Lett.,* 1997, **78**, 1667–1670.

16. S. Weiss, *Science,* 1999, **283**, 1676–1683.

17. P. Anger, P. Bharadwaj and L. Novotny, *Phys. Rev. Lett.,* 2006, **96**, 113002.

18. Y. F. Cheng and N. J. Dovichi, *Science,* 1988, **242**, 562–564.

19. X. H. C. Huang, M. A. Quesada and R. A. Mathies, *Anal. Chem.,* 1992, **64**, 2149–2154.

20. C. A. Monnig and R. T. Kennedy, *Anal. Chem.,* 1994, **66**, R280–R314.

21. A. L. Washburn and R. C. Bailey, *Analyst,* 2011, **136**, 227–236.

22. F. Ceyssens, D. Witters, T. Van Grimbergen, K. Knez, J. Lammertyn and R. Puers, *Sens. Actuators B,* 2013, **181**, 166–171.

23. D. Ho, M. O. Noor, U. J. Krull, G. Gulak and R. Genov, *IEEE Trans. Bio-Med. Circuits Syst.,* 2013, **7**, 643–654.

24. D. Ho, M. O. Noor, U. J. Krull, G. Gulak and R. Genov, *IEEE Trans. Circuits Syst.,* 2013, **60**, 2116–2129.

25. E. A. Jares-Erijman and T. M. Jovin, *Nature Biotechnol.,* 2003, **21**, 1387–1395.

26. J. R. Lakowicz, *Principles of Fluorescence Spectroscopy*, Springer, New York, 3rd edn, 2006.

27. W. H. Hu, Z. S. Lu, Y. S. Liu, T. Chen, X. Q. Zhou and C. M. Li, *Lab Chip,* 2013, **13**, 1797–1802.

28. Y. Gao, W. L. Stanford and W. C. W. Chan, *Small,* 2011, **7**, 137–146.

29. S. Giri, E. A. Sykes, T. L. Jennings and W. C. W. Chan, *ACS Nano,* 2011, **5**, 1580–1587.

30. M. Bruchez, M. Moronne, P. Gin, S. Weiss and A. P. Alivisatos, *Science,* 1998, **281**, 2013–2016.

31. W. C. W. Chan and S. M. Nie, *Science,* 1998, **281**, 2016–2018.

32. Y. L. Gao, A. W. Y. Lam and W. C. W. Chan, *ACS Appl. Mater. Interfaces,* 2013, **5**, 2853–2860.

33. S. Sun, M. Ossandon, Y. Kostov and A. Rasooly, *Lab Chip,* 2009, **9**, 3275–3281.
34. L. L. Cao, L. W. Cheng, Z. Y. Zhang, Y. Wang, X. X. Zhang, H. Chen, B. H. Liu, S. Zhang and J. L. Kong, *Lab Chip,* 2012, **12**, 4864–4869.
35. M. O. Noor, E. Petryayeva, A. J. Tavares, U. Uddayasankar, W. R. Algar and U. J. Krull, *Coord. Chem. Rev.,* 2014, **263–264**, 25–52.
36. W. R. Algar and U. J. Krull, *Langmuir,* 2006, **22**, 11346–11352.
37. W. R. Algar and U. J. Krull, *Anal. Chim. Acta,* 2007, **581**, 193–201.
38. W. R. Algar and U. J. Krull, *Anal. Chem.,* 2009, **81**, 4113–4120.
39. W. R. Algar and U. J. Krull, *Langmuir,* 2009, **25**, 633–638.
40. W. R. Algar and U. J. Krull, *Langmuir,* 2008, **24**, 5514–5520.
41. W. R. Algar and U. J. Krull, *Sensors,* 2011, **11**, 6214–6236.
42. W. R. Algar and U. J. Krull, *Langmuir,* 2010, **26**, 6041–6047.
43. L. Chen, W. R. Algar, A. J. Tavares and U. J. Krull, *Anal. Bioanal. Chem.,* 2011, **399**, 133–141.
44. A. J. Tavares, M. O. Noor, C. H. Vannoy, W. R. Algar and U. J. Krull, *Anal. Chem.,* 2012, **84**, 312–319.
45. M. O. Noor, A. J. Tavares and U. J. Krull, *Anal. Chim. Acta,* 2013, **788**, 148–157.
46. M. O. Noor and U. J. Krull, *Anal. Chem.,* 2013, **85**, 7502–7511.
47. M. O. Noor, A. Shahmuradyan and U. J. Krull, *Anal. Chem.,* 2013, **85**, 1860–1867.
48. I. L. Medintz, L. Berti, T. Pons, A. F. Grimes, D. S. English, A. Alessandrini, P. Facci and H. Mattoussi, *Nano Lett.,* 2007, 7, 1741–1748.
49. E. Petryayeva, W. R. Algar and U. J. Krull, *Langmuir,* 2013, **29**, 977–987.
50. E. Petryayeva and U. J. Krull, *Langmuir,* 2012, **28**, 13943–13951.
51. K. A. Willets and R. P. Van Duyne, in *Annu. Rev. Phys. Chem.,* 2007, vol. 58, pp. 267–297.
52. K. L. Kelly, E. Coronado, L. L. Zhao and G. C. Schatz, *J. Phys. Chem. B,* 2003, **107**, 668–677.
53. S. Link and M. A. El-Sayed, *Int. Rev. Phys. Chem.,* 2000, **19**, 409–453.
54. J. N. Anker, W. P. Hall, O. Lyandres, N. C. Shah, J. Zhao and R. P. Van Duyne, *Nature Mater.,* 2008, 7, 442–453.
55. R. Elghanian, J. J. Storhoff, R. C. Mucic, R. L. Letsinger and C. A. Mirkin, *Science,* 1997, **277**, 1078–1081.
56. C. Escobedo, *Lab Chip,* 2013, **13**, 2445–2463.
57. T. W. Ebbesen, H. J. Lezec, H. F. Ghaemi, T. Thio and P. A. Wolff, *Nature,* 1998, **391**, 667–669.
58. R. Gordon, D. Sinton, K. L. Kavanagh and A. G. Brolo, *Acc. Chem. Res.,* 2008, **41**, 1049–1057.
59. H. SadAbadi, S. Badilescu, M. Packirisamy and R. Wuthrich, *Biosens. Bioelectron.,* 2013, **44**, 77–84.
60. L. H. Guo, Y. C. Yin, R. Huang, B. Qiu, Z. Y. Lin, H. H. Yang, J. R. Li and G. N. Chen, *Lab Chip,* 2012, **12**, 3901–3906.
61. C. Escobedo, Y. W. Chou, M. Rahman, X. B. Duan, R. Gordon, D. Sinton, A. G. Brolo and J. Ferreira, *Analyst,* 2013, **138**, 1450–1458.

62. R. Kurita, Y. Yokota, Y. Sato, F. Mizutani and O. Niwa, *Anal. Chem.*, 2006, **78**, 5525–5531.
63. L. Malic, T. Veres and M. Tabrizian, *Biosens. Bioelectron.*, 2009, **24**, 2218–2224.
64. T. Springer, M. Piliarik and J. Homola, *Sens. Actuators B*, 2010, **145**, 588–591.
65. P. L. Stiles, J. A. Dieringer, N. C. Shah and R. R. Van Duyne, in *Annu. Rev. Anal. Chem.*, 2008, vol. 1, pp. 601–626.
66. R. Gao, N. Choi, S. I. Chang, S. H. Kang, J. M. Song, S. I. Cho, D. W. Lim and J. Choo, *Anal. Chim. Acta*, 2010, **681**, 87–91.
67. P. B. Monaghan, K. M. McCarney, A. Ricketts, R. E. Littleford, F. Docherty, W. E. Smith, D. Graham and J. M. Cooper, *Anal. Chem.*, 2007, **79**, 2844–2849.
68. G. Wang, C. Lim, L. Chen, H. Chon, J. Choo, J. Hong and A. J. deMello, *Anal. Bioanal. Chem.*, 2009, **394**, 1827–1832.
69. A. Walter, A. Marz, W. Schumacher, P. Rosch and J. Popp, *Lab Chip*, 2011, **11**, 1013–1021.
70. A. Marz, B. Monch, P. Rosch, M. Kiehntopf, T. Henkel and J. Popp, *Anal. Bioanal. Chem.*, 2011, **400**, 2755–2761.
71. X. N. Lu, D. R. Samuelson, Y. H. Xu, H. W. Zhang, S. Wang, B. A. Rasco, J. Xu and M. E. Konkel, *Anal. Chem.*, 2013, **85**, 2320–2327.
72. J. Kim, J. Elsnab, C. Gehrke, J. Li and B. K. Gale, *Sens. Actuators B*, 2013, **185**, 370–376.
73. K. Yunus and A. C. Fisher, *Electroanalysis*, 2003, **15**, 1782–1786.
74. V. N. Goral, N. V. Zaytseva and A. J. Baeumner, *Lab Chip*, 2006, **6**, 414–421.
75. S. Lindsay, T. Vázquez, A. Egatz-Gómez, S. Loyprasert, A. A. Garcia and J. Wang, *Analyst*, 2007, **132**, 412–416.
76. N. Gharib Naseri, S. J. Baldock, A. Economou, N. J. Goddard and P. R. Fielden, *Electroanalysis*, 2008, **20**, 448–454.
77. T. Yasukawa, K. Nagamine, Y. Horiguchi, H. Shiku, M. Koide, T. Itayama, F. Shiraishi and T. Matsue, *Anal. Chem.*, 2008, **80**, 3722–3727.
78. W. A. El-Said, C. H. Yea, H. Kim, B. K. Oh and J. W. Choi, *Biosens. Bioelectron.*, 2009, **24**, 1259–1265.
79. W. Dungchai, O. Chailapakul and C. S. Henry, *Anal. Chem.*, 2009, **81**, 5821–5826.
80. R. Ge, R. W. K. Allen, L. Aldous, M. R. Bown, N. Doy, C. Hardacre, J. M. MacInnes, G. McHale and M. I. Newton, *Anal. Chem.*, 2009, **81**, 1628–1637.
81. Y. Liu, H. Wang, J. Huang, J. Yang, B. Liu and P. Yang, *Anal. Chim. Acta*, 2009, **650**, 77–82.
82. W. Jung, A. Jang, P. L. Bishop and C. H. Ahn, *Sens. Actuators B*, 2011, **155**, 145–153.
83. N. Triroj, P. Jaroenapibal, H. Shi, J. I. Yeh and R. Beresford, *Biosens. Bioelectron.*, 2011, **26**, 2927–2933.

84. D. P. Wasalathanthri, V. Mani, C. K. Tang and J. F. Rusling, *Anal. Chem.,* 2011, **83**, 9499–9506.

85. Y. Liu, J. Yan, M. C. Howland, T. Kwa and A. Revzin, *Anal. Chem.,* 2011, **83**, 8286–8292.

86. M. Brun, J. F. Chateaux, A. L. Deman, P. Pittet and R. Ferrigno, *Electroanalysis,* 2011, **23**, 321–324.

87. D. Jiang, G. Xiang, C. Liu, J. Yu, L. Liu and X. Pu, *Int. J. Electrochem. Sci.,* 2012, **7**, 10607–10619.

88. E. Sinkala, J. E. McCutcheon, M. J. Schuck, E. Schmidt, M. F. Roitman and D. T. Eddington, *Lab Chip,* 2012, **12**, 2403–2408.

89. K. Islam, S. K. Jha, R. Chand, D. Han and Y. S. Kim, *Microelectron. Eng.,* 2012, **97**, 391–395.

90. A. Vasudev, A. Kaushik, Y. Tomizawa, N. Norena and S. Bhansali, *Sens. Actuators B,* 2013, **182**, 139–146.

91. J. H. Park, Y. S. Song, J. G. Ha, Y. K. Kim, S. K. Lee and S. J. Bai, *Sens. Actuators B,* 2013, **188**, 1300–1305.

92. N. Triroj, P. Jaroenapibal and R. Beresford, *Sens. Actuators B,* 2013, **187**, 455–460.

93. M. D. M. Dryden, D. D. G. Rackus, M. H. Shamsi and A. R. Wheeler, *Anal. Chem.,* 2013, **85**, 8809–8816.

94. L. Krejcova, L. Nejdl, M. A. M. Rodrigo, M. Zurek, M. Matousek, D. Hynek, O. Zitka, P. Kopel, V. Adam and R. Kizek, *Biosens. Bioelectron.,* 2014, **54**, 421–427.

95. S. Gu, Y. Lu, Y. Ding, L. Li, H. Song, J. Wang and Q. Wu, *Biosens. Bioelectron.,* 2014, **55**, 106–112.

96. M. Medina-Sánchez, S. Miserere, E. Morales-Narváez and A. Merkoçi, *Biosens. Bioelectron.,* 2014, **54**, 279–284.

97. P. Wang, L. Ge, M. Yan, X. Song, S. Ge and J. Yu, *Biosens. Bioelectron.,* 2012, **32**, 238–243.

98. J. Wang, R. Polsky, B. Tian and M. P. Chatrathi, *Anal. Chem.,* 2000, **72**, 5285–5289.

99. Y. J. Kim, J. E. Jones, H. Li, H. Yampara-Iquise, G. Zheng, C. A. Carson, M. Cooperstock, M. Sherman and Q. Yu, *J. Electroanal. Chem.,* 2013, **702**, 72–78.

100. J. Wang, *Analytical Electrochemistry*, Wiley-VCH, Hoboken, 3rd edn, 2006.

101. P. J. A. Kenis, R. F. Ismagilov and G. M. Whitesides, *Science,* 1999, **285**, 83–85.

102. C. Karuwan, K. Sukthang, A. Wisitsoraat, D. Phokharatkul, V. Patthanasettakul, W. Wechsatol and A. Tuantranont, *Talanta,* 2011, **84**, 1384–1389.

103. P. Dubois, G. Marchand, Y. Fouillet, J. Berthier, T. Douki, F. Hassine, S. Gmouh and M. Vaultier, *Anal. Chem.,* 2006, **78**, 4909–4917.

104. K. Choi, A. H. C. Ng, R. Fobel and A. R. Wheeler, in *Annu. Rev. Anal. Chem.*, 2012, vol. 5, pp. 413–440.

105. A. Bhimji, A. A. Zaragoza, L. S. Live and S. O. Kelley, *Anal. Chem.,* 2013, **85**, 6813–6819.

106. I. Ivanov, J. Stojcic, A. Stanimirovic, E. Sargent, R. K. Nam and S. O. Kelley, *Anal. Chem.,* 2013, **85**, 398–403.
107. B. Lam, Z. C. Fang, E. H. Sargent and S. O. Kelley, *Anal. Chem.,* 2012, **84**, 21–25.
108. A. Apilux, W. Dungchai, W. Siangproh, N. Praphairaksit, C. S. Henry and O. Chailapakul, *Anal. Chem.,* 2010, **82**, 1727–1732.
109. R. F. Carvalhal, M. S. Kfouri, M. H. O. De Piazetta, A. L. Gobbi and L. T. Kubota, *Anal. Chem.,* 2010, **82**, 1162–1165.
110. Z. Nie, F. Deiss, X. Liu, O. Akbulut and G. M. Whitesides, *Lab Chip,* 2010, **10**, 3163–3169.
111. Z. Nie, C. A. Nijhuis, J. Gong, X. Chen, A. Kumachev, A. W. Martinez, M. Narovlyansky and G. M. Whitesides, *Lab Chip,* 2010, **10**, 477–483.
112. D. Tobjörk and R. Österbacka, *Adv. Mater.,* 2011, **23**, 1935–1961.
113. F. Patolsky, G. F. Zheng and C. M. Lieber, *Anal. Chem.,* 2006, **78**, 4260–4269.
114. K.-I. Chen, B.-R. Li and Y.-T. Chen, *Nano Today,* 2011, **6**, 131–154.
115. F. Patolsky, G. F. Zheng, O. Hayden, M. Lakadamyali, X. W. Zhuang and C. M. Lieber, *Proc. Natl. Acad. Sci. U. S. A.,* 2004, **101**, 14017–14022.
116. B. L. Allen, P. D. Kichambare and A. Star, *Adv. Mater.,* 2007, **19**, 1439–1451.
117. A. K. Wanekaya, W. Chen, N. V. Myung and A. Mulchandani, *Electroanalysis,* 2006, **18**, 533–550.
118. Y. Cui, Q. Q. Wei, H. K. Park and C. M. Lieber, *Science,* 2001, **293**, 1289–1292.
119. G. F. Zheng, F. Patolsky, Y. Cui, W. U. Wang and C. M. Lieber, *Nature Biotechnol.,* 2005, **23**, 1294–1301.
120. F. N. Ishikawa, H. K. Chang, M. Curreli, H. I. Liao, C. A. Olson, P. C. Chen, R. Zhang, R. W. Roberts, R. Sun, R. J. Cote, M. E. Thompson and C. Zhou, *ACS Nano,* 2009, **3**, 1219–1224.
121. C. Li, M. Curreli, H. Lin, B. Lei, F. N. Ishikawa, R. Datar, R. J. Cote, M. E. Thompson and C. Zhou, *J. Am. Chem. Soc.,* 2005, **127**, 12484–12485.
122. A. Choi, K. Kim, H.-I. Jung and S. Y. Lee, *Sens. Actuators B,* 2010, **148**, 577–582.
123. F. Patolsky, G. Zheng and C. M. Lieber, *Nanomedicine,* 2006, **1**, 51–65.
124. B. Z. Tian, P. Xie, T. J. Kempa, D. C. Bell and C. M. Lieber, *Nature Nanotechnol.,* 2009, **4**, 824–829.
125. J. T. Hu, T. W. Odom and C. M. Lieber, *Acc. Chem. Res.,* 1999, **32**, 435–445.
126. R. M. Penner, *Annu. Rev. Anal. Chem.,* 2012, **5**, 461–485.
127. F. Patolsky and C. M. Lieber, *Mater. Today,* 2005, **8**, 20–28.
128. L. B. Luo, J. S. Jie, W. F. Zhang, Z. B. He, J. X. Wang, G. D. Yuan, W. J. Zhang, L. C. M. Wu and S. T. Lee, *Appl. Phys. Lett.,* 2009, **94**, 123103.
129. M. Wipf, R. L. Stoop, A. Tarasov, K. Bedner, W. Fu, I. A. Wright, C. J. Martin, E. C. Constable, M. Calame and C. Schönenberger, *ACS Nano,* 2013, **7**, 5978–5983.

130. J. Hahm and C. M. Lieber, *Nano Lett.,* 2004, **4**, 51–54.

131. Z. Q. Gao, A. Agarwal, A. D. Trigg, N. Singh, C. Fang, C. H. Tung, Y. Fan, K. D. Buddharaju and J. M. Kong, *Anal. Chem.,* 2007, **79**, 3291–3297.

132. G. J. Zhang, J. H. Chua, R. E. Chee, A. Agarwal and S. M. Wong, *Biosens. Bioelectron.,* 2009, **24**, 2504–2508.

133. T. W. Lin, P. J. Hsieh, C. L. Lin, Y. Y. Fang, J. X. Yang, C. C. Tsai, P. L. Chiang, C. Y. Pan and Y. T. Chen, *Proc. Natl. Acad. Sci. U. S. A.,* 2010, **107**, 1047–1052.

134. T. Y. Lin, B. R. Li, S. T. Tsai, C. W. Chen, C. H. Chen, Y. T. Chen and C. Y. Pan, *Lab Chip,* 2013, **13**, 676–684.

135. W. U. Wang, C. Chen, K. H. Lin, Y. Fang and C. M. Lieber, *Proc. Natl. Acad. Sci. U. S. A.,* 2005, **102**, 3208–3212.

136. E. Stern, J. F. Klemic, D. A. Routenberg, P. N. Wyrembak, D. B. Turner-Evans, A. D. Hamilton, D. A. LaVan, T. M. Fahmy and M. A. Reed, *Nature,* 2007, **445**, 519–522.

137. E. Stern, A. Vacic, N. K. Rajan, J. M. Criscione, J. Park, B. R. Ilic, D. J. Mooney, M. A. Reed and T. M. Fahmy, *Nature Nanotechnol.,* 2010, **5**, 138–142.

138. M. Curreli, Z. Rui, F. N. Ishikawa, H.-K. Chang, R. J. Cote, Z. Chongwu and M. E. Thompson, *IEEE Trans. Nanotechnol.,* 2008, **7**, 651–667.

139. Y. L. Bunimovich, Y. S. Shin, W.-S. Yeo, M. Amori, G. Kwong and J. R. Heath, *J. Am. Chem. Soc.,* 2006, **128**, 16323–16331.

140. W. J. Royea, A. Juang and N. S. Lewis, *Appl. Phys. Lett.,* 2000, **77**, 1988–1990.

141. G.-J. Zhang, J. H. Chua, R.-E. Chee, A. Agarwal, S. M. Wong, K. D. Buddharaju and N. Balasubramanian, *Biosens. Bioelectron.,* 2008, **23**, 1701–1707.

142. C. J. Chu, C. S. Yeh, C. K. Liao, L. C. Tsai, C. M. Huang, H. Y. Lin, J. J. Shyue, Y. T. Chen and C. D. Chen, *Nano Lett.,* 2013, **13**, 2564–2569.

143. A. R. Gao, N. Lu, Y. C. Wang, P. F. Dai, T. Li, X. L. Gao, Y. L. Wang and C. H. Fan, *Nano Lett.,* 2012, **12**, 5262–5268.

144. A. Gao, N. Zou, P. Dai, N. Lu, T. Li, Y. Wang, J. Zhao and H. Mao, *Nano Lett.,* 2013, **13**, 4123–4130.

145. R. Elnathan, M. Kwiat, A. Pevzner, Y. Engel, L. Burstein, A. Khatchtourints, A. Lichtenstein, R. Kantaev and F. Patolsky, *Nano Lett.,* 2012, **12**, 5245–5254.

146. S. Hideshima, H. Hinou, D. Ebihara, R. Sato, S. Kuroiwa, T. Nakanishi, S. I. Nishimura and T. Osaka, *Anal. Chem.,* 2013, **85**, 5641–5644.

147. Y. Q. Fu, J. K. Luo, X. Y. Du, A. J. Flewitt, Y. Li, G. H. Markx, A. J. Walton and W. I. Milne, *Sens. Actuators B,* 2010, **143**, 606–619.

148. F. R. Baxter, C. R. Bowen, I. G. Turner and A. C. E. Dent, *Ann. Biomed. Eng.,* 2010, **38**, 2079–2092.

149. G. L. Smith, J. S. Pulskamp, L. M. Sanchez, D. M. Potrepka, R. M. Proie, T. G. Ivanov, R. Q. Rudy, W. D. Nothwang, S. S. Bedair, C. D. Meyer and R. G. Polcawich, *J. Am. Ceram. Soc.,* 2012, **95**, 1777–1792.

150. S. Zhengguo, L. Dongling, W. Zhiyu and Z. Xingqiang, *J. Semicond.,* 2013, **34**, 114013.
151. I. Sayago, M. J. Fernández, J. L. Fontecha, M. C. Horrillo, C. Vera, I. Obieta and I. Bustero, *Sens. Actuators B,* 2012, **175**, 67–72.
152. L. García-Gancedo, W. I. Milne, J. K. Luo and A. J. Flewitt, *Proc. SPIE,* 2013, 879308.
153. K. Keiji Kanazawa and J. G. Gordon Ii, *Anal. Chem.,* 1985, **57**, 1770–1771.
154. V. L. Strashilov and V. M. Yantchev, *IEEE Trans. Sonics Ultrason.,* 2005, **52**, 812–821.
155. H. Zhang, M. S. Marma, S. K. Bahl, E. S. Kim and C. E. McKenna, *IEEE Sens. J.,* 2007, **7**, 1587–1588.
156. H. Zhang, M. S. Marma, E. S. Kim, C. E. McKenna and M. E. Thompson, *17th, IEEE Int. Conf. Micro Electro Mech. Syst.*, Maastricht, Netherlands, 2004.
157. P. C. H. Li, W. Wang and M. Parameswaran, *Analyst,* 2003, **128**, 225–231.
158. M. Michalzik, R. Wilke and S. Büttgenbach, *Sens. Actuators B,* 2005, **111–112**, 410–415.
159. K. Zhang, L. B. Zhao, S. S. Guo, B. X. Shi, T. L. Lam, Y. C. Leung, Y. Chen, X. Z. Zhao, H. L. W. Chan and Y. Wang, *Biosens. Bioelectron.,* 2010, **26**, 935–939.
160. C. Li, P. M. Wu, A. Browne, S. Lee and C. H. Ahn, *Proc. IEEE Sens.*, 2007, 4388436, 462–465.
161. K. Zhang, S. H. Choy, L. Zhao, H. Luo, H. L. W. Chan and Y. Wang, *Microelectron. Eng.,* 2011, **88**, 1028–1032.
162. Y. H. Hsu and W. C. Tang, *Microfluid. Nanofluid.,* 2011, **11**, 459–468.
163. Y. H. Hsu and W. C. Tang, *Smart Mater. Struct.,* 2009, **18**, 129901.
164. P. S. Dittrich, K. Tachikawa and A. Manz, *Anal. Chem.,* 2006, **78**, 3887–3907.
165. D. R. Reyes, D. Iossifidis, P. A. Auroux and A. Manz, *Anal. Chem.,* 2002, **74**, 2623–2636.
166. T. Vilkner, D. Janasek and A. Manz, *Anal. Chem.,* 2004, **76**, 3373–3385.
167. C. J. Easley, J. M. Karlinsey, J. M. Bienvenue, L. A. Legendre, M. G. Roper, S. H. Feldman, M. A. Hughes, E. L. Hewlett, T. J. Merkel, J. P. Ferrance and J. P. Landers, *Proc. Natl. Acad. Sci. U. S. A.,* 2006, **103**, 19272–19277.
168. J. Khandurina, T. E. McKnight, S. C. Jacobson, L. C. Waters, R. S. Foote and J. M. Ramsey, *Anal. Chem.,* 2000, **72**, 2995–3000.
169. V. Srinivasan, V. K. Pamula and R. B. Fair, *Lab Chip,* 2004, **4**, 310–315.
170. E. Petryayeva and W. R. Algar, *Anal. Chem.,* 2013, **85**, 8817–8825.
171. J. L. Delaney, E. H. Doeven, A. J. Harsant and C. F. Hogan, *Anal. Chim. Acta,* 2013, **790**, 56–60.
172. M. Arciuli, G. Palazzo, A. Gallone and A. Mallardi, *Sens. Actuators B,* 2013, **186**, 557–562.

Subject Index

Illustrations and figures are in **bold**. Tables are in *italics*.